Essentials of Physical Geography Today

SECOND EDITION

Theodore M. Oberlander
University of California, Berkeley

Robert A. Muller
Louisiana State University

RANDOM HOUSE *New York*

Second Edition
987654321
Copyright © 1987 by Random House, Inc.

Library of Congress Cataloging-in-Publication Data

Oberlander, Theodore.
  Essentials of physical geography today.
  Includes index.
  1. Physical geography. I. Muller, Robert A.
II. Title.
GB55.023 1987   910'.02   86-29711
ISBN 0-394-36280-2

Cover Illustration: *Landscape, Hudson Valley* by Frederick E. Church. Courtesy of the Cooper-Hewitt Museum, Smithsonian Institution/Art Resource.

Graphic Supervision and Cover Design: Lorraine Hohman

Production Manager: Laura Lamorte

Manufactured in the United States of America

Preface

This introduction to physical geography is concerned with the natural world as it is seen from the human perspective. Physical geography is the geography of the human environment, and is more than a mere composite of other physical sciences such as meteorology, climatology, biology, pedology, and geology. Physical geography not only describes the natural phenomena at the earth's surface but also, and more importantly, seeks explanations of how and why physical processes act as they do, how their effects cascade through the earth's physical and biotic systems, and how they are significant to life on our planet. The subject matter of physical geography centers upon the surface region where land, sea, and air meet and interact, and where life flourishes. Human impacts on the natural systems operating at this interface are an essential consideration, tending to disturb the equilibria established among natural physical systems over long periods of time and triggering rapid change in nearly all of the earth's environments.

The processes that influence surface phenomena involve both energy and materials. Energy continuously cascades through physical systems ranging in scale from the global biosphere to single microbes. Materials such as carbon, water, and oxygen flow through these systems in never-ending cycles that involve inputs, storages, outputs, stability thresholds, and quasiequilibrium states. The study of physical geography emphasizes the ways in which the various physical systems interact, constantly exchanging energy and materials.

While our goal is to organize the complexity of our planet by using the concepts of energy systems, we have no desire that these concepts diminish the fascination derived from consideration of the individual wonders of the real world. Despite our use of generalizing theory, our focus is on the world itself, in all its awesome but comprehensible complexity.

This second edition of *Essentials of Physical Geography Today* combines the general organization, perspective, and style of its predecessor with many new elements: some from the third edition of *Physical Geography Today* and others presented here for the first time. Those familiar with the first edition of *Essentials* will detect many important changes in the text. Biogeography receives greater recognition in the form of a chapter concerned with the biosphere and ecological energetics, modified from a similar chapter in *Physical Geography Today*. Discussion of geological structures is advanced to the chapter on the earth's lithosphere; and climatic influences on landforms, omitted from the preceding edition, are surveyed in the introductory chapter on landforms.

New "Environmental Issues" boxes have been added to highlight a wide variety of current topics featured in the news and of concern to both geographers and the general public. Our new "Case Studies in Physical Geography" supplement includes 27 case studies, 11 more than in the last edition. Sixteen of these case studies are totally new. The case studies have been moved from the text to a supplement to provide instructors with more flexibility and choice in the selection of topics for special emphasis.

For the first time, third-order headings have been utilized. They clarify the text for both the instructor and the student. The lengthy figure captions of former editions of *Essentials* and *Physical Geography Today* have been abbreviated, with portions of the former captions incorporated into the text. This more effectively integrates the text and figures. The chapter summaries of the former edition have been replaced by lists of key words. Review questions and class applications of the text material, suitable for discussion or laboratory work, are appended to each chapter. We have also included a glossary of terms, which most students find helpful for quick reference and review.

Our aim has not been to introduce and define every term used by physical geographers in their professional work, nor to touch on every phenomenon that enters into the realm of physical geography. Rather than lightly skimming a very large surface, we have sought to develop an understanding of the most essential facts and relationships by extended treatment of the topics that seem to be of paramount importance. Some of these have not previously been explored in beginning textbooks in physical geography.

The text serves as the basis for a semester or a two-semester sequence. Much attention has been devoted to continued improvement of the book's graphics, which have been a popular feature of past editions of *Physical Geography Today* and *Essentials*. Many new illustrations have been introduced in this new edition in the hope of enhancing student understanding of processes and interactions, and of the overall perception of physical landscapes.

Acknowledgments

We are grateful for the assistance of the reviewers whose comments and suggestions helped us shape the third edition of *Physical Geography Today* and thereby the second edition of *Essentials,* which builds upon the foundation of the former book. A preliminary draft of the manuscript of *Physical Geography Today* was reviewed by Robert B. Batchelder, Boston University; Vernon Meentemeyer, University of Georgia; Laurence S. Kalkstein, University of Delaware; Wayne N. Engstrom, California State University, Fullerton; Patricia F. McDowell, University of Oregon; Patrick J. Bartlein, University of Oregon; the late Jack R. Villmow, Northern Illinois University; Marlyn L. Shelton, University of California, Davis; Kenneth L. White, Texas A. & M. University; and Jay R. Harman, Michigan State University. Detailed comments on each chapter of *Physical Geography Today* were provided by Robert B. Batchelder, Boston University; John J. Alford, Western Illinios University; David McArthur, San Diego State University; Carl L. Johannessen, University of Oregon; John Lier, State University of California at Hayward; and John Street, University of Hawaii. We are also grateful for the work of Harry Spector, who collated reviewers' comments and edited the text.

In preparing the second edition of *Essentials of Physical Geography Today* we are indebted to Donald W. Ash, Indiana State University; Anthony J. Brazel, Arizona State University; Ronald H. Isaac, Ohio University; Patricia F. McDowell, University of Oregon; and Kathleen C. Parker, University of Georgia, who contributed detailed reviews of the first edition of *Essentials.* Their comments and suggestions have been extremely helpful to us. We gratefully acknowledge the assistance of many people at Random House in the preparation of this edition: notably Barry Fetterolf, Suzanne Thibodeau, Sylvia Shepard, Holly Gordon, Carolyn Viola-John, Laura Lamorte, Lorraine Hohman, and Kathy Bendo. We appreciate their encouragement as well as all of their direct contributions. Mary Moulton and Lucille Oberlander deserve thanks for their time-consuming work on this book's Index. As always, our wives, Lucille and Jeanne, did many of the things that were most crucial in bringing this undertaking to fruition. And, of course, we must acknowledge the contributions of our students, our teaching assistants, and the countless persons whose paths we have crossed to our benefit in the preparation of this book—as well as those whose needs have had to wait while project-related deadlines and emergencies were being met.

THEODORE M. OBERLANDER
ROBERT A. MULLER

Overview

Contents

Essentials of Physical Geography Today

The Mystery, by T. M. Oberlander, 1966

The origin of the universe and our solar system with its nine planets remains a matter of scientific speculation. What is clear is that the planet earth has changed astonishingly since its formation about 5 billion years ago. Physical geography is the study of the processes that continue to modify our planetary environment.

1
Evolution of the Earth

In every country of the world there are individuals who identify themselves as geographers. Everyone is aware that geography is concerned with *where things are,* but beyond this most people are unsure about what the study of geography involves. Some have encountered a type of "geography" that merely describes the earth and the natural processes and human activities occurring on its surface, and others are aware that geography is a field that seems to have no clear boundaries. Everything in the universe has some spatial properties. Once these are described, what remains to be done? The answer is that description is only the beginning. Having determined the spatial pattern of any phenomenon, one naturally asks further questions: How was the pattern achieved? Why did it develop in such a way? What are its current tendencies? How does it influence other phenomena? What is its future? These questions are the essence of geographical inquiry and have been for more than 2,000 years.

PHYSICAL AND HUMAN GEOGRAPHY

The two broadest subdivisions within the field of geography are *physical geography,* which focuses on the natural processes that create physical diversity on the earth, and *human geography,* which is concerned with human activity on our planet. However, human activities clearly alter natural processes and the physical environment, and natu-

3

Climatology Meteorology Hydrology Pedology Botany Ecology

Astronomy

Geology

Geodesy

Geomorphology

Cartography

Physical Geography

Physical Oceanography

Figure 1.1 (opposite)
Physical geography draws upon the specialized
knowledge of many disciplines. The unique
contribution of physical geography is its focus on the
interactions of the varying phenomena that combine to
give each place its particular character. (Tom Lewis)

ral processes and features exert many influences on human activities. Thus physical and
human geography are intimately interwoven.
While they cannot be entirely separated, it is
possible to focus upon either the human phenomena played out on the physical stage or the
physical stage on which humans must perform.
This book does the latter.

The earth's surface is a constantly changing
arena in which energy from the sun and from
the earth's interior act upon air, rock, soil,
water, and a host of living organisms. All are
intricate systems linked by and to the physical
processes that shape our natural surroundings.
Because physical geography analyzes the complex natural processes that determine the human environment, its subject matter must be
extremely diverse (Figure 1.1). Its focus is on
interactions, such as the effect of solar energy
on atmospheric motion, the role of water in the
development of soil, or the influence of vegetation on erosion processes. Similarly, human
geography studies the interplay of physical,
cultural, historical, and economic influences
and their effect on human activities throughout the world. Geography's emphasis on the interactions among various physical and human
systems provides a unique point of view, one
that no other single field of science offers.

Physical geography is the original environmental science—traditionally concerned with
the interaction between human beings and the
physical environment. The relentless intensifi-
cation of our extraction of water, food, fuel, and
raw materials from the earth is increasingly
affecting our environment and the natural
processes that maintain it. In many cases
human activities have unintentionally triggered changes in natural systems, affecting

even the earth's capacity to support life.
Knowledge of the subtle linkages within and
between environmental systems can help us
reduce the danger of negative side effects resulting from our activities on the surface of the
earth.

As this chapter stresses, the present is part
of a continuum of change that goes back some 5
billion years. The changes have left a variety of
traces. These include the deposits made by geological processes, which speak to us of changing climates, upheavals of the land, and rises
and falls in the level of the seas; fossilized
plants and animals, which record the history of
life on our planet; datable materials of various
types, which enable us to develop a reliable
chronology of events; and even a record of the
earth's changing magnetic field, which has
contributed critical information about long-
term movements of the earth's crust. Such
clues have led to a generally accepted reconstruction of the principal events in the history
of the earth. This chapter outlines those
events, which have created the major physical
systems that are the subjects of later chapters.

FORMATION OF THE EARTH

Birth of the Universe

Before the earth could be formed, the universe had to come into being. Scientific evidence suggests that the universe did not always exist, but originated at a definite point in
time. Most scientists believe that before this
event all the matter and energy in the universe
had been squeezed into a single nucleus, or
"cosmic egg."

The Big Bang

Between 10 and 20 billion years ago, its internal energy caused this nucleus to explode in
what is called the *Big Bang*, which threw matter outward in all directions. During this expansion, the elements were formed by the fu-

Figure 1.2
Stars cluster by the billions in galaxies separated by vast expanses of empty space. Galaxies take many forms, the more common of which are illustrated in these telescope photographs. At the top are a barred spiral galaxy (left) and an elliptical galaxy (right). Below these are a well-structured spiral galaxy (left), much like our own Milky Way, and an irregular spiral galaxy (right).

The lower photograph shows our own galaxy's nearest neighbor in space, the great spiral galaxy M31, which is visible to the naked eye as a hazy patch of light in the constellation Andromeda. The Andromeda galaxy M31 is virtually a twin to our Milky Way galaxy, but the distance between the two is so great that it takes more than 2 million years for the light from one to reach the other. (Hale Observatories, California Institute of Technology and Carnegie Institution of Washington, and U.S. Naval Observatory)

sion of atomic protons, electrons, and neutrons. The first stars did not form until millions of years later, when they condensed out of vast clouds of gases and dust that were drawn together by the gravitational attraction of neighboring matter. During consolidations the squeezed clouds of atoms became so heated that self-sustaining nuclear reactions commenced, emitting energy and causing them to glow. In this way they became stars—like our sun. The stars cluster by the billions in enormous rotating galaxies, many of them resembling pinwheels (Figure 1.2). The galaxies are so far apart that their light takes millions of years to pass across the voids between them.

Expansion of the Universe

The present apparent rate of expansion of the universe suggests that the Big Bang occurred some 10 to 20 billion years ago. To realize how long ago that was in relation to later events, think of 10 billion years as a 24-hour day, with the universe originating at midnight, 24 "hours" ago. Each second of such a day would represent 100,000 actual years! Our sun's energy output suggests that it became active some 6 billion years ago. Using the most conservative estimate of the age of the universe, this would be almost 10 hours after the beginning of the 24-hour "day" initiated by the Big Bang.

Formation of the Planets

Although there are billions of stars, the only planets we can detect are the nine that are circling our own sun. Since our sun seems to be an average star, it is assumed that many stars have similar planetary systems. However, they are invisible, even through the most powerful telescopes, because planets give off only reflected light, which is extremely faint compared with light emitted by the furnace-like stars.

The most widely accepted explanation of the birth of our planetary system is that of Harold Urey, an American chemist and winner of a Nobel Prize. Urey proposed that our solar system began as a rotating, disk-shaped cloud of gases and dust. As the center of the cloud condensed to form the star we call the sun, the outer portions broke into separate eddies that themselves condensed into spinning swarms of solid matter in the form of *planetesimals*. Gravitational attraction caused the separate swarms of planetesimals to condense individually to form the existing planets. The moons of the planets are regarded as left-over planetesimal masses. Several types of evidence indicate that the earth formed in this manner about 4.6 billion years ago, or about 1 P.M. (one hour after noon in our imaginary 24-hour day).

Urey's theory is supported by two simple facts. First, all the planets circle the sun in the same direction, presumably the direction of spin of the original cloud of matter. Second, the orbits of all the planets lie in approximately the same plane, thought to be the plane of the disk-like cloud (Figure 1.3).

Structure of the Earth

According to Urey's model, the earth was formed by the condensation of planetesimals at moderate temperatures. Our planet's hot interior is thought to have developed later by heating resulting from the *radioactive decay* of elements in the earth's interior. The temperature of the earth's outer layers seems to have been stable for at least 2 billion years. Radioactive decay affects the atoms of certain unstable heavy elements (such as uranium) that change by casting off protons from their atomic nuclei. This releases energy that heats the surrounding material. As the early earth's interior was strongly heated by radioactive decay of unstable elements, solid material began to melt and move. Over millions of years light material slowly rose toward the surface, while heavier

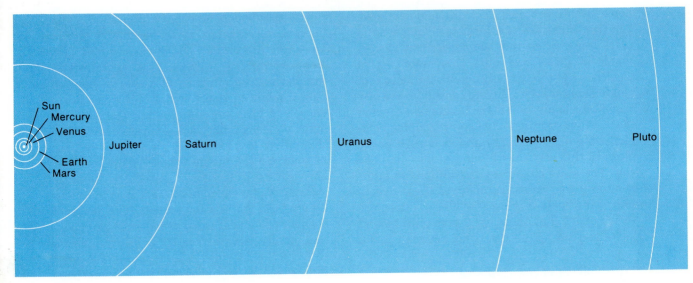

Sun, Mercury, Venus, Earth, Mars, Jupiter, Saturn, Uranus, Neptune, Pluto

(a)

material gradually sank toward the center. Eventually three concentric shells evolved (Figure 1.4). These can be detected by observing the behavior of earthquake shock waves as they pass through the earth.

The Core

Extending about half way to the surface from the earth's center is the *core,* which has a radius of 3,400 kilometers (km), or 2,100 miles. The core has such a high density that it is thought to be dominated by a mixture of iron and nickel. The physical state of this material is uncertain, since the combination of high pressure and high temperature at the center of the earth cannot be duplicated in laboratories.

The Mantle

Surrounding the core is the *mantle,* which is some 2,900 km (1,800 miles) thick. It is composed of high-density rock material dominated by silica, magnesium, and iron. Temperatures in the mantle are high enough to melt this material, but pressures are so great at these depths that most of the mantle is rigid. However, in the upper mantle there is a zone, ex-

tending from about 70 to 240 km (40 to 150 miles) beneath the earth's surface, in which molten mantle material can move at a very slow rate. This zone is known as the *asthenosphere.*

The Lithosphere

Above the asthenosphere is the outer zone of solid rock termed the *lithosphere* (Figure 1.4). This includes both the uppermost portion of the mantle and the chemically distinct outermost shell of the earth, which is called the *crust.* The boundary between the crust and upper mantle is referred to as the Mohorovičić discontinuity (after its discoverer) or *Moho.* The crust varies in thickness from about 10 to 40 km (6 to 25 miles) and is the only portion of the earth to which humans have had access. It is composed of the solid rock that forms the continents and ocean floors. As we shall see in Chapter 12, the continents are raft-like slabs of low-density rock, solidly embedded in a continuous layer of higher-density rock that underlies them as well as the ocean floor. The rock of the continental areas is diverse in origin and in many places has been severely disturbed by vertical and horizontal crustal movements. The rock

(b)

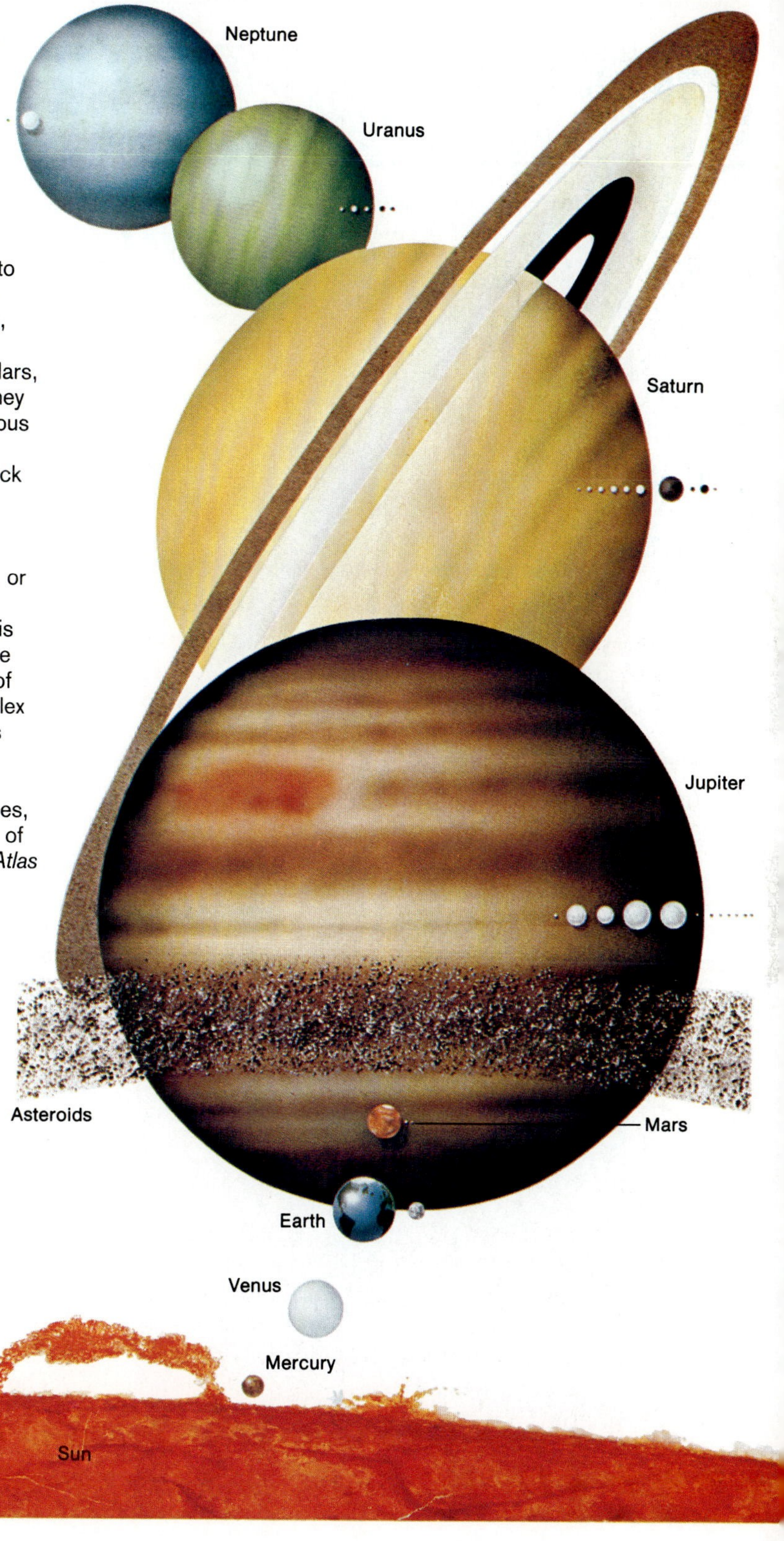

Figure 1.3 (opposite)

(a) The distances of the planets from the sun are shown in correct scale. The planets are too small to be shown at the scale of their orbits. Between the orbits of Mars and Jupiter are swarms of asteroids, which may be fragments of earlier planets.

The inner planets, Mercury, Venus, Earth, and Mars, together with Pluto, are called *terrestrial planets.* They are solid bodies surrounded by relatively thin gaseous atmospheres. The outer planets, Jupiter, Saturn, Uranus, and Neptune, are largely gaseous, with thick hydrogen-rich atmospheres. (Doug Armstrong)

Figure 1.3 (right)

(b) The relative sizes of the planets, their satellites, or moons, and part of the solar sphere, which at this scale would have a diameter of 1.5 meters. There is still much to be learned about the solar system. The Voyager 2 spacecraft sent back stunning pictures of Uranus during late January 1986, revealing a complex and almost bizarre terrain on the moon Miranda, as well as 10 additional small moons and more ring fragments that encircle Uranus.

Shown on the sun's surface are solar prominences, or "flares," which are jets of gases that spurt tens of thousands of kilometers above the solar surface. (*Atlas of the Earth,* Mitchell Beazley, Ltd., 1971)

Pluto

Neptune

Uranus

Saturn

Jupiter

Asteroids

Mars

Earth

Venus

Mercury

Sun

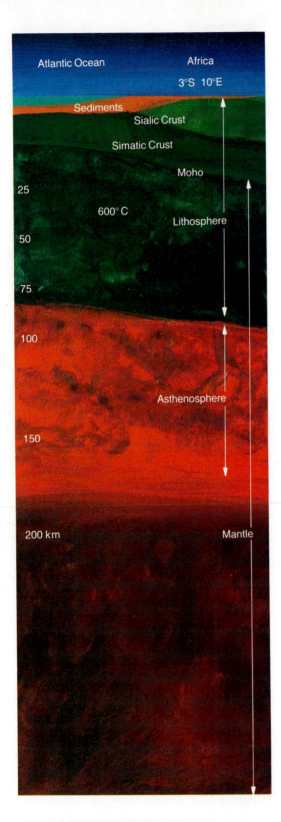

Figure 1.4
The earth in cross section. Our knowledge of the inner properties of the earth has come primarily from a study of the shock waves generated by earthquakes. These waves, called *seismic waves*, travel through the earth at speeds that vary according to the properties of the material through which they pass. (Robert Kinyon/Millsap & Kinyon after R. Phinney)

underlying the deep sea floors is mainly solidified lava produced by submarine volcanic eruptions.

The Early Atmosphere

Since the earth and sun are thought to have formed from the same cloud of matter, they should initially have had similar compositions. But the gaseous element neon, which is abundant in the sun and other stars, is now absent from the film of gases, known as the *atmosphere,* which surrounds the earth. This indicates that the earth lost its original gaseous atmosphere. The loss may have been caused by heating when the planetesimals condensed, even before the planets were formed.

The absence of an atmosphere was only temporary, for volcanic eruptions on the land began feeding new gases to the surface, gradually producing a new atmosphere. At first its composition would have been similar to that of volcanic gases, being dominated by carbon dioxide, with some water vapor, nitrogen, and sulfurous gases, and only traces of free oxygen. The planet Venus has such an atmosphere at present—one that is quite incapable of supporting life. The earth's initial atmosphere was eventually altered, as we shall see shortly.

The Early Landscape

The earth's surface 3 or 4 billion years ago must have resembled the volcanic landscape in

Figure 1.5
Because the earth's early landscapes lacked any form of vegetative protection, areas of erodible materials would have been deeply gashed by the effects of water running off the land surface, as is this area in California's Death Valley. (T. M. O.)

Figure 1.5. Flows of molten rock and blankets of volcanic ash covered the land. The ground was completely barren, and no living organisms were present. Water (condensed from clouds of volcanic steam) fell as rain, helping to decay rock by chemical processes. Large and small rock fragments and mineral grains littered the ground. There was no soil, and rainwater streamed off, cutting deep gullies and carrying rock particles and dissolved minerals to the growing seas. Between rains, windstorms swept sand and dust across the naked landscape.

With no vegetative protection, erosional forms in rainy regions must have been spectacular. Hillslopes were steep, rocky, rutted with gullies, and bordered by piles of rock debris. Sand dunes were no doubt common. The combination of much rainfall and no covering vegetation and soil gave these early landscapes an appearance seen nowhere on earth today. The naked scenery lacked the warm colors that characterize our present deserts, for there was little or no free oxygen to combine with iron-bearing substances. Iron oxides color both rocks and soils the rusty hues to which we are accustomed.

LIFE ON THE EARTH

About 3.5 billion years ago—with only 8 or 9 hours left in our imaginary 24-hour day—the history of the earth took a unique turn. Life appeared! How this occurred remains a puzzle. Experiments show that when mixtures of the simple volcanic gases present in the earth's primitive atmosphere are subjected to electric discharges, complex molecules are formed. These include amino acids, which are the building blocks of the proteins that are found in every living cell. In the early atmosphere, lightning and ultraviolet radiation from the sun could have provided the energy to break apart and recombine molecules. These new, more complex molecules would have been transported to the oceans during rainfalls. In time the oceans may have become an "organic broth," rich with complex molecules that were not "living" but had the potential to be organic building blocks. Whatever the source, an unusual combination of molecules eventually appeared that was able to absorb energy from its surroundings and to reproduce itself. This combination of molecules was the first "living" organism.

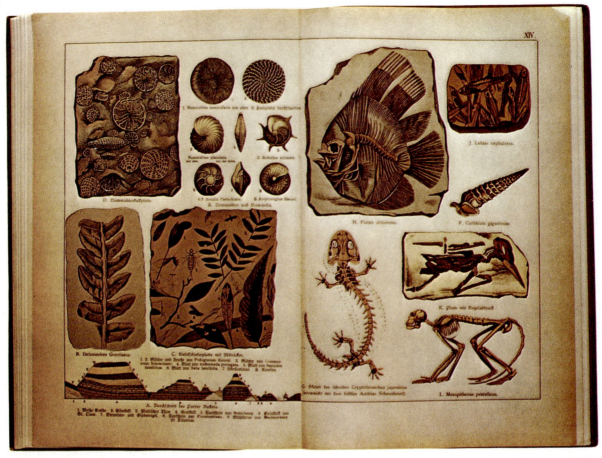

Figure 1.6
This nineteenth-century German scientific treatise illustrates the diagnostic fossils of rocks of different ages found in the Paris Basin of France, which is portrayed in a geological cross section at the lower left. The wide range of fossil types present in the rocks of this region is clearly evident. (Photo by Werner Kalber, from the John Sinkankas Collection)

The Fossil Record

We can only guess about the origin of life on earth, for there are no known traces of the first life forms. Our understanding of life in the remote past is based on the *fossil record. Fossils* are mineral replacements of the remains of plants and animals that were buried by mud, silt, or sand, before they could decompose (Fig-ure 1.6). As the covering deposits accumulated to great depths, they slowly became compacted and cemented to form layers of rock. Such a process requires hundreds of thousands of years. Over such spans of time the remains of plants and animals within the deposits are replaced by more resistant minerals that become hard casts of the original forms, thus preserving them indefinitely. In undisturbed sedimentary rocks, the oldest life forms are of course found as fossils in the lowest layers of rock. The fossils of more recent life forms are in higher layers. Thus an erosional trench through a thick mass of sedimentary rock, as at the Grand Canyon of the Colorado River, gives us a catalogue of the life forms that have inhabited the region through hundreds of millions of years.

The catalogue is not complete, however, for most fossils represent only hard bones, shells, and woody plant parts. The soft tissues of organisms usually disappear before burial and fossilization. Furthermore, if sediment is not accumulating at a location, there is no possibility of natural burial, and there is a gap in the fossil record for organisms of that region. Finally, much of the fossil record at a locality may have been removed by natural erosion of fossil-bearing rocks.

Organic Evolution

The earliest living organisms preserved in the fossil record had already evolved much beyond the earth's initial life forms. The earliest fossilized organisms resemble the blue-green algae seen on the surfaces of ponds today. These early plants obtained energy by *photosynthesis,* a process that uses the visible light of solar radiation as an energy source to produce carbohydrates, such as sugars and starches, from carbon dioxide and water, with oxygen given off as a by-product (see Chapter 9). This brings us to an important question: how did more complex plants and animals arise from the earliest simple forms of life?

An answer came from the British naturalists Charles Darwin and Alfred Russel Wallace, both of whom published theories of *organic evolution* in 1859. According to the various theories of evolution, most organisms do not reproduce exact copies of themselves. In fact, the offspring of any one pair of parents may vary considerably. In a given environment some variations have survival value. They may increase an organism's ability to gather food, escape predators that would use *it* for food, or withstand environmental stress. Such a favored organism has an above-average chance to survive and transmit the helpful variation to its offspring. This affects the evolution of the

species as a whole, which thus slowly changes by the process of *natural selection.*

Natural Selection

If an organism finds an environment to which it is extremely well adapted, additional changes offer no new benefits and the species may not evolve further. An example is the opossum, which is virtually unchanged over the past 60 million years. On the other hand, natural selection may result in alteration in a species in just a few decades. A widely cited example is the English peppered moth. Once a light-colored moth with dark speckles, it became predominantly dark in color between 1850 and 1900 as industrialization altered its physical environment. The darker coloration apparently helped the moth hide from predatory birds in the newly soot-covered towns and nearby woodlands.

In time, the changes in species resulting from the process of natural selection may produce entirely new species—groups of mutually fertile individuals that differ from similar groups in some constant way. According to Darwin and Wallace, and most other biologists, this is the process that has produced the amazing number of plant and animal species that populate the earth today. All of these are genetically related to the first primitive organisms that formed in the earth's ancient seas.

Interaction of Life with the Environment

The appearance of microscopic green plants in the oceans 3.5 billion years ago—at 3:36 P.M. in our 24-hour day—set in motion new processes that changed the face of the earth. For a long time plants were confined to the sea, where water shielded the organisms from the sun's cell-destroying ultraviolet radiation. But the simple marine plants had to remain in the sunlit upper portions of the seas to carry on the vital photosynthesis process. During photosynthesis water molecules are split into hydrogen and oxygen atoms. Plant cells use some of the oxygen and release the rest. Little of the oxygen released by the first plants remained in the atmosphere. Most of it combined with iron-bearing minerals in decomposed rock, coloring the rock various shades of red, brown, or yellow.

Atmospheric Alteration

It was only after the reactive minerals exposed at the earth's surface had been oxidized that the free oxygen released by photosynthesis began to accumulate in the atmosphere. The oxidation process appears to have required some 2 billion years—about 5 hours in our 24-hour day. Then, at last, the atmosphere began to absorb free oxygen generated by biotic activity. Today's atmosphere contains more than 20 percent free oxygen, most of it the special gift of the green plants that have lived and died through the ages.

Appearance of Animals

At 10:30 P.M. in our 24-hour day the fossil record reveals the sudden appearance of complex, multicellular marine animals with shells. How they developed from marine plant life is still unknown. What is clear is that these animals evolved the ability to make use of the oxygen that was being released by marine plants, as well as the ability to use plant material for food. Once these adaptations appeared, genetic variation and the natural selection process produced explosive expansion in the diversity of oxygen-breathing animal species.

Carbon Fixation

Soon marine animals and plants were working together to alter the earth's environment. Aquatic plants extracted carbon from carbon dioxide dissolved in water to produce carbohydrates, as part of the photosynthesis process. Tiny aquatic animals that fed on these plants built hard shells by combining the carbon with calcium weathered from rocks and washed from the land. When these animals died, their remains drifted to the ocean floor to form a growing deposit of calcium carbonate sediment, which was gradually transformed into the rock known as *limestone* (Figure 1.7). Limestone, one of the world's most abundant rock types (Chapter 12), is the chief storehouse of the carbon that once dominated the earth's atmosphere. In many places, carbon from the remains of marine organisms combined with hydrogen to form the *hydrocarbons:* petroleum and natural gas.

The removal of carbon from seawater by marine organisms gave the oceans greater capacity to gain carbon from other sources—mainly from atmospheric carbon dioxide. Thus, while biological processes were adding oxygen to the atmosphere, they were also removing carbon dioxide from it. This changed the atmosphere very significantly, reversing the initial proportions of oxygen and carbon dioxide. Today carbon dioxide is present in the atmosphere in only minor amounts, about 335 parts per million (ppm).

Formation of Ozone

The growing quantities of oxygen in the earth's atmosphere led to the formation of *ozone,* a molecule composed of 3 atoms of oxygen, as compared to the oxygen molecule's 2 atoms. Ozone is created in the upper atmosphere by high-energy solar radiation that splits apart oxygen molecules, the atoms of which recombine as ozone molecules. Ozone is of crucial importance to life on the earth because it absorbs much of the biologically destructive ultraviolet radiation streaming from the sun. The formation, high in the atmosphere, of an ozone layer that could screen out lethal radiation was required before the earth's land surfaces could become habitable to living organisms. Not until about 400 million years ago—only one hour before midnight in our 24-hour day—were plants able to begin colonizing

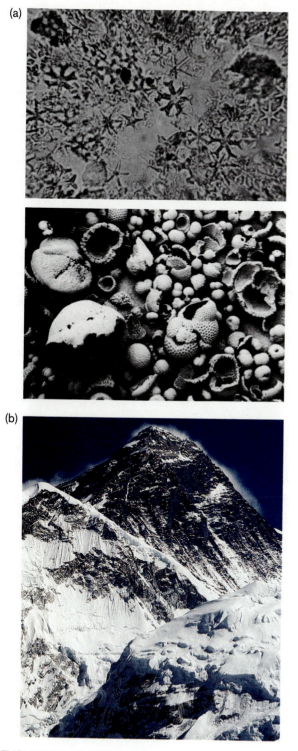

(a)

(b)

Figure 1.7

(a) Photomicrographs of biologically produced (biogenic) deep-sea sediment. At the top is an accumulation of the calcareous (calcium-rich) skeletons of star-shaped *discoasters,* which are now extinct. Their disappearance from the seas is thought to be an indication of the onset of the Ice Ages. Below is an accumulation of *foraminifera* fragments, which are an important component of present marine sediments. (Lamont-Doherty Geological Observatory/CORE Laboratory)

(b) The summit of Mount Everest, on the border of Nepal and Tibet, is the highest point on our planet. This giant peak in the Himalaya Range rises 8,848 meters (29,028 ft) above sea level and is composed of limestone that originated on the sea floor about 100 million years ago. (Barbara Brower)

the land surfaces. Small invertebrate animals, such as insects and scorpion-like arachnids, appeared soon afterward, with larger invertebrates moving onto the land some 50 million years after the arrival of the first plants.

Effect of Land Plants

When plants appeared on the land, the process of rock decay, erosion, and movement of sediment all were altered. The chemical action of plant acids helped decay rock surfaces. Later, the prying action of the roots of more highly evolved plants assisted in the breakup of rock masses. The activities of terrestrial plant and animal organisms began to create the first true soils, which thickened under vegetative protection, softening the previously angular outlines of landscapes.

The widespread growth of land vegetation had the effect of removing carbon from the atmosphere on a massive scale. Here and there vigorous plant growth in swampy coastal areas led to the accumulation of thick masses of undecomposed plant remains. These became buried by sediment and were eventually transformed into coal. An enormous amount of carbon that was taken from the atmosphere is now fixed in mineral form in the world's coal fields.

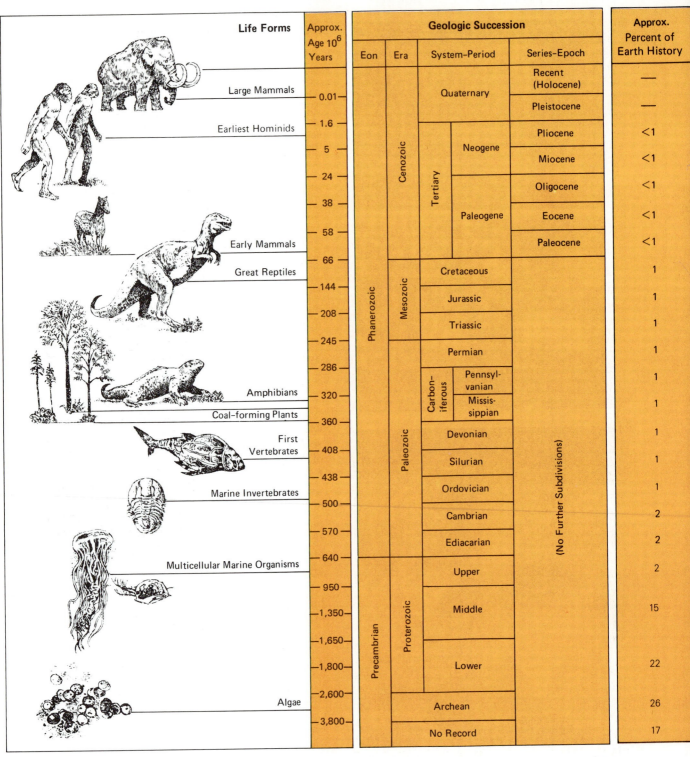

Life Forms	Approx. Age 10^6 Years	Eon	Era	System-Period	Series-Epoch	Approx. Percent of Earth History
		Phanerozoic	Cenozoic	Quaternary	Recent (Holocene)	—
Large Mammals	0.01				Pleistocene	—
Earliest Hominids	1.6			Tertiary / Neogene	Pliocene	<1
	5				Miocene	<1
	24				Oligocene	<1
	38			Tertiary / Paleogene	Eocene	<1
	58				Paleocene	<1
Early Mammals	66		Mesozoic	Cretaceous		1
Great Reptiles	144			Jurassic		1
	208			Triassic		1
	245		Paleozoic	Permian		1
	286			Carboniferous / Pennsylvanian		1
Amphibians	320			Carboniferous / Mississippian		1
Coal-forming Plants	360			Devonian		1
First Vertebrates	408			Silurian		1
	438			Ordovician		1
Marine Invertebrates	500			Cambrian		2
	570			Ediacarian		2
Multicellular Marine Organisms	640	Precambrian	Proterozoic	Upper	(No Further Subdivisions)	2
	950			Middle		15
	1,350					
	1,650			Lower		22
	1,800					
Algae	2,600			Archean		26
	3,800			No Record		17

Geologic Succession

Figure 1.8 (opposite)
The geologic time scale. To the left are shown selected life forms and the approximate times when they first appeared or were dominant. A general trend from less complex to more complex life forms reflects the processes of organic evolution. Note that the time scale in the diagram emphasizes the most recent 600 million years. (John Dawson after *Adventures in Earth History,* edited by Preston Cloud, W. H. Freeman Company. Copyright © 1970 and the Geological Society of America, 1983)

appear, presumably our own ancestors. Although we are newcomers on the earth, we are certainly not intruders here. Like all our fellow organisms we have evolved with the earth and are adapted to it. Our respiration depends on the gases in the current atmosphere, and our eyes respond to those wavelengths of solar radiation that reach the earth's surface in greatest abundance. We are a product of the earth's environment as a whole, and our interactions with its complex individual environments will continue to affect our progress as a species.

Evolution of Terrestrial Life

While plant and animal life has completely transformed our planet, the nonliving environment has at times taken its toll on the living world. Climatic changes due to astronomical or other causes have affected the evolutionary development of plants and animals. So also has such geological activity as the uplift of mountain ranges and changes in the size and position of continents and ocean basins (see Chapter 12).

The early small amphibians evolved into giant reptiles known as dinosaurs (Greek: "terrible lizards"), which dominated the earth some 130 million years ago (Figure 1.8). These huge creatures suddenly disappeared, along with many other life forms, about 65 million years ago—just 10 minutes before midnight in our 24-hour day. Various causes have been proposed. One hypothesis involves a major asteroid impact that shrouded the earth in dust for some months, reducing the penetration of sunlight and causing the collapse of plant-dependent food chains. Chemical evidence for such an event has recently been uncovered in rocks dating from the time of the most recent mass extinctions of species some 65 million years ago (see the Case Study on species extinction).

The giant reptiles were replaced by the ancestors of modern mammals. Not until about 4 million years ago—only half a minute before midnight—did the first human-like creatures

DATING THE EARTH

Before the year 1800 it was generally believed that the earth's landscape had been formed by a series of violent catastrophes that had lasted only a few thousand years. Toward the end of the eighteenth century, James Hutton, a Scottish naturalist, took a fresh look at the mountains and streams of his native land and interpreted what he saw as the products of erosion and mountain uplift. He theorized that these slow processes, given enough time, could have produced the landscapes of the British Isles, and concluded that the eras of geologic time must be much longer than previously thought. Not long afterward, Charles Darwin concluded that evolution required time spans on the order of hundreds of millions of years.

Radiometric Dating

To check the ideas of Hutton and Darwin, scientists attempted to develop an accurate way to measure geologic time. The accurate natural timekeeper that scientists sought was found early in the twentieth century when it was discovered that the passage of time could be measured by using the decay rates of *radioactive isotopes*. An isotope is one of a number of different forms of a single chemical element, such as the isotopes uranium 234 and uranium

238 (the number representing the mass of each of its atoms). A radioactive isotope decays spontaneously into another element, at a certain predictable rate that is unaffected by changes in temperature, pressure, or other external factors. The rate of decay is specified in terms of the isotope's *half-life,* which is the time required for half of an initial amount of radioactive atoms to decay.

Suppose a rock initially contained a certain amount of uranium 238 when it solidified from the molten state. Uranium 238 decays with a half-life of 4.5×10^9 years through a chain of steps to the stable end product, lead 206. Each uranium atom that decays eventually becomes one atom of lead. Since the rate of decay is known, measurement of the relative numbers of uranium 238 and lead 206 atoms in the rock allows us to compute the elapsed time.

If the rock initially contained some lead 206, or is younger than about 60 million years, the uranium 238/lead 206 method will not be accurate. The most useful radiometric clocks for geologic time are listed in Table 1.1. The decay of potassium 40 to argon 40 has been particularly useful as an age indicator because potassium occurs in several of the minerals composing common rock types formed by volcanic activity. Perhaps the most widely noted uses of potassium/argon age determinations have

been the mapping of the ages of ocean floors (Chapter 12) and the dating of fossils of early human-like creatures (hominids) discovered in the volcanic regions of East Africa (see the Case Study on hominid evolution).

Using radiometric dating methods, it is possible to make a fairly accurate determination of the age of the earth. The oldest rocks presently identified on the earth are in Greenland. Their radiometric age is about 3.8×10^9 years, meaning that the earth is at least 3.8 billion years old. However, since erosion may have destroyed still earlier rocks, it is probable that the earth is even older. (Rocks from the moon have been dated at 4.2 billion years.) According to currently accepted models of the creation of the solar system, meteorites were formed at the same time as the earth. Radiometric dating of meteorites should therefore be an accurate method of determining the earth's true age. This method results in an age of 4.6 billion years. Direct measurements of the total uranium and lead content of the earth's crust also point to an age of 4.6 billion years.

Carbon 14 is a radioactive isotope that has been important for dating events occurring within the past 40,000 years or so. Carbon is a useful substance for age determination because it is a constituent of virtually all plant and animal matter. Carbon 14—a radioactive form of carbon that is continuously formed in the earth's atmosphere by the action of cosmic rays on nitrogen 14—makes up a definite proportion of all the carbon ingested by plants and animals. When an organism dies, it ceases to take in carbon. The amount of carbon 14 in its cells steadily declines due to radioactive decay, but the amount of ordinary carbon remains constant. Therefore, measuring the ratio of radioactive carbon to ordinary carbon allows the age of the sample to be calculated. Carbon 14 dating has provided highly accurate data about relatively recent events in the earth's history. For example, by dating wood from trees that were killed by the advance of glacial ice, it has been possible to fix the date of the ice's advance through a particular area. Car-

Table 1.1
Half-Lives of Radioactive Elements

RADIOMETRIC CLOCK	HALF-LIFE (YEARS)
Rubidium 87/Strontium 87	5.0×10^{10}
Uranium 238/Lead 206	4.5×10^9
Potassium 40/Argon 40	1.3×10^9
Uranium 235/Lead 207	0.7×10^9
Uranium 234/Thorium 230	248×10^3
Carbon 14	5730

Source: Dwight E. Gray, ed. *American Institute of Physics Handbook,* 3rd ed. New York: McGraw-Hill, 1972.

bon 14 dating can also be applied to charcoal, shells, bone, and the calcium carbonate in soils. Thus it is of great assistance in archaeological work.

Other Dating Methods

A number of other dating techniques are useful for measuring relatively short time spans of thousands of years. One method uses *varves,* which are annual sediment layers deposited in lakes and on the seafloor. Coarse sediment is deposited seasonally during periods of stream flow, with fine sediment settling out during the dry season or when water surfaces are frozen. In Scandinavia varve analysis has been extended more than 10,000 years into the past, and correlation of varve patterns from different localities has enabled scientists to reconstruct the history of glacial recession in northern Europe.

Tree-ring analysis, or *dendrochronology,* is similar to varve analysis and has been used both in age determination and to reconstruct climatic conditions in the recent geologic past. It involves counting the annual growth rings of trees and measuring their relative widths to estimate how moisture and temperature have fluctuated through time. In this way, trees more than 4,000 years old have been found among the bristlecone pines of California and Nevada. Logs used in the construction of ancient Indian pueblos in the Southwestern United States have been dated by correlation of their rings with ring sequences of known age. The ages of many kinds of topographic surfaces of recent origin can be determined by counting the rings of the largest trees growing on them.

Other dating techniques are applicable in special circumstances, and the number of chronological indicators will surely increase with further research.

KEY TERMS

physical geography
human geography
Big Bang
planetesimals
radioactive decay
core
mantle
asthenosphere
lithosphere
crust
Moho
atmosphere

fossils
photosynthesis
organic evolution
natural selection
limestone
hydrocarbons
ozone
radioactive isotopes
half-life
varves
dendrochronology

REVIEW QUESTIONS

1. In what way does the field of geography differ from all other sciences?
2. How has the age of the earth been determined?
3. What is the source of the earth's internal energy?

4. How was the initial atmosphere of the earth different from the earth's present atmosphere?
5. During what proportion of the earth's history were the surfaces of the continents devoid of any form of life?

6. How did the appearance of life in the seas eventually alter the global environment?
7. Indicate two reasons why animal life developed later than plant life on the earth.
8. What two factors are thought to be the most important influences on organic evolution and the development of new species?

9. What has happened to the carbon that was once a dominant component of the earth's atmosphere?
10. What is the physical basis of radiometric age determination techniques?

APPLICATIONS

1. Where within North America would you look for landscapes most similar to those of 500 million years ago? To those of 3 billion or more years ago?
2. What are the ages of the rocks present in the area of your campus? Around your home? Do these rocks contain fossils? If so, what life forms are represented? Where in your area is a good collection of local fossils on display?

3. For what purely physical reasons would life not be likely on either Venus or Mars?
4. What are some clear examples of "natural selection," by which specific life forms have become adapted to particular environments? Is the human species, *Homo sapiens,* adapted to a particular environment?

FURTHER READING

Abell, George O. *Exploration of the Universe,* 3rd ed. New York: Holt, Rinehart & Winston (1975), 738 pp. This introductory text often used in astronomy courses is especially useful for recent ideas about the formation of the universe and solar system.

Beatty, J. K., B. O'Leary, and **A. Chaikin, eds.** *The New Solar System,* 2nd ed. Cambridge, Mass.: Sky Publishing Corp. (1982), 240 pp. An exciting up-to-date treatment of all aspects of the solar system. Includes colored maps and diagrams, and includes spectacular images from satellites and planetary probes.

Cloud, Preston. *Cosmos, Earth, and Man.* New Haven: Yale University Press (1978), 372 pp. A lucid account of the history of the universe and planet earth, emphasizing the major changes that have occurred due to the origin and evolution of life, by a leading earth scientist.

Dott, Robert H., Jr., and **Roger L. Batten.** *Evolution of the Earth.* New York: McGraw-Hill (1971), 649 pp. This introductory text traces the evolution of crustal features, flora, and fauna from a geological perspective.

Eicher, Donald L. *Geologic Time.* Englewood Cliffs, N.J.: Prentice-Hall (1968), 150 pp. This brief monograph, part of the Foundations of Earth Science series, surveys each of the methods used in dating crustal materials and features.

Poirier, Frank E. *Fossil Evidence: The Human Evolutionary Journey,* 3rd ed. St Louis: C. V. Mosby (1981), 428 pp. A full account of evolutionary systematics and the hominid fossil remains that may represent the ancestors of the human race. It is well illustrated and highly informative.

Sagan, Carl. *Cosmos.* New York: Random House (1980), 365 pp. A brilliantly written and copiously illustrated anecdotal account of cosmic evolution, from the birth of the universe to the present. The book is based on the 13-part television series produced by the author, a well-known astronomer.

Ceaseless flows of energy cause continuous change on the earth. Energy streaming from the sun and from the earth's interior powers the earth's dynamic physical systems, which constantly interact to maintain the environments we inhabit.

The Starry Night by Vincent Van Gogh, 1889. (Collection, the Museum of Modern Art, New York; acquired through the Lillie P. Bliss bequest)

2
Energy and the Earth's Systems

Energy is involved in every process and is the source of all change. But energy cannot produce change unless there is something that can react to it. The moon receives energy from the sun, but having no atmosphere, no water, and no life—systems that can react to energy—the moon has changed little over billions of years. The earth, however, is a complex of active physical and biological systems—the atmosphere, the oceans, landforms, soils, and the plant and animal realms—all changing restlessly in response to energy in various forms. The ways in which these systems respond to energy and interact are major themes of physical geography. To set the stage for later discussions, this chapter deals with energy—its sources and forms and its general effects on the earth's physical systems.

SOURCES OF ENERGY

The most important source of energy for the earth is the sun. The sun's radiation provides the power to set in motion most of the physical processes important to life, including the movements of the atmosphere that redistribute energy, produce weather, and drive the oceanic circulation. Solar energy also causes the growth of plants, which, in turn, sustain all animal life on our planet.

Two other sources of energy reside within the earth itself. One is gravitational force, which exerts a toward-the-center-of-the-earth pull on all objects, near and far. The pull is strongest on or near the earth's surface. This causes flows of air and water and falls of soil and rock. The second internal source of energy is the heat generated within the earth by radioactive decay of unstable atomic isotopes. The sudden shock of an earthquake or the eruption of a volcano reminds us of unseen forces below the earth's surface. These forces work over long spans of time to lift mountain ranges, to open new oceans and close old ones, and to change the positions of the continents.

ENERGY CONSERVATION

The total amount of energy in the universe remains constant. Energy cannot be created, nor can it be destroyed. It may seem that matter is being converted to energy in a nuclear reaction, but such reactions merely release the energy binding atomic nuclei, energy that is required for matter to exist. However, energy can be converted from one form to another.

Forms of Energy

While the energy binding atomic nuclei, or *nuclear energy,* can produce titanic effects and can even be harnessed for human use, the forms of energy most significant to natural processes on the earth's surface are solar radiant energy, heat energy, gravitational energy, kinetic energy, and chemical energy (Figure 2.1).

Solar Radiant Energy

Radiant energy from the sun heats the atmosphere and the earth's surface. It does this both directly and indirectly, as we shall see in Chapter 3. Solar radiation reaches the earth in about 8⅓ minutes, traveling at a speed of about 300,000 km (186,000 miles) per second. The largest portion of the sun's energy output is in the wavelengths of visible light, which our eyes are adapted to sense. The invisible longer-wavelength thermal radiation is felt as heat, and the invisible shorter-wavelength ultraviolet radiation is what causes sunburn. On the earth, solar radiation varies in intensity from place to place and from time to time. It is the seasonal geographical variation of solar radiation that creates the earth's weather systems and climates (Chapter 3).

Heat Energy

Heat energy is the energy resulting from the random motion of the atoms and molecules of substances. The hotter a substance is, the more vigorous is the motion of its atoms. This motion must be generated by an input of energy, often in another form, such as solar radiation.

There is a distinction between temperature, which is a measure of average molecular motion in a substance, and heat energy, which refers to the total energy in the entire volume of a substance. A cup of hot coffee at 50°C (122°F) has a higher temperature than a bathtub of warm water at 35°C (95°F), but there is actually much more heat energy in the bathtub water because of its larger volume.

As we shall see later in this chapter, there is also an important difference between sensible heat and latent heat. *Sensible heat* is the ordinary heat, created by molecular motion, which we can feel and which can be measured directly with a thermometer. *Latent heat* is a type of stored energy that cannot be measured directly. It becomes sensible heat only when released by a change of state of a substance, as when water vapor condenses to liquid water or liquid water freezes to ice.

Gravitational Energy

Gravitational energy is the attractive force exerted by the mass of an object, and is proportional to the mass and inversely proportional to the square of the distance from the center of the object. On or near the surface of the earth gravitational energy is the *potential energy* an object has due to its elevation, or distance from the earth's center. A large boulder on a hillside or a parcel of air aloft in the atmosphere has the potential to move downward if its support, or the force holding it aloft, is lost. Gravitational energy is proportional to altitude and mass. A large boulder near the top of a mountain has more potential energy—and will make a bigger splash in a lake below it—than a smaller rock at the same elevation or a somewhat larger boulder much farther downhill.

Kinetic Energy

Kinetic energy is energy of motion. The faster the speed at which an object is moving, the more energy it has. The size of the splash produced by a boulder that falls into a lake is actually a consequence of the kinetic energy it gains as it loses its potential energy. For a given velocity, kinetic energy is proportional to the mass of the object. If equal volumes of air and water are moving at the same speed, the volume of water will have much greater kinetic energy because of its vastly greater mass. This is most apparent along coasts, where the destructive force of wind-driven ocean waves is obvious. The winds that generate storm waves blow with much greater velocity than water waves can attain. But the wind can barely displace coarse sand, while the impact of a wave can move masses of rock weighing many thousands of kilograms (tens of tons).

Chemical Energy

Chemical energy is energy stored in the electrical bonds that bind together the molecules or atoms of substances. When substances react chemically, energy is released, absorbed, or converted to other forms of energy. During combustion, as when one strikes a match, chemical energy is transformed into heat and light. In plant photosynthesis (Chapter 9) solar radiation is used to generate carbohydrates. These are stored in plant tissues and may later be consumed by animals and subsequently oxidized (combined with oxygen), releasing the chemical energy animals require. Chemical energy is not the same as nuclear energy, which results from the separation and recombination of the protons and neutrons in atomic nuclei. The energy produced by a nuclear reactor is a result of the splitting of heavy uranium-235 atoms, which produces isotopes of lighter elements plus an explosive release of the energy that previously bound the uranium nucleus together. Stars, including the sun, are natural nuclear reactors in which hydrogen is being converted to helium with energy as a byproduct.

Energy Transformations

Standing in the sunlight, you feel warm because solar radiant energy excites the molecules of your skin, creating heat energy there. Running uses chemical energy stored in the cells of your body to produce energy of motion. Coasting downhill on a bicycle, your speed increases as your initial potential energy is converted to kinetic energy.

As noted earlier, energy is never destroyed but is constantly being converted from one form to another. Some of these forms are of little use. When a snowball hits a wall, its kinetic energy is abruptly transformed. Its atoms, and some of those of the wall, are violently jostled, producing heat energy that diffuses into the air. In fact, some heat is generated in every process that involves the transfer of energy. This heat—or random motion of molecules—merely warms the surrounding environment and is not otherwise usable. Since energy transfers always generate a certain amount of heat that is quickly dissipated, usable energy can never be transferred with 100 percent efficiency.

Figure 2.1
This painting interprets some of the forms of energy important in physical geography and illustrates ways in which energy interacts with systems on the earth.

Most change on the earth is caused by two kinds of energy: solar radiation (1) and the earth's internal heat (14). Solar energy powers the motion of the atmosphere (2) by causing temperature differences on the earth's surface. The wind in turn transfers energy to the ocean, producing ocean currents and waves (3). Ocean currents and the moving atmosphere help distribute energy from the hot equatorial regions of the earth to the cool polar regions. The energy of waves actively shapes coastlines (4) by eroding rocks and by transporting sand to and from beaches.

On the seas and on the land, energy from the sun frees water molecules to enter the atmosphere by evaporation (5). Water vapor is transported by the moving atmosphere and returns to the earth's surface as precipitation (6). Some of the water falling on the land runs into streams (7). As gravitational energy causes streams to flow toward lower altitudes, their kinetic energy causes erosion of the land and transportation and deposition of sediment. In cold climates or at high altitudes, winter precipitation takes the form of snow (8). Accumulated snow produces glaciers (9), which are moving masses of ice that erode the land beneath.

Vegetation (10) absorbs solar radiant energy and transforms it to stored chemical energy by the process of photosynthesis. Some of the stored energy is passed on to animals (11), which ultimately depend on green plants as their source of food energy. Vegetation, moisture, and rock materials interact to form soils (12). The chemical energy in some vegetation and tiny marine animals becomes stored as fossil fuels, which constitute the main power source for modern industrial society (13).

The energy released by radioactive decay (15) creates heat within the earth. This geothermal energy can be tapped and may become an important source of power. Uneven heating below the earth's crust causes motion within the asthenosphere. At "hot spots" (17), currents rise and spread. This moves the continents and ocean floors, causing the crust to be wrinkled in some places and broken by faults (16) in others. Where large segments of the crust are forced together (18 and white arrows), the heavier segment, usually an oceanic plate, descends into the mantle and melts. The molten material rises back toward the surface to create volcanic eruptions (19). (John Dawson)

Energy and Work

All processes in the physical world can be viewed in terms of energy and work. Energy is the capacity for doing work, which often means producing change. Heat energy within the earth creates volcanoes, warps the earth's crust, lifts mountains, and moves the continents and ocean floors. Chemical energy stored in plants causes their growth, supports animal life, and helps to form soils. The energy in some vegetation is stored as fossil fuel, the main source of power for industrial society.

Radiant energy from the sun is what directly or indirectly produces most change on the earth (Figure 2.1). In one important sequence of events, it causes molecules of water to move from the earth to the atmosphere, where they have potential energy. When this water falls as rain, it trades potential energy for kinetic energy, which on impact with the ground causes soil particles to break up or move to new locations. Some of the water falling on the land runs into streams. As streams flow toward lower altitudes, kinetic energy is used in the work of altering the landscape—eroding the land and transporting and depositing sediment. In cold climates and at high altitudes, precipitation may take the form of snow. Accumulated snow can become the moving ice of a glacier that also alters the landscape by erosion and the deposition of sediment.

By producing temperature differences on the earth's surface, radiant energy also causes motion of the atmosphere. The resulting wind, in turn, transfers energy to the ocean, giving rise to ocean currents and waves. The kinetic energy of waves modifies shorelines by eroding rocks and transporting sand to and from beaches. More important still, water in both the seas and the atmosphere continually transports energy from warmer to cooler regions of the earth, helping to equalize the earth's energy distribution. Solar radiant energy is also important in many other ways that will be apparent in later chapters.

WATER: AN ENERGY CONVERTER

Water plays a critical role in converting energy from one form to another, in moving energy from place to place, and in energy storage. Let us now look at the two properties of water that enable it to play such an important part in planetary processes.

Phase Changes

All substances can take three physical forms—solid, liquid, and gaseous. These are called *phases* of the substance. The phase in which a particular substance occurs depends on its temperature and pressure environment. For example, we ordinarily see iron as a metallic solid. But we know that foundries melt iron to a liquid that can be poured to make castings. Subjected to even more energy, such as a high-intensity beam of atomic particles, iron can be vaporized. Water, however, is the only common chemical compound that occurs in all three phases under the natural conditions prevailing at the earth's surface. Moreover, our planet is the only one in the solar system that has the proper range of temperatures to permit water to appear in all three phases at the planetary surface.

For a substance to change phase, heat energy must be either absorbed or released by its molecules or atoms. A change to a phase having tighter binding of molecules or atoms releases latent energy. This results in sensible heat that can be measured. Such phase changes can be seen when water vapor condenses to liquid water, when liquid water freezes, and when water vapor changes directly to ice. A change to a phase that has looser binding of molecules, as when water or ice evaporates or ice melts, absorbs energy, which becomes latent heat stored in the resulting substance.

Energy is commonly measured in units called *gram calories,* or simply calories. One gram calorie (with a small "c") is the amount of energy required to raise the temperature of a gram of water from 14.5°C to 15.5°C. (The Calories, spelled with a capital "C," used in relation to food energy in diets are kilocalories, equal to 1,000 gram calories.) Table 2.1 and Figure 2.2 summarize the caloric values of the energy absorbed or released by the phase changes of water and identify the various terms.

Latent heat is an important factor in the earth's temperature balance. In moist climates and over warm seas, much of the incoming solar energy is utilized for evaporation rather than heating, resulting in significant local cooling. The atmospheric circulation transports water vapor created by evaporation to other regions where its latent energy is released by condensation that forms clouds. Where there is little water to be evaporated, as in desert regions or on city streets, most incoming solar radiation is converted directly to sensible heat, causing local temperatures to be higher than those in moist areas.

Heat Capacity

In addition to its readiness to change phase, water also has an unusual capacity to act as a

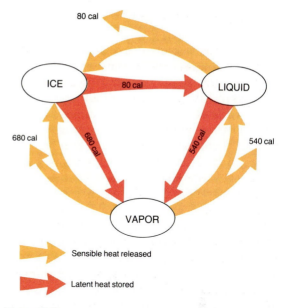

Figure 2.2
Significant energy changes occur when water changes from one physical phase to another. Energy must be added to change water from a phase in which the water molecules are tightly bound to a phase in which they are more loosely bound. To change 1 gram of ice directly into gaseous water vapor requires adding approximately 680 calories of energy.

Energy is released when a substance changes from a phase in which its molecules are loosely bound to a phase in which they are more tightly bound. For example, 680 calories are released from 1 gram of water vapor at 15°C to form 1 gram of ice at the same temperature.

Table 2.1
Heat Transfers Associated with Phase Changes of Water

PHASE CHANGE	HEAT TRANSFER	TYPE OF HEAT
Liquid water to water vapor	540–590 calories absorbed	Latent heat of vaporization
Ice to liquid water	80 calories absorbed	Latent heat of fusion
Ice to water vapor	680 calories absorbed	Latent heat of sublimation
Water vapor to liquid water	540–590 calories released	Latent heat of condensation
Liquid water to ice	80 calories released	Latent heat of fusion
Water vapor to ice	680 calories released	Latent heat of sublimation

reservoir of heat energy. Water requires a large input of energy to become warmed, and once warmed it cools more slowly than most other substances. The amount of heat energy required to raise the temperature of 1 gram of a substance by 1°C at normal atmospheric pressure is known as the substance's *specific heat*. The specific heat of water (1 calorie per gram of water per degree Celsius) is about five times that of soil or air. This means that 1 gram of water must absorb five times more heat energy than 1 gram of soil or air to increase in temperature an equal amount.

The capacity of a substance to absorb heat in relation to its volume is its *heat capacity*, defined as the amount of heat required to raise the temperature of a unit volume of the substance by 1°C. Because of the very high density of water relative to air, the heat capacity of water is many times that of air, and is two to three times that of dry soil. Its high heat capacity means that water can absorb and store large amounts of energy without changing temperature greatly. By contrast, small amounts of energy will produce large changes in the temperatures of land surfaces or air. This helps explain why temperatures vary much less in the oceans than on the land.

THE ATMOSPHERE: AN ENERGY TRANSPORTER

The earth's atmosphere is a major topic in physical geography because it is the chief means of transporting energy and moisture over the earth's surface. To understand how the mixture of gases forming in the atmosphere behaves, it is necessary to know something about the behavior of gases in general.

Atmospheric Pressure

Unlike the rigid structures of solids and the loosely bound molecules of liquids, the molecules of gases are free to move independently of one another. Gas molecules frequently collide with solid surfaces and with each other. Each collision exerts a push on the molecule or surface that is struck. The total force that molecular collisions exert on a given area of surface in contact with the gas is called the *pressure* of the gas.

The pressure exerted by the earth's atmosphere decreases rapidly as altitude increases. At an altitude of 5 to 6 km (3 to 3.6 miles) the atmospheric pressure is only half the pressure at sea level, as shown in Figure 2.3. The graph shows the standard value of atmospheric pressure at each altitude; actual atmospheric pressure at a particular location on a particular

Figure 2.3
The pressure exerted by air in the atmosphere decreases rapidly with increased altitude. As the graph indicates, average pressure at sea level is approximately 1,000 millibars, but at an altitude of 5 km (3 miles) it falls to about 540 millibars. Note that the pressure is shown on a *logarithmic scale.* (Doug Armstrong adapted from *Handbook of Geophysics and Space Environment,* edited by Shea L. Valley, Air Force Cambridge Research Laboratories, U.S. Air Force, 1965)

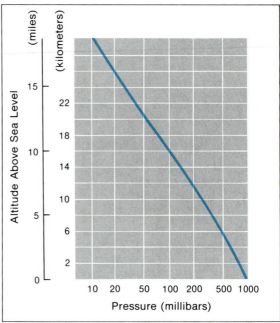

CHAPTER 2

day can be a few percent higher or lower than the standard value. If the air is denser than usual, the pressure is higher than normal, and if the air is less dense, the pressure is lower than normal.

A *barometer* measures atmospheric pressure in terms of the height of a mercury column that exerts the same downward pressure as the atmosphere. Under average conditions at sea level, this is 760 millimeters (mm), or 29.9 inches of mercury. Atmospheric pressure is expressed also in *millibars* (mb)—a unit of pressure equal to 1,000 dynes per square centimeter. (A dyne is the force required to cause a mass of 1 gram to accelerate by 1 centimeter per second in each second of time.) Standard atmospheric pressure at sea level is about 1,000 millibars.

The pressure, temperature, and volume of a given amount of gas are dependent on one another. A change in one always causes a change in either or both of the others (Figure 2.4a and b). A rise in temperature will increase the vol-

ume of an unconfined gas. But if a gas is sealed in a container so that it cannot expand, a rise in temperature will cause the pressure of the gas to rise.

Figure 2.4
(a) The pressure, volume, and temperature of a fixed amount of gas are interdependent; a change in one of the three quantities is always accompanied by a change in one or both of the others. If a gas is confined in a sealed box so that its volume is constant, changing the temperature of the gas changes the pressure the gas exerts on the walls of the box.

(b) A parcel of air in the atmosphere is unconfined. If the air is heated, it must expand its volume to maintain a constant pressure. The greater the volume occupied by the gas molecules in the parcel, the lower the density of the gas.

(c) The transfer of energy to and from parcels of air in the atmosphere can produce motion. A parcel of air that is hotter than the surrounding atmosphere is less dense than the atmosphere; the upward buoyant force on the parcel exceeds the downward gravitational force, and the parcel rises. Conversely, if a parcel of air is cooler than the surrounding atmosphere, it is denser. The downward gravitational force in this case exceeds the upward buoyant force, and the parcel descends. (Tom Lewis)

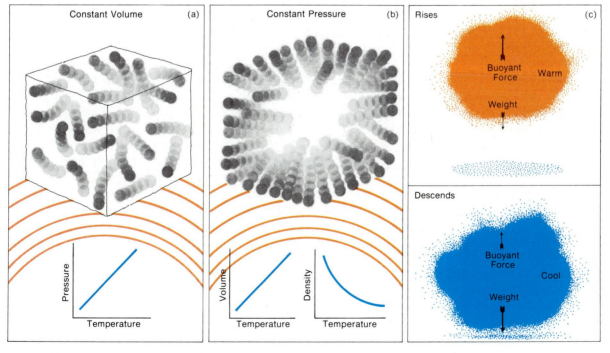

ENERGY AND THE EARTH'S SYSTEMS

Pressure and Wind

If a volume of unconfined air close to the earth's surface is heated, it will expand without changing its pressure. This expanded air is less dense than the surrounding cooler air because it has fewer molecules per unit of space. The heated air is pushed upward by a buoyant force that exceeds the downward gravitational force on the air, and the warm air rises like a hot-air balloon (Figure 2.4c). Differences in the density of air thus cause sensible heat to be transferred upward in the lower atmosphere. This vertical transfer of air and of heat energy in turn sets air in motion horizontally, producing wind.

The general effect of the earth's wind systems is to transfer warm air from regions that receive greater amounts of solar radiation to regions that receive lesser amounts. Winds also carry the latent heat present in water vapor from regions where evaporation has occurred to far-off places, where the vapor condenses back to water or ice.

THE EARTH'S PHYSICAL SYSTEMS

A system is any collection of interacting objects. The earth as a whole can be thought of as a single physical system. However, the earth contains so many phenomena interacting in so many ways that to understand them it is necessary to subdivide the world into smaller systems. None of these is truly independent; all interact to some degree.

The environment at the earth's surface is divisible into four major systems. These are the *atmosphere,* which is the envelope of gases surrounding the solid surface of the earth; the *hydrosphere,* comprising all the waters of the earth; the *lithosphere,* composed of the solid material forming the outer shell of the earth; and the *biosphere,* which encompasses all life forms present on our planet.

Open Systems

All physical systems have some common features. All have energy sources and contain materials of some type. Most are *open systems,* with inputs of energy and materials from outside the system, transformations of energy and materials within the system, storage capacity, and outputs of energy and materials that become the inputs for other systems (Figure 2.5). In a *closed system* there would be no external inputs and no outputs to other systems. Closed systems would be entirely self-contained, like a battery-powered wristwatch. When its internal energy source exhausts itself, a closed system would cease to function. In actuality it is hard to conceive of a completely closed system, since some of its energy would be lost as heat during every energy transformation within it.

When we look at the inputs and outputs of any system, we become aware of the interdependence of different systems. Think for a moment of waves breaking on a beach. The inputs to the beach system are wave energy and sand particles. The outputs are relocated sand and the form of the beach itself.

Daily changes in sand flow and beach form are the direct result of wave action against the shore. However, by tracing the inputs and outputs farther, we can see that the beach system is part of a much larger picture. The kinetic energy of the waves was originally produced by the force of wind against the water surface hundreds or thousands of kilometers out to sea. The wind resulted from geographical variations in atmospheric pressure. The pressure variations are a consequence of place-to-place variations in solar energy input to the atmosphere. Solar energy itself is an output of atomic reactions within the sun. The material input of the beach system—sand—is an output of a complex erosional system on the land and was probably delivered to the shore by a river. Finally, one output of the beach system—the changing form of the sand deposit—has important effects on the living organisms that in-

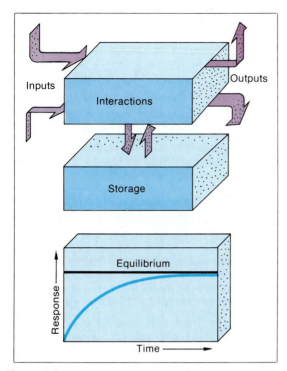

Figure 2.5
A system is any collection of interacting objects. The inputs to a system represent energy and material received from outside the system's boundaries. The inputs are transformed by system interactions to new forms that are either placed in storage for a time or transferred to other systems as outputs of material and energy.

The lower diagram illustrates schematically one of the ways that systems respond in time to an input. In this model of systems behavior, the output does not reach full strength immediately upon application of an input. Time is required for a steady condition to be achieved. Some systems, such as the atmosphere, respond rapidly to new inputs; other systems, such as soils, may require hundreds of thousands of years to attain equilibrium. (Tom Lewis)

habit the beach. This is but one simple example of the complexity of almost any open system.

Systems can store energy or materials for varying amounts of time. Rocks that are being stretched or squeezed in the earth's crust are storing energy. They are behaving like a compressed or stretched spring. Eventually, when their deformation, or strain, exceeds their strength, they snap. This releases the stored energy in the form of an earthquake. The slow bending of rocks in western California has stored so much energy that a major destructive earthquake is expected within the decade.

System Equilibrium

Many systems tend toward a stable condition in which continued inputs of energy and materials produce no long-term changes in the system or its outputs. Such a system is said to be in *equilibrium*. Short-term changes may occur, yet the long-term behavior is constant. For example, a flood may cause a stream channel to be enlarged and deepened by erosion of its bed and banks. But when the flood subsides, the channel is restored to its previous state by the deposition of sediments that fill in sections enlarged by erosion. Although there are short-term fluctuations, most stream channels remain about the same from year to year.

The maintenance of equilibrium in a physical system requires that inputs of energy and materials maintain an average balance with outputs over the period of time considered. When the balance is upset, the system reacts by changing in such a way that a new balance can be achieved.

In general, systems do not react to every small change in inputs; rather they incorporate a certain amount of inertia. They resist change until the degree of imbalance produced by changed inputs is too great to withstand. This inertia minimizes work—the expenditure of energy—in the system.

Thresholds

The point at which a system becomes so unbalanced that it begins to change to restore equilibrium is known as a system *threshold*. Think again of a boulder, with all its stored potential energy, resting on a steep slope.

Imagine scratching away some of the soil at the foot of the boulder on its downhill side. At first nothing happens. But no sensible person would continue digging incautiously at the downhill base of the boulder. Even without thinking of potential energy or thresholds, anyone would realize that at some point the digging would undermine the boulder, which would topple over and go crashing down the hillside. The digging would steepen the slope under the boulder to the threshold value at which the boulder would be moved by the force of gravity.

There are different types of thresholds in most systems. Some are externally controlled, involving changes in inputs of energy or materials; others are internal, having to do with storage of materials or energy within the system. Systems change when certain parameters either rise above or fall below threshold values. A current of water begins to pick up and carry solid particles of a certain size when its velocity increases to a certain value. Similarly, a flow of water that is carrying particles begins to drop them when flow velocity decreases to a certain point. Thresholds can be reached by either gradual or sudden changes, which may be either natural in origin or a direct result of human activities.

Response and Relaxation Times

The time required for an equilibrium system to begin to change as a result of a change in system parameters is the *response time*. The time required for a system to reestablish equilibrium after a change has begun is the *relaxation time*. These times vary greatly in different systems. A river channel responds in hours to a change in inputs and reestablishes equilibrium quickly, within days. Other systems, such as hillslopes, have response and relaxation times of years, centuries, or even longer.

When a system threshold is reached, the resulting change may be violent, as in the case of an earthquake or a hurricane, each of which releases stored energy as suddenly as the toppling boulder described previously. Other threshold phenomena can cause a system to assume a new form. Damming a river reduces the load of sediment it carries to the sea. This reduces the supply of sand to coastal beaches. Within a few years or decades after a dam is built, coastal beaches may shrink or may disappear altogether. In this case the supply of sand has dropped below the threshold value required to sustain the beaches.

Cycles of Materials

As long as the earth and its atmosphere do not experience a long-term change in temperature, we can assume that the earth's input and output of energy are in balance. But this energy is used and transformed in many ways between the time solar radiation, the major energy input, arrives on our planet and is returned to space in a different form. These intervening energy exchanges are the driving force for most physical systems on the earth.

The materials composing these systems are fixed in amount. Except for rare meteorite falls and the escape of some gaseous molecules, the earth is composed of the same matter that came together when the planet formed 4.6 billion years ago. Since the earth's materials are finite, they must be constantly recycled to permit the earth's systems to continue functioning.

The rate at which materials are recycled varies greatly from system to system. Water can evaporate from the ocean surface and enter the atmosphere as vapor, then recycle to the ocean as precipitation, all in a few hours. However, if the water were to fall as snow in a polar region, it could become part of a glacier, where it might be held as ice for hundreds or even thousands of years before returning to the sea. Some of the ice in Antarctic glaciers is more than 120,000 years old.

Other materials participate in even longer cycles. Oceanic salts have been stored for hun-

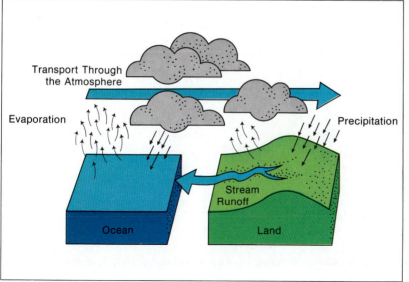

Figure 2.6
The hydrologic cycle is a key element in physical geography because water has important interactions with systems such as vegetation and landforms. (Tom Lewis)

dreds of millions of years in beds of rock salt extending from New York to Ohio and Michigan—the bed of a shallow sea 400 million years ago. Eventually the slow erosion of the North American continent will return this salt to the oceans; the earth's crust is itself being recycled ever so slowly by geological processes.

Hydrologic Cycle

The earth's hydrosphere is a dynamic physical system that is dependent on the *hydrologic cycle,* illustrated in Figure 2.6. Streams carry water endlessly from the continents to the oceans. If there were no "return flow" back to the continents, the landmasses would eventually become waterless. The hydrologic cycle in its simplest form begins with evaporation of seawater, which enters the atmosphere as water vapor. The atmospheric circulation transports much of this water vapor to the continents. Over the land, water vapor condenses into clouds composed of water droplets. The droplets coalesce and fall as rain or snow. Some of the rain water or melted snow runs off into streams that carry the water back to the oceans to complete the cycle.

Actually this is an oversimplification. Much precipitation also falls on the oceans themselves, producing a short subcycle within the larger cycle. And evaporation takes place on the continents—from various bodies of water and from plant leaves, which transpire vast amounts of moisture. Most rainfall does not run off directly into streams but enters the soil and becomes soil moisture. Some of it percolates deeper to become groundwater, which nourishes springs and wells and sustains streamflow between rains. Even though it is a simplified model, the basic concept of the cycle is correct—it involves circulation and changes of state of material, with periods of residence in different forms.

Carbon Cycle

Another material cycle of great importance to life on the earth is the *carbon cycle,* depicted in Figure 2.7. The most critical substance in the carbon cycle is gaseous carbon dioxide (CO_2). As noted in Chapter 1, green plants require atmospheric carbon dioxide, along with water and solar energy, to manufacture carbohydrates in the process of photosynthesis. Al-

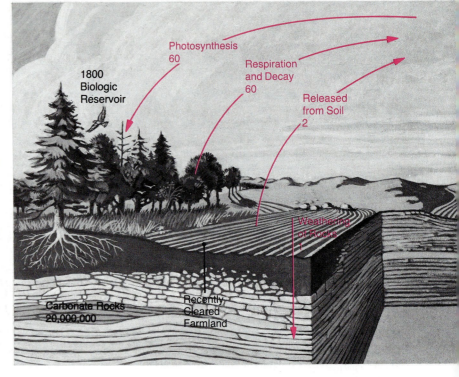

Figure 2.7
This diagram illustrates the large number of systems through which carbon can pass during its cycle. The numbers are estimated values in billions of tons of carbon released or absorbed annually in each process (red) or the total amount stored in each reservoir (black).

The transfer of CO_2 to and from organisms is believed to be nearly in balance over the year. The CO_2 content of the atmosphere, however, is increasing by several percent each decade. This increase is the result of human industrial activities; it would be even greater if it were not for the CO_2 taken up by the oceans. (John Dawson after Gilbert and Plass, "Carbon Dioxide and Climate," *Scientific American,* 1959)

though the carbon dioxide in the atmosphere is continually being depleted by plant activity, it is also continually released at the surface by plant respiration and decay, as well as by animal respiration, which expels CO_2 in exhaled breath as a waste product of the process by which animals oxidize nutrients to obtain energy.

Only a small percentage of the earth's carbon is present as CO_2 in the atmosphere. Far larger amounts of carbon compounds are dissolved in the seas and stored in vegetation, limestone bedrock, and deposits of fossil fuels (coal, oil, and natural gas). However, since the beginning of the Industrial Revolution in the nineteenth century, the burning of fossil fuels has increased the CO_2 content of the atmosphere by 10 to 15 percent, as shown in Figure 2.8. This is a matter of concern because CO_2 is a strong absorber of the thermal energy that is radiated from the earth's surface.

The increase in atmospheric CO_2 over the past 100 years may have produced a 0.3°C rise in the temperature of the lower atmosphere. It is estimated that the level of CO_2 in the atmosphere will double in 50 to 70 years, causing a 3°C global temperature increase, with the greatest warming in polar regions. Continuation of the current tendency toward global warming would eventually reduce the size of the polar icecaps in Antarctica and Greenland. The change of phase removing water from glacial storage would gradually raise the level of the seas, causing heavily populated coastal regions to become submerged, as discussed further in Chapters 16 and 17.

Budgets of Energy and Materials

Inputs, outputs, storage, and balance can be treated as parts of *energy and material budgets*

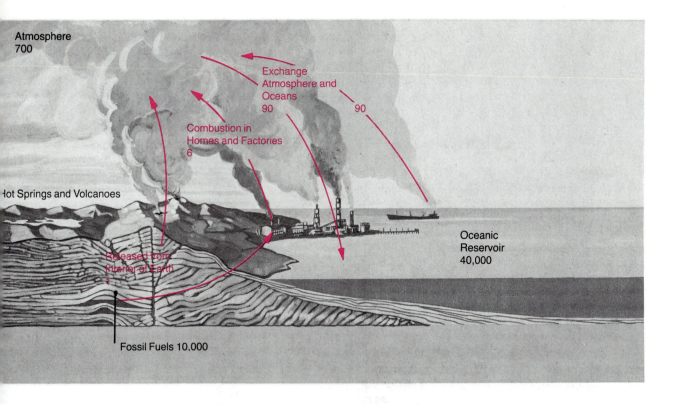

Atmosphere
700

Exchange
Atmosphere and
Oceans
90

90

Combustion in
Homes and Factories
6

Hot Springs and Volcanoes

Released from
Interior of Earth
1

Oceanic
Reservoir
40,000

Fossil Fuels 10,000

of a system. As the term suggests, the budget concept is similar to the financial concept of a budget that balances income and savings against expenses.

The example of a concrete swimming pool can serve to illustrate the budget concept. The pool gains water from rainfall and the city water supply. It loses water by evaporation and splashing. But to account for all the water passing through the pool, we must understand all interactions in the system. If there are cracks in the concrete, leaks must be included in the output side of the budget. So must the films of water clinging to persons leaving the pool. Similarly, any runoff of rainwater into the pool must be added to the input.

We could also develop an energy budget for the pool, in which energy inputs and outputs are reflected in the changing temperature of the pool water. Temperature changes indicate changes in the energy budget in much the

same way that water-level changes reflect the materials budget.

The accounting system for energy and materials sometimes reveals that things are occurring in the system that we do not understand or have failed to measure. For example, the global CO_2 budget shows that the increase in atmospheric CO_2 produced by the burning of fossil fuels over the past century is only about half the expected amount. The missing CO_2 is believed to have gone into solution in the oceans or to have been absorbed by vegetation on the land. The CO_2 budget shows that either ocean water or land vegetation has a higher CO_2 storage capacity than expected, somewhat reducing the harmful effect of fossil fuel use.

One of the tasks of physical geography is to identify the significant similarities and differences among the systems at work in the various regions of the earth. Systems that are subject to similar inputs make similar responses.

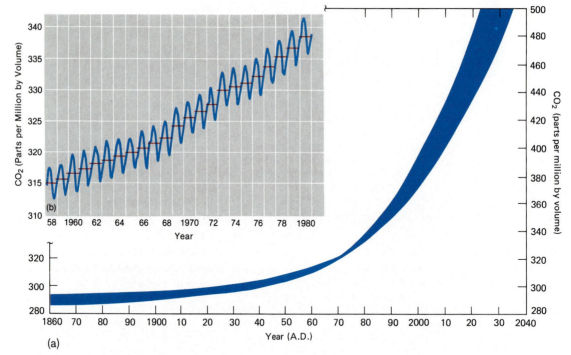

Figure 2.8
(a) Estimates of the global increase in atmospheric CO_2 since the Industrial Revolution and continuing into the twenty-first century. Estimates take into account uptake of CO_2 by the oceans and the biosphere and assume a continued exponential increase in the burning of fossil fuels (especially coal), with about half the released CO_2 remaining in the atmosphere, where it contributes to global warming. (After W. W. Kellogg and R. S. Schware, *Climate Change and Society*, Westview Press, 1981)
  (b) The monthly and average annual CO_2 concentrations measured at Mauna Loa Observatory, Hawaii, show an increase of almost 8 percent between 1958 and 1981. The seasonal variation is due to the annual cycle of vegetation growth in the northern hemisphere. (After McCracken and Moses, *Bulletin of the American Meteorological Society*, 1982)

Thus the problem met, the lessons learned, and the solutions developed in one area can be applied to others of a similar type. This can help us avoid the costly errors that have already led to environmental degradation in many parts of our planetary home.

KEY TERMS

nuclear energy
solar radiant energy
heat energy
gravitational energy
potential energy
kinetic energy
chemical energy

phases of water
sensible heat
latent heat
phase changes
gram calories
latent heat of sublimation, fusion,
  vaporization, condensation

specific heat
heat capacity
pressure
barometer
millibars
atmosphere
hydrosphere
lithosphere
biosphere

open system
closed system
system equilibrium
system threshold
response time
relaxation time
hydrologic cycle
carbon cycle
energy and material budgets

REVIEW QUESTIONS

1. What are the principal energy sources that produce change on the face of the earth?
2. Why is energy resulting from the force of gravity known as *potential* energy?
3. How is energy related to the existence of matter?
4. What energy transformations are involved in the initial formation and later human utilization of coal and petroleum?
5. How are energy transfers at the earth's surface and in the atmosphere related to changes in phase by water?
6. What is the difference between specific heat and heat capacity?
7. Considering the heat capacity of water, how might the climate of a coastal region be expected to differ from that of a continental interior?
8. What is the difference in meaning between the response time and the relaxation time of a physical system?
9. What are the similarities and differences between the inputs, forms of residence, and outputs of the hydrological cycle and the carbon cycle?
10. The system budget for a swimming pool is analogous to that of a natural lake. What are some other systems for which energy and material budgets can be formulated?

APPLICATIONS

1. Diversion of "wasted" solar energy to controlled uses will cause decreases in solar energy input to other systems. Will this have noticeable consequences?
2. Latent heat is a type of energy that might be compared to gravitational energy. What are the similarities and differences between latent heat and gravitational energy?
3. How is national policy for energy production of concern to climatologists?
4. What is the source of the water you use each day? What artificial systems have been installed to bring this water to you? How has the simple hydrologic cycle described in the text been altered or "shortcircuited" in your area? Has this produced any unexpected consequences?
5. What is the largest-scale closed system you can think of? What is the smallest-scale open system imaginable?
6. What sort of artificial disturbances of natural systems have occurred and are occurring in your area? What are the response times of these disturbances? Are the relaxation times known?

FURTHER READING

Asimov, Isaac. *Life and Energy.* New York: Bantam (1965), 378 pp. An account of energy principles and their applications to living organisms by one of the most adept of all writers on scientific topics.

Chorley, Richard J., ed. *Water, Earth, and Man.* London: Methuen (1969), 588 pp. This is a most useful collection of 38 selections, mostly by British authors, dealing with various aspects of the hydrologic cycle. The selections are at introductory and intermediate levels and complement a number of chapters in this text.

Energy and Power, A Scientific American Book. San Francisco: W. H. Freeman (1971), 144 pp. A collection of articles on energy printed originally in the periodical, *Scientific American.* Energy sources, flows, and transformations are all treated in a lively, highly readable manner, with many excellent illustrations.

Odum, Howard T. *Environment, Power, and Society.* New York: Wiley-Interscience (1971), 331 pp. This book attempts to show how natural and engineered systems are related in terms of energy flows and basic system structures. Odum uses the principles of ecology to offer solutions to current problems of global importance. A particular contribution is the symbolic language used to depict energy flows and transformations.

Weyl, Peter K. *Oceanography: An Introduction to the Marine Environment.* New York: John Wiley (1970), 535 pp. This book complements *Essentials of Physical Geography Today* very well. The chapters on oceanic salts and geochemical cycles are especially pertinent to Chapter 2 of this text.

Number 8 by Mark Rothko, 1952. (From the collection of Mr. and Mrs. Burton Tremaine, Meriden, Connecticut)

All life processes are driven by complex exchanges of visible and invisible forms of radiant energy from the sun. Desert heat and the cold of the polar zones are consequences of the earth-sun relation in space and of energy transfers on the earth.

3
Energy and Temperature

Solar radiation is the principal source of energy for the natural processes that create diversity and change on the earth. However, if the earth continually received energy from the sun without returning an equal amount to space, the oceans would boil and the land would be scorched. Since the average temperature of the atmosphere remains nearly the same from one year to the next, the earth must be returning about as much energy to space as it receives from the sun.

Of course, not all locations on earth have equal energy gains and losses. Each year, tropical regions receive a greater amount of energy than they radiate back into space. Polar regions, on the other hand, annually lose more energy to space than they receive from the sun. We know that the tropical regions are not progressively heating up nor are the polar regions cooling off. This means that there must be a flow, or *flux,* of energy from areas of excess to areas of deficiency. The atmosphere and oceans circulate the energy that the earth receives, transporting warm air and water from the tropics toward the poles while moving cool air and water back toward the equator. In this chapter we shall examine the gains and losses of energy that maintain the earth's temperature balance.

SOLAR RADIATION

Solar energy is transferred through space as *electromagnetic radiation*. Such radiation travels in waves and can be classified according to wavelength—the distance between similar points on successive waves. Radiation wavelengths are measured in units called *micrometers,* which are equal to 1 millionth of a meter. Figure 3.1 presents the *electromagnetic spectrum,* in which various types of electromagnetic radiation are identified according to wavelength. Visible light includes wavelengths from 0.4 to 0.7 micrometers. Our eyes sense the various wavelengths of visible light as different colors. When visible light is bent, or refracted, and then reflected by water droplets in the atmosphere, we see the various wavelengths as a rainbow. Solar radiation also includes wavelengths both longer and shorter than those of visible light. Ultraviolet radiation, X-rays, gamma rays, and cosmic rays are all emitted at wavelengths shorter than those of visible light, while infrared radiation and radio waves have longer wavelengths than those that are visible.

Every solid, liquid, and gas, whether warm or cold, emits electromagnetic radiation as a result of the motion of its molecules. The temperature of the radiating substance determines the wavelengths of emission. The hotter the object, the shorter the wavelengths of the radiation emitted. When we turn on the heating element of an electric stove, the coil remains dull black while warming up. During this time it emits infrared radiation, which we can feel but not see. As it grows hotter, the wavelength of maximum emission becomes shorter, shifting over into the visible portion of the spectrum. The coil glows dull red, then bright red. If it could be heated further without melting, it would eventually glow yellow-white, like the sun.

At the earth's surface and in the lower atmosphere, temperatures normally range between −40° and +40°C (−40° and +104°F). Substances at such temperatures emit electromagnetic radiation with wavelengths of 4 micrometers or more. The surface of the sun, on the other hand, has a temperature of more than 6,000°C (10,800°F). At this temperature,

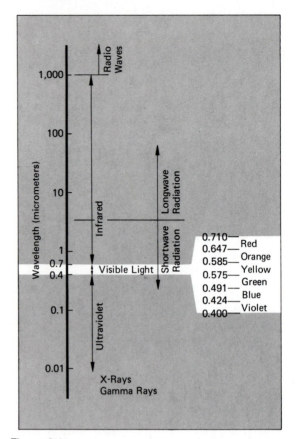

Figure 3.1
The electromagnetic spectrum is conventionally divided on the basis of wavelength. (Doug Armstrong)

most radiation is emitted at wavelengths of less than 4 micrometers. Hence we can think of most solar radiation as *shortwave radiation,* and all radiation emitted by terrestrial sources (the earth's surface and atmosphere) as *longwave radiation* (Figure 3.2).

One of the laws of physics is that a surface emits an amount of radiation per unit of time that is in direct proportion to the surface temperature. A very hot object, such as the sun, not only emits shorter wavelength radiation than a cooler object like the earth, but it also emits radiation at a far greater rate. Accordingly, solar radiation is shortwave radiation emitted at a high intensity, and any radiation

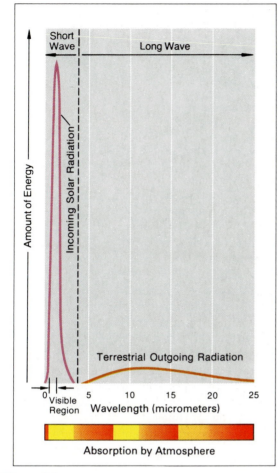

Figure 3.2
The high surface temperature of the sun causes energy to be emitted primarily at wavelengths shorter than 4 micrometers, with much of the energy concentrated in the visible region of the spectrum. In contrast, the longwave radiation emitted by the earth is confined to wavelengths longer than 4 micrometers.

The lower portion of the figure indicates the degree to which atmospheric gases, primarily carbon dioxide and water vapor, absorb electromagnetic energy near the earth's surface. Wavelength bands of strong absorption are shown in red, and bands of relative transparency are indicated by yellow shades. (Doug Armstrong after G. M. Dobson, *Exploring the Atmosphere,* 1963, by permission of the Clarendon Press, Oxford)

emitted by the earth and its atmosphere must be longwave radiation emitted at a low intensity.

The lower portion of Figure 3.2 indicates that the lower atmosphere is relatively opaque to longwave radiation, so that much of the radiant energy emitted by the earth's surface is absorbed by gases there. The atmosphere is relatively transparent to electromagnetic radiation in the band from 8.5 to 11 micrometers, and radiant energy in this band can escape to space if the sky is clear.

SOLAR ENERGY INPUT TO THE EARTH

The sun radiates energy equally in all directions. The earth, which in relation to the sun is like a grain of sand to a football 100 yards away, intercepts only a tiny fraction of the total energy emitted by the sun. But this small fraction is an enormous quantity of energy, amounting to 2.6×10^{18} calories per minute (2.6×10 multiplied by itself 18 times). The solar energy intercepted by the earth in one minute is about equal to the total electrical energy artificially generated on earth in one year.

Not all this radiant energy reaches the earth's surface because, as we shall see later in this section, the earth's atmosphere modifies the solar radiation that strikes it. Nor do equal amounts of radiant energy strike all parts of the upper atmosphere. This is because the distribution of solar radiation reaching the top of the atmosphere is controlled by the length of the daylight period and the elevation of the sun above the horizon. Both of these factors vary with latitude and are determined by the earth-sun relationships.

Earth-Sun Relationships

Like every planet in the solar system, the earth follows a fixed path, or orbit, around the sun. The orbit is slightly elliptical, so that the earth's distance from the sun varies a small amount through the year. However, this distance never varies more than 1.8 percent from the average distance of 149.5 million km (92.9 million miles). As a result, the monthly income of solar energy over the entire earth varies from the 12-month average by no more than 7 percent. The earth intercepts the greatest amount of solar radiation in early January, when it is closest to the sun (*perihelion*), and receives the least in July, when it is farthest from the sun (*aphelion*).

As the earth rotates on its axis, the circle of illumination appears to sweep around the earth from east to west over a period of 24 hours. Of course, the circle actually remains fixed, facing the sun, as the earth turns through it from west to east. This introduces an additional daily cycle of solar energy income related to the times of sunrise and sunset. If the earth's axis of rotation were perpendicular to the plane of its orbit, every place on the earth would have 12 hours of daylight and 12 hours of darkness every day of the year. In the middle and high latitudes, the seasonal change from long daylight periods in summer to short daylight periods in winter would not occur.

The Earth's Tilted Axis

Figure 3.3 shows that as the earth circles around the sun, the earth's axis always remains tilted in the same direction. This phenomenon is the cause of the seasons and the variation in daylight periods from the equator to each of the poles. The earth's axis tilts at an angle of about 23½° from the perpendicular to the plane of the orbit. This causes the duration of daylight and darkness to vary seasonally everywhere except at the equator, which is always cut exactly in half by the *circle of illumination*.

The inclination of the earth's axis also affects the angle of the sun above the horizon. On December 21 or 22, the sun appears to be directly overhead at noon at latitude 23½°S, close to Rio de Janeiro. This latitude, called the *tropic of Capricorn*, marks the southernmost position at which an observer on the ground would see the midday sun directly overhead.

Winter Solstice

The day on which the noon sun is directly overhead at the tropic of Capricorn is known in the northern hemisphere as the *winter solstice*. In the northern hemisphere this is the day of the year that has the fewest hours of daylight, with the sun appearing low in the southern sky even at noon. In areas north of the *Arctic Circle* (latitude 66½°N), the sun is not visible at all at the winter solstice. Areas situated within the Arctic Circle, such as Greenland and northern Scandinavia, experience 24 hours of darkness on this date. Moving south from the Arctic Circle, an increasing proportion of each parallel lies within the circle of illumination, so that the days become longer. Still, the period of darkness continues to be longer than the period of daylight until one arrives at the equator (0° latitude), where day and night are of equal duration at all times.

As one proceeds south of the equator, entering the summer hemisphere, proportionately more of each parallel lies within the circle of illumination, so that the daylight period becomes progressively longer than the period of darkness. At the *Antarctic Circle* (latitude 66½°S) and beyond it to the South Pole, the sun remains above the horizon for 24 hours, so that there is no night at all. Of course, while it is the winter solstice in the northern hemisphere, it is the summer solstice in the southern hemisphere. Table 3.1 indicates the varying lengths of the daylight period at different latitudes at the northern hemisphere winter solstice.

Summer Solstice

As the earth circles the sun, the tilt of the earth's axis causes the noon sun to be directly

Table 3.1

Length of Daylight During Winter Solstice*

LATITUDE	DAYLIGHT
90°N	0
80°N	0
70°N	0
60°N	5 hr 52 min
50°N	8 hr 4 min
40°N	9 hr 20 min
30°N	10 hr 12 min
20°N	10 hr 55 min
10°N	11 hr 32 min
0°	12 hr 7 min
10°S	12 hr 28 min
20°S	13 hr 5 min
30°S	13 hr 48 min
40°S	14 hr 40 min
50°S	15 hr 56 min
60°S	18 hr 8 min
70°S	24 hr
80°S	24 hr
90°S	24 hr

*Not counting twilight. Daylight includes time when at least the upper edge of the sun's disk is above the horizon.

Source: Adapted from Robert J. List, Ed. *Smithsonian Meteorological Tables,* 6th rev. ed. Washington, D.C.: Smithsonian Institution Press, 1951.

poles and coincides with the meridians of longitude. This causes the periods of daylight and darkness to be of equal duration everywhere. Thus these dates are known as the equinoxes— the *vernal equinox* in March, and the *autumnal equinox* in September, using the perspective of the northern hemisphere. The exact dates of the solstices and equinoxes vary slightly from year to year because the astronomical relationships between the earth and sun do not coincide exactly with the days determined by the earth's rotation.

Solar Elevation

The tilt of the earth's axis not only produces solstices and equinoxes; it also affects the intensity of solar radiation. Imagine holding a flashlight close to a square of cardboard representing the earth and its atmosphere. With the flashlight perpendicular to the cardboard, the circle of light is small but bright. If the cardboard is tilted to make an angle with the beam of light, the illuminated area of the cardboard becomes larger but also dimmer. The same amount of light is emitted by the flashlight, but the light intensity received per unit area of the cardboard surface has decreased.

The geometric relationship between the elevation of the sun above the horizon (beam angle) and the intensity of the solar beam at the earth's surface is shown in Figure 3.3 (bottom left). When the sun is directly overhead, the solar elevation is at the maximum of 90° and the intensity of the solar beam per unit of surface area is greatest. As the solar angle above the horizon decreases from the maximum, the intensity of the solar beam per unit of surface area decreases, reaching zero at sunset, when the solar elevation is 0°.

The amount of solar energy received per unit of surface area thus depends primarily on the angle at which the incoming radiation strikes the surface. Winter sunbathers in Miami recognize this, propping themselves in lounge chairs in order to be perpendicular to the win-

overhead at different locations at different times of the year (Figure 3.3). Six months after the northern hemisphere winter solstice, the noon sun is directly overhead north of the equator at latitude 23½°, known as the *tropic of Cancer.* The moment the vertical rays of the sun fall on the tropic of Cancer, about June 21, is the northern hemisphere *summer solstice.* At the summer solstice, the area north of the Arctic Circle has 24 hours of daylight, while in latitudes south of the Antarctic Circle the sun does not appear above the horizon.

Equinoxes

Midway between the solstices—about March 21 and September 21—the noon sun is directly overhead at the equator. At these two times the circle of illumination cuts through the

Autumnal
Equinox
September 21

Winter Solstice
December 21

Vernal
Equinox
March

N

Sun's Rays

Circle of
Illumination

N

Arctic
Circle

Tropic of
Cancer

23½

Equator

Tropic of
Capricorn

Antarctic
Circle

Winter Solstice; December 2

Summer Solstice
June 21

N
90
80
70
60
50
40
30
20
10

Latitude

Parallels

Meridians

Longitude

W 90 80 70 60 50 40 30 20 10 0 E

Prime
Meridian

Figure 3.3

(top) The amount of solar energy that reaches a given location at the top of the atmosphere depends on the distance between the earth and the sun and on the orientation of the earth. The top half of the diagram shows the earth at 12 different times during the year. The maximum distance of the earth from the sun is 152 million km (94.4 million miles) and occurs early in July. The minimum distance is 147 million km (91.4 million miles) and occurs early in January.

The earth's axis of rotation is $23\frac{1}{2}°$ from the perpendicular to the plane of the earth's orbit. As the earth circles the sun, the orientation of the axis remains the same.

(bottom left) The amount of solar radiant energy that a unit area on the earth's surface intercepts depends on the angle between the sun's rays and the plane of the surface. As the sketch shows, the solar beam is spread over a wider area where it meets the surface obliquely, reducing the energy received per unit of area. A surface intercepts the greatest amount of radiant energy when the surface is perpendicular to solar rays.

(bottom center) The solar radiant energy input to a location on the earth during a given 24-hour period depends partly on the duration of daylight. At any single moment, one-half of the earth is illuminated by the sun. The *circle of illumination* is the boundary between the light and dark regions of the earth. Because of the tilt of the earth's axis, the duration of daylight varies at different locations. The diagram illustrates the situation at winter solstice.

(bottom right) Points on the spherical surface of the earth can be located in terms of two sets of intersecting lines. The *parallels of latitude* are circles parallel to the plane of the equator. These circles connect points having the same angular distance north or south of the plane of the equator, the angle being formed at the center of the earth.

The *meridians of longitude* are north-south lines connecting points of equal angular distance, east or west of the *prime meridian*, which passes through the astronomical observatory in Greenwich, England. Longitude can vary from 0° to 180°.

A degree of latitude or longitude is divided into 60 minutes of angular measure, and a minute is divided into 60 seconds. The latitude and longitude of Washington, D.C., for example, can be written 38 degrees 54 minutes North, 77 degrees 2 minutes West, or in abbreviated form as 38°54′N, 77°02′W. (John Dawson)

ter sun's rays. Vineyards are planted on steep south-facing slopes along the Rhine and Mosel rivers in western Germany for the same reason. When the sun is 60° above the horizon, a field on a 30° slope facing toward the sun receives nearly 15 percent more solar energy than a horizontal field of the same size. It is the constant tilt of the earth's axis that helps produce summer in the northern hemisphere when the earth is farthest from the sun in its orbit. Solar elevations are higher in summer, and the hours of daylight longer. Hence, the northern hemisphere receives more solar radiation per unit area than during the other seasons.

Solar Radiation at the Top of the Atmosphere

The rate at which perpendicular rays of solar radiation strike the top of the earth's atmosphere is known as the *solar constant*. The average value over the year, based on satellite, rocket, and high mountain data, is about 1.94 calories per sq cm per minute. Scientists express amounts of solar radiation by a unit known as the *langley*. One langley is equal to 1 calorie per sq cm. The solar constant, therefore, can be expressed as 1.94 langleys per minute.

The amount of solar radiation striking an area at the top of the atmosphere during one day depends on the value of the solar constant, the true distance between the earth and the sun, the duration of sunlight, and minute-by-minute changes in the angle of the solar beam. Geographic variations in solar radiation received at the top of the atmosphere result from the last two factors, which are related to latitude.

The total daily solar radiation input to horizontal surfaces at the top of the atmosphere at different latitudes is shown in Figure 3.4. Each curve shows solar radiation input in langleys per 24 hours. For example, the energy input at the latitude of Philadelphia or Denver (about 40°N) on September 1 is about 800 langleys.

This decreases to only about 325 langleys at the winter solstice. The shaded areas represent periods of continuous darkness in the polar regions.

Figure 3.4 shows that through the year the solar radiation input varies much more at the poles than at the equator. At the equator, the variation is from about 780 langleys per day at the summer solstice to about 880 during late February and early November. At the equator the hours of daylight and darkness are always about equal, and the solar altitude at noon is always high. The most extreme annual variation is at the South Pole, ranging from more than 1,100 langleys in mid-December, when the earth is at perihelion, to 0 from mid-March to early September. That is because the South Pole receives continuous sunlight between September and March, while between March and September the sun is always below the horizon. Averaged over a full year, however, locations at the equator receive nearly two and one-half times as much solar radiation as the South Pole.

The energy input into the northern and southern hemispheres is not perfectly symmetrical. At summer solstice in the northern hemisphere (June 21), locations at latitude 15°N receive about 900 langleys per 24 hours, but at summer solstice in the southern hemisphere (December 21), locations at latitude 15°S receive close to 1,000 langleys per 24 hours. The reason is that the earth is nearer the sun in January than in July, and the energy input to the earth is 7 percent higher in January than in July.

Solar Radiation and the Atmosphere

Before solar radiation can reach the surface of the earth, it must pass through the earth's atmosphere, which is composed of gases, particles, and clouds that respond differently to solar radiation at various wavelengths. The atmosphere therefore exerts a strong influence

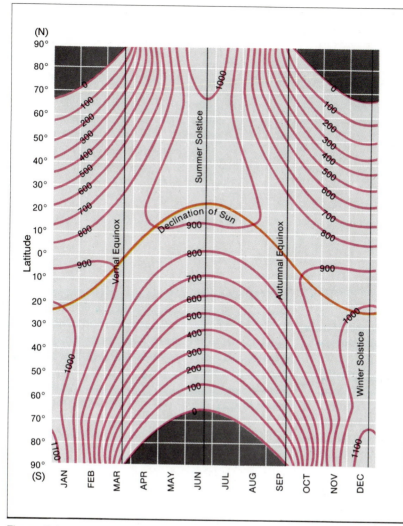

Figure 3.4
This graph shows solar radiant energy input to the top of the atmosphere at varying latitudes throughout the year. The curved contour lines give the solar radiant energy input to the top of the atmosphere in units of langleys per 24 hours. The shaded areas of the diagram poleward of latitude 66½° represent times of perpetual darkness when there is no solar radiant energy input.

The orange curve in the graph represents the latitude where the solar elevation at noon is 90°. (Doug Armstrong after *Smithsonian Meteorological Tables,* 6th ed., edited by Robert J. List, 1971, by permission of the Smithsonian Institution)

on the amounts and types of solar radiation that reach the earth's surface.

The chemical makeup of the atmosphere is given in Table 3.2. In Chapter 1 we saw how events during the early history of the earth caused nitrogen, oxygen, and carbon dioxide to be introduced into the atmosphere. The inert gases neon, helium, and krypton are probably remnants from the earth's original atmosphere. Argon, a more abundant inert gas, seems to have been produced largely by the

Table 3.2

Principal Gases in the Earth's Lower Atmosphere

GAS	MOLECULAR FORMULA	MOLECULES OF GAS PER MILLION MOLECULES OF AIR	PROPORTION (PERCENT)
Nitrogen	N_2	7.809×10^5	78.09
Oxygen	O_2	2.095×10^5	20.95
Water Vapor	H_2O	variable	variable
Argon	Ar	9.3×10^3	0.93
Carbon Dioxide	CO_2	330.0	0.03
Neon	Ne	18.0	1.8×10^{-3}
Helium	He	5.0	5.0×10^{-4}
Krypton	Kr	1.0	1.0×10^{-4}

Source: Adapted from Robert J. List, Ed. *Smithsonian Meteorological Tables*, 6th rev. ed. Washington, D.C.: Smithsonian Institution Press, 1951.

decay of radioactive potassium in the earth's crust.

The nitrogen, oxygen, and inert gases in the atmosphere are present in the same relative proportions everywhere. The amounts of other gases, such as carbon dioxide, water vapor, and ozone, vary with time, location, and altitude. Water vapor, which enters the atmosphere from the earth's surface through evaporation and plant transpiration, rarely appears above an altitude of 10 km (6 miles). Ozone occurs primarily at an altitude of about 25 km (15 miles), where it is produced by reactions between solar radiation and oxygen molecules.

Structure of the Atmosphere

One can think of the atmosphere as being divided into "spheres" related to temperature changes with altitude. Figure 3.5 shows that in the *troposphere*—the portion of the atmosphere nearest the earth's surface—the temperature decreases with altitude. The average vertical temperature decrease in the troposphere is 6.5°C per kilometer (3.6°F per 1,000 ft). The troposphere is warmest at the earth's surface, on the average, because it is heated primarily from below by the transfer of heat energy from the surface. The troposphere contains water

vapor, and most clouds and weather phenomena are confined to this layer. Temperature increases with increased altitude in the *stratosphere*, largely because atmospheric gases such as ozone absorb a portion of the radiant energy incident from the sun. The stratosphere is nearly devoid of water vapor, so clouds seldom form there. Above the stratosphere, molecules of the gases forming the atmosphere become very widely dispersed and atmospheric pressure is negligible.

Absorption and Scattering

Solar radiation interacts with gas molecules in the atmosphere through the processes of absorption and scattering. In the process of absorption, a molecule takes up radiant energy and converts it to heat. Scattering refers to the process in which gas molecules, dust particles, and water droplets deflect incoming solar radiation from its original path. Because of scattering, the earth's surface receives solar radiation from the sky as well as directly from the sun. Shorter wavelengths, especially the blue portion of visible light, are scattered more effectively than longer wavelengths, causing the sky to appear blue. The water droplets that form clouds also reflect some solar radiation

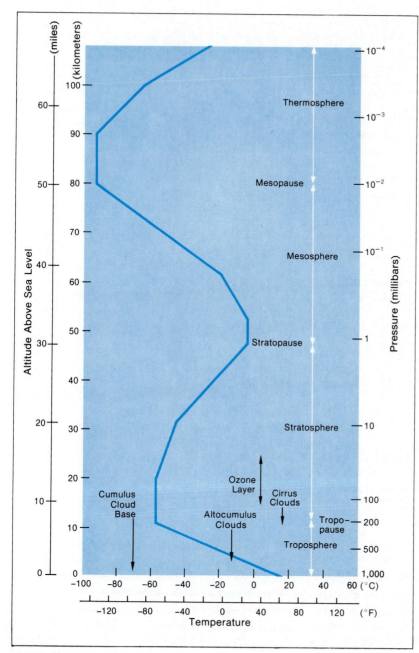

Figure 3.5
The atmosphere is conventionally divided into layers according to the variation of temperature with altitude. (Doug Armstrong after the U.S. Air Force, 1965)

ENERGY AND TEMPERATURE

Putting Solar Energy to Work

Soldiers Grove, Wisconsin, is a small village along the Kickapoo River that has been plagued by flooding of the local business district, located within a loop of the river. In 1976 the U.S. Army Corps of Engineers decided against a flood control reservoir and protective levees, and after another disruptive flood in 1978, local citizens decided instead to seek federal aid to relocate the business district to high ground above the floodplain.

The local business economy was suffering from the very high energy costs of the middle 1970s. Most village officials saw the relocation plan, in which the federal government would put up a little more than 50 percent of the costs as a flood abatement project, as a once-in-a-lifetime opportunity to reduce winter-season heating costs significantly by utilizing solar energy for heating. In the very cold winter climate of Wisconsin, heating is normally necessary for eight to nine months, with additional brief periods of modest demands during cool spells in summer. With costs of heating fuel so high, everyone was aware that dollars spent on heating meant dollars not available for anything else—the energy dollars were needed to remain in the local community.

In planning the new business district, professional architects advised that the sun could provide as much as 75 percent of the annual heating needs, even in this climatic region where runs of stormy weather can persist for three or more days. However, the siting and design of each building would have to be carefully controlled and very high levels of insulation installed. The overall costs of construction would be significantly greater than with conventional heating systems, despite federal tax credits of 15 percent of the costs of solar heating systems for business buildings at that time.

Nevertheless, the solar heating technology gained support, and the village council passed an ordinance requiring each business building to obtain a minimum of 50 percent of its energy needs for heating directly from the sun. Soldiers Grove became the first municipality in the United States to require solar heating.

Construction began in 1979 and most was completed by 1983. Month-by-month outlays for heating have been greatly reduced, and in some cases almost eliminated. Solar heating is destined to become increasingly important in the future, but the sharply lower energy costs in the mid-1980s together with the withdrawal of federal solar energy subsidies will hinder and delay the orderly development of the technology for some years.

---

back out to space as well as absorbing and scattering incoming radiation.

Gas molecules absorb most strongly within the longwave portion of the spectrum. Since most solar radiation is in the range of visible light, the majority passes through the atmosphere. More incoming radiation is reflected back to space by clouds than is absorbed by the atmosphere. As we shall see later in this chapter, nearly all the radiation absorbed by the earth's lower atmosphere is longwave radiation from the earth's surface, not incoming solar radiation.

Ultra-shortwave radiation, such as X-rays, gamma rays, and ultraviolet radiation, requires special mention because its energy can destroy the complex molecules required for life processes. Fortunately, little ultra-shortwave radiation actually reaches the earth's surface; most of it is absorbed in the upper atmosphere. Molecular oxygen in the outer layers of the earth's atmosphere and the ozone layer in the stratosphere strongly absorb ultra-shortwave radiation, causing these portions of the atmosphere to be heated to temperatures close to those at the earth's surface.

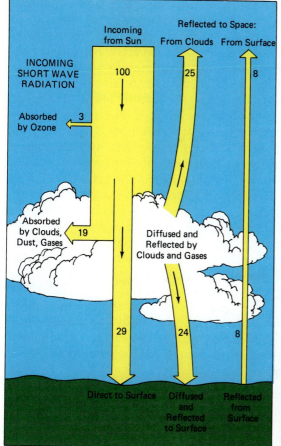

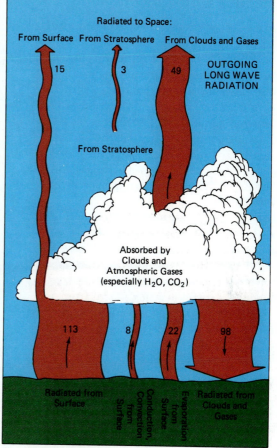

THE ENERGY BALANCE OF THE EARTH-ATMOSPHERE SYSTEM

Because the earth is neither heating up nor cooling off, the amount of energy the earth receives from the sun and the amount it radiates back into space must be in balance. To understand this we must examine the ways in which the earth and its atmosphere lose energy and thereby maintain an energy balance. It is important to keep in mind that there is both seasonal and geographical variation in energy exchanges within the earth-atmosphere system, but we shall be considering an average for the entire globe over a year. For this purpose,

Figure 3.6

This diagram traces 100 units of shortwave solar radiation that arrive at the top of the earth's atmosphere and the interactions of this radiation with the atmosphere and the surface. The numbers represent average global values. The interactions of incoming shortwave radiation are shown on the left, and the interactions of outgoing longwave radiation are shown on the right. All the indicated interactions occur at a given location during daylight hours, but at night, when there is no solar radiant energy input, only the longwave interactions occur. (After Herbert Riehl, *Introduction to the Atmosphere*, 3rd ed., © 1978 by McGraw-Hill Book Company)

we shall assume, as in Figure 3.6, that the total solar radiation intercepted by the earth and its atmosphere over a year is equivalent to 100 units of energy.

Reflection and Albedo

Airline passengers are often startled by the dazzling brightness when their plane emerges from thick clouds after taking off under a heavy overcast. Thick clouds are very effective reflectors of shortwave radiation, throwing back as much as 90 percent of the solar energy that falls on their upper surfaces. This reflected radiation combines with the incoming solar beam to produce the extremely bright zone just above the clouds. The proportion of incoming solar radiation that an object reflects is its *albedo*.

Clouds and fresh snow reflect 55 to 95 percent of the incoming solar beam, and the albedo of a good mirror approaches 100 percent. Most land surfaces, except those covered by snow, have albedos between 10 and 30 percent. The albedo at a given location varies from season to season as snow appears and disappears and as bare fields become covered with crops (Table 3.3). It is particularly important to take albedo into account in analyzing the radiation balance, because in the process of reflection no radiation is absorbed; it is simply redirected—generally outward to space. The reflected energy is lost to the earth-atmosphere system.

Figure 3.6 shows that the albedo of the whole earth-atmosphere system is about 33 percent. About 25 percent of incoming radiation is reflected by clouds, and another 8 percent by the earth's surface. Absorption by atmospheric gases and dust accounts for an additional 22 percent. Of this, 3 percent is ultraviolet radiation absorbed by ozone in the stratosphere. Ultimately, 45 percent of the solar radiation is absorbed by the land and water areas of the earth. Figure 3.7 summarizes the average annual latitudinal distribution of solar radiation in the earth-atmosphere system. In polar regions the combined effects of low solar altitudes, clouds, snow cover, and icecaps are evident. In the tropics seasonal cloudiness reduces the radiation gain at the surface. The highest gains are in the cloud-free subtropical desert regions.

Insolation

Figure 3.8 (see pp. 58–59) shows the average solar radiation input, or *insolation,* at the surface during winter and summer for the United States and for the entire earth. At the top of the atmosphere the amount of solar energy input varies only with day length and the angle of the sun's rays. But at the earth's surface differences in cloudiness can cause irregularities in the pattern of energy input. The southwestern deserts in the United States are farther north than Florida but receive more energy because their skies are generally clear. In North America the lowest energy inputs at the surface are in the east and northwest, where fog and clouds are common. The world maps show strong north-south differences in energy received at the surface in the winter hemisphere. The highest values occur over deserts in the summer hemisphere.

Table 3.3
Albedo of Various Surfaces

SURFACE	ALBEDO (PERCENT)
Fresh Snow	80–95
Dense Stratus Clouds	55–80
Ocean (Sun Near Horizon)	40
Ocean (Sun Halfway Up Sky)	5
Bare Dark Soil	5–15
Bare Sandy Soil	25–45
Desert	25–30
Dry Steppe	20–30
Meadow	15–25
Tundra	15–20
Green Deciduous Forest	15–20
Green Fields of Crops	10–25
Coniferous Forest	10–15

Sources: Robert J. List, Ed. *Smithsonian Meteorological Tables,* 6th rev. ed. Washington, D.C.: Smithsonian Institution Press, 1951. M.I. Budyko. *Climate and Life.* Trans. D. H. Miller. New York: Academic Press, 1974, pp. 54–55.

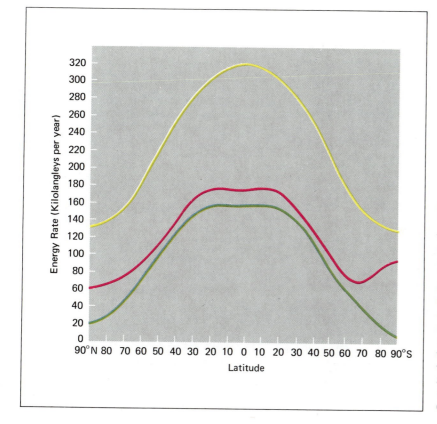

Figure 3.7
This graph shows the average annual distribution of solar radiation in the earth-atmosphere system by latitudes. The yellow curve represents incoming solar radiation at the top of the atmosphere, the red curve the incoming solar radiation just above the earth's surface, and the green curve the net solar radiation gain of the surface. (After William D. Sellers, *Physical Climatology*, © 1965, The University of Chicago Press)

Longwave Radiation: The Greenhouse Effect

Because the earth's surface is warm relative to its atmosphere and the space around it, having an average temperature of about 15°C (59°F), it emits longwave radiation upward to the atmosphere and space. If the earth had no atmosphere, all this longwave radiation would be lost to space, and the surface would cool very rapidly after each sunset. This does not occur, however, because the atmosphere acts like a greenhouse, allowing shortwave solar radiation to pass through to the surface, but trapping the longwave radiation emitted by the earth itself. Thus the atmosphere acts as a blanket that keeps the earth warm.

Water vapor and carbon dioxide molecules are primarily responsible for the absorption of longwave radiation in the lower atmosphere. The absorbed radiation heats the lower atmosphere, especially in humid regions where there is abundant water vapor. The atmosphere, in turn, reradiates longwave radiation. Some of this is directed upward and is lost to space, but much more is returned downward to the surface of the earth. On the average, such longwave radiation exchanges result in a net loss of longwave radiation at the surface and a net gain in the troposphere. Hence the surface is cooled and the troposphere is warmed by longwave radiation exchanges.

Surface cooling by longwave radiation is dramatically reduced by the presence of clouds. Clouds absorb most of the longwave radiation emitted from the surface, severely reducing losses to space. Like the atmospheric gases, clouds also emit longwave radiation, much of

Figure 3.8
The average radiation received at the surface in the United States during January and July and globally during December and June is shown in units of langleys per day.

During December and January, when the sun is overhead in the southern hemisphere, the solar radiation received in the northern hemisphere decreases rapidly with increased latitude.

During the summer months (June to September in the northern hemisphere, December to March in the southern hemisphere), the solar radiation input exhibits little variation with latitude. The increased duration of daylight with increased latitude compensates for the lower altitude of the sun toward the pole. The variation in solar radiation input between different locations during the summer is caused primarily by differences in cloudiness. (*The National Atlas of the United States of America,* 1970; and Löf, Duffie, and Smith, *World Distribution of Solar Radiation,* Report No. 21, University of Wisconsin, 1966)

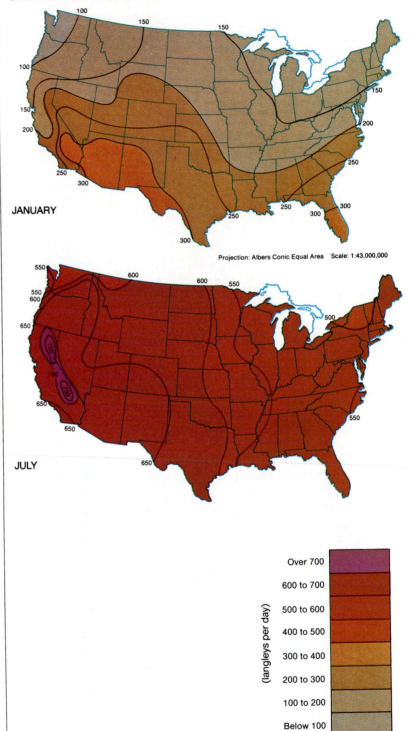

United States Mean Solar Radiation

Projection: Albers Conic Equal Area   Scale: 1:43,000,000

JANUARY

JULY

(langleys per day)

Over 700

600 to 700

500 to 600

400 to 500

300 to 400

200 to 300

100 to 200

Below 100

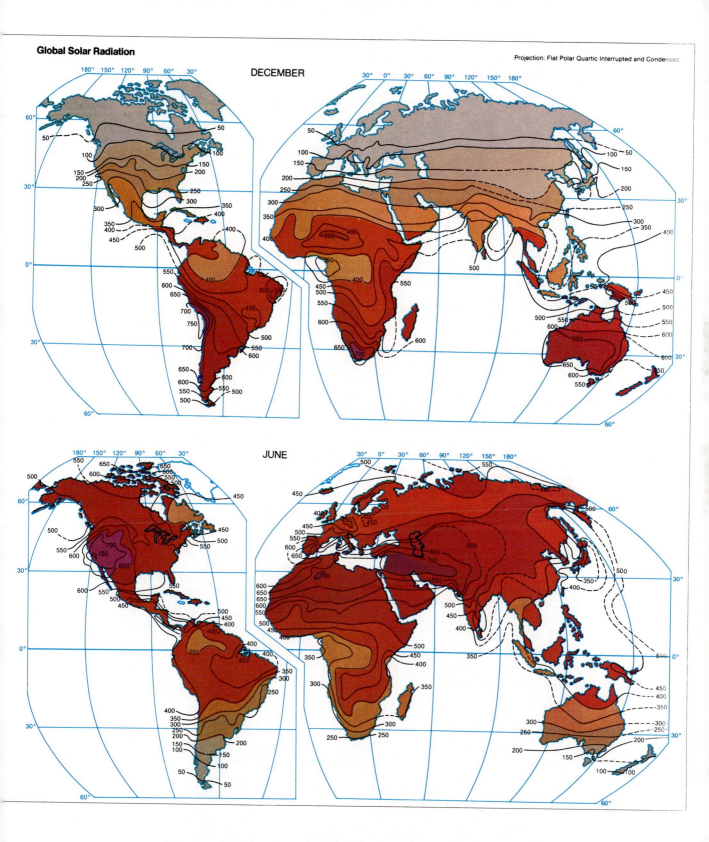

Global Solar Radiation

Projection: Flat Polar Quartic Interrupted and Condensed

DECEMBER

JUNE

which is received by the ground. That is why cloudy nights tend to be much warmer than clear nights with starry skies.

Radiation Exchanges in the Earth-Atmosphere System

The right-hand side of Figure 3.6 traces the exchanges of longwave radiation within the energy balance of the earth-atmosphere system. The figure shows that an average of 113 units of longwave radiation is emitted by the earth's surface. Most of this (98 units) is absorbed by clouds, water vapor, and carbon dioxide in the troposphere, and only a small proportion (15 units) escapes to space. The troposphere, in turn, radiates the absorbed longwave radiation (98 units) back to the surface, and an additional 49 units upward to space. The warm stratosphere also loses some longwave radiation (3 units) to space.

The longwave radiation emitted by the troposphere back downward to the surface, plus the emissions upwards to space, exceeds the longwave radiation received from the surface. This is because it is supplemented by 19 units of solar radiation absorbed by clouds, dust, and gases, and converted to longwave radiation in the troposphere. It is also supplemented by 30 units of energy transferred from the surface by evapotranspiration, conduction, and convection, which are explained further in the following section.

We have considered the solar energy intercepted by the earth and atmosphere as 100 units of shortwave radiation. As Figure 3.6 shows, returned to space are 33 units of reflected shortwave radiation and 67 units of longwave radiation emitted from the troposphere, the surface, and the stratosphere. Therefore, outgoing shortwave and longwave radiation (33 + 67 = 100) is equal to the incoming solar radiation (100 units), and the earth-atmosphere system as a whole is in balance.

Conduction, Convection, and Evapotranspiration

Considered separately, however, the atmosphere and the earth's surface do not show balanced radiation budgets. Figure 3.6 shows that radiative losses in the troposphere exceed gains by 30 units. Similarly, the earth's surface gains 143 units and radiates away only 113 units of longwave radiation. Consequently, the surface experiences an annual net gain of 30 units—an amount just equal to the net loss in the troposphere. We know that the surface is not becoming progressively warmer, nor is the troposphere becoming cooler. Therefore, other energy exchanges must compensate the imbalance in the troposphere and at the earth's surface. These energy exchanges occur through conduction, convection, and evapotranspiration.

Conduction

When two objects at different temperatures are in contact, heat flows from the warmer to the colder object by the process of *conduction*. The same process plays a role in the transfer of heat energy at the surface of the earth. By midday, solar radiation causes the top layer of the ground to become warmer than the deeper layers. As a result, heat is transferred deeper into the ground by conduction. At night, when radiation losses have cooled the surface layer, heat flows from the deeper layers back toward the surface.

Heat is also conducted upward from the earth's surface to the atmosphere when land and water surfaces are warmer than the air in contact with them. This helps use up some of the net radiation gain of the surface and offsets some of the net radiation loss of the troposphere.

Convection

The efficiency of the conduction process is increased by the process of *convection*—the vertical rise of parcels of low-density warm air.

Nuclear Winter

In recent years research in the United States, the United Kingdom, West Germany, and the Soviet Union has suggested that even a medium exchange of nuclear weapons between two warring nations could possibly lead to greatly modified climates in both the northern and southern hemispheres. The primary agent of climatic change would be enormous clouds of thick smoke and soot that would rise from massive fires devastating metropolitan regions, forests, and even some agricultural lands for days and perhaps even weeks after a nuclear exchange. The smoke would rise into the upper troposphere and be swept rather quickly eastward around the northern hemisphere in middle and higher latitudes.

The black soot from petrochemical fires and even the smoke would absorb almost all of the incoming solar radiation high in the troposphere, leaving the surfaces below in an unheated dank twilight, if not in total darkness. Some of the computer models predict that surface temperatures would fall even more rapidly because the soot particles would allow more longwave radiation from the surface to escape out to space than would natural clouds, known for their greenhouse characteristics. The models predict that if the initial nuclear exchange took place in sum-

mer, surface temperatures over the continental interiors could fall to as low as −25°C (−13°F) within two weeks, destroying most plants and crops. This grim scenario has been termed "nuclear winter."

Climate patterns over the earth would also be very different from today. Because of the enormous storage of sensible heat in the oceans, the water surfaces would cool little more than 2°C (4°F), setting up more frequent storms off the coasts. A winter monsoon circulation pattern, somewhat similar to the winter monsoon of eastern and southern Asia at the present, would prevail over all the continents. The circulation aloft would also transport smoke over the southern hemisphere, where temperatures might fall as much as 5°C (9°F).

The models suggest that it might take as long as two years for most of the smoke and soot to settle and for air temperatures to return to normal. However, there are many uncertainties among the interactions within the global circulation models used for these analyses, and much more research will be needed to "prove" beyond a shadow of a doubt that the clouds of smoke and soot would cool the earth's surface so rapidly.

Convection exchanges warmer surface air with cooler air aloft. At times, however, warm air moves over cooler land or water surfaces. At such times, conduction causes heat to flow from the atmosphere downward to the surface. This also occurs at night when the surface cools by longwave radiation. Conduction in this direction chills the lower atmosphere and removes any possibility of upward convection. As shown in Figure 3.6, an average of 8 units of energy is transferred from the surface to the atmosphere by conduction and convection.

Evapotranspiration

A most important energy exchange between the earth's surface and the troposphere involves the change of liquid water and solid ice to water vapor. Recall from Chapter 2 that every gram of water that is vaporized under average conditions transfers heat energy from the surface to the atmosphere. Evaporation from water surfaces of all types is a constant process that transfers heat to the atmosphere. Less obvious is the process by which plants absorb liquid water from the soil and emit water vapor to the atmosphere through pores in their leaves. This process is called *transpiration*. Since transpiration and evaporation involve similar energy exchanges and are difficult to measure separately, they are often considered jointly as *evapotranspiration*. Figure 3.6 shows that, on the average, 22 units of energy are transferred from the earth's surface to the troposphere by evapotranspiration. This energy is stored in the water vapor in the form of latent heat. It becomes sensible heat, capable of warming the surrounding air, only when the water vapor condenses back into water droplets or ice crystals elsewhere in the atmosphere.

Heating of the atmosphere by evapotranspiration, conduction, and convection offsets the net radiation gain of the earth's surface and the net radiation loss of the atmosphere, balancing the global energy budget. A final component, plant photosynthesis, also consumes solar energy, but utilizes too small a fraction to be of concern in an analysis of the energy balance of the earth-atmosphere system.

Latitudinal Differences in Solar Income and Radiational Losses

The energy balance of the earth-atmosphere system shown in Figure 3.6 represents an annual average for the system as a whole. If we look at specific locations on the earth, however, we discover that energy exchanges are not in balance. Figure 3.9 shows that equatorial and tropical regions receive much more solar energy than they return to space through reflection and longwave radiation. Polar regions, on the other hand, lose more energy through reflection and longwave radiation than they gain from solar radiation.

These latitudinal imbalances in energy gains and losses are the cause of the atmospheric and oceanic circulations that are discussed in Chapter 4. Flows in the atmosphere and oceans transport warm air, latent heat in

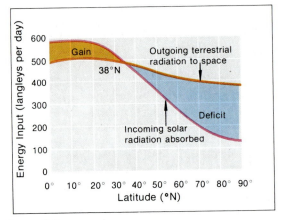

Figure 3.9
The graph shows average annual global values of absorbed solar radiant energy and emitted longwave radiation at each latitude in the northern hemisphere for the earth's surface and the atmosphere together. (Doug Armstrong after F. K. Hare, *The Restless Atmosphere,* 1966, Hutchinson Publishing Group, Ltd.)

water vapor, and warm water into high latitudes, and colder air and water toward the equator. The temperature in the lower latitudes therefore rises until the rate of heat flow toward the poles is sufficient to carry away the excess energy. As Figure 3.10 indicates, there is a strong seasonality to the poleward flux of energy. In the winter hemisphere only the zone extending about 20° beyond the equator receives more energy from radiation than it loses, but in the summer hemisphere all regions gain more radiational energy than they lose. The temperature difference between low and high latitudes is greater during the winter than during the summer. The temperature differential helps to power atmospheric motion and a poleward flow of energy from the tropics.

ENERGY BALANCES AT THE EARTH'S SURFACE

The large-scale energy exchanges between the sun, the earth-atmosphere system, and

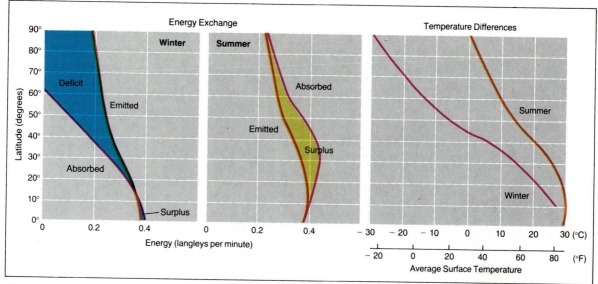

Figure 3.10
Energy exchange and temperature in the northern hemisphere. The diagrams on the left show the average rates at which the earth and the atmosphere absorb solar radiant energy and emit longwave radiation to space during the winter and summer seasons.

The diagram on the right shows the average temperature of the atmosphere during winter and summer at a height of approximately 2 meters above the surface. (Doug Armstrong after Herbert Riehl, *Introduction to the Atmosphere,* 3rd ed., © 1978 by McGraw-Hill Book Company)

space have been measured with increasing accuracy in recent years by a number of specialized satellites. Our knowledge of local energy balances at the earth's surface is based on more routine measurements at various experimental stations. By interpreting these data according to the principles of energy exchange discussed earlier in this chapter, scientists are gaining a better understanding of how the energy balances at particular locations are affected by such factors as cloud forms, vegetation cover, local water bodies, and human activities.

A study of surface energy balances usually includes evaluation of the radiation budget—the income and outgo of shortwave and long-

wave radiation at the surface. The net gain or loss of radiative energy at the earth's surface is generally called the *net radiation,* which is normally positive during the daytime (heating) and negative at night (cooling). The all-inclusive energy balance contains, in addition to the radiation budget, the other energy balance components: latent heat (evapotranspiration) and sensible heat (measurable heating of the air and soil).

Local Energy Balances

Figure 3.11 provides examples of hourly patterns of energy balance factors at Hancock, Wisconsin, representative of humid midlatitude locations. During the sunny daytime hours, more shortwave and longwave radiation is received at the surface than is reflected or radiated back to the atmosphere or into space. Following sunrise, the rate of incoming solar radiation rises rapidly, resulting in a net radiation gain that peaks near solar noon, and returns to zero a few minutes before sunset. Thus, during the day, the radiation budget at Hancock is said to be positive. The daytime net

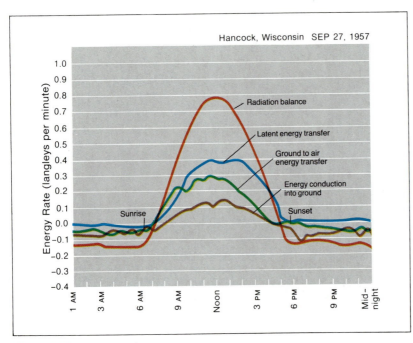

Figure 3.11

The local energy balance and its principal components during a 24-hour period are shown for Hancock, Wisconsin, a moist midlatitude location. For net radiation, the sign convention is positive (+) for a radiational gain and negative (−) for a radiational loss.

Energy cannot accumulate at the surface, and the excess or deficiency of radiant energy can be accounted for as the sum of three components: the latent heat removed by evapotranspiration of water; the sensible heat flowing from the surface into the air by conduction and convection; and sensible heat that flows into the soil from the surface by conduction. For these three components, however, the sign convention is just the opposite of that for radiation. Energy flows away from the surface are treated as positive (+), and energy flows to the surface as negative (−).

At Hancock at midday, latent heat transfer is positive because of evaporation and transpiration from plants. The small negative value of latent heat transfer before sunrise indicates condensation of water vapor on the ground as dew. Heat transfer from the ground to the air is positive at midday, which indicates that surface heat is warming the air. Overnight, the conduction of heat from the soil to the surface is negative, indicating that heat is flowing by conduction from the warmer soil to the cooler surface. (Doug Armstrong after William D. Sellers, *Physical Climatology,* © 1965, The University of Chicago Press)

radiation gain is used for evapotranspiration, atmospheric heating, and heating of the soil. The portion used in plant photosynthesis is negligible.

At night, however, somewhat more long-wave radiation is lost to the atmosphere than is received from it, so the radiation budget of the cooling surface is negative. Small amounts of energy are conducted downward from the warmer atmosphere and upward from the warmer subsoil to the cooling surface. Thus the lower atmosphere is heated by the surface during the day, when the surface radiation budget is positive, and cooled by the surface at night, when the surface radiation budget is negative.

In Figure 3.11, evapotranspiration is also seen to be closely controlled by the net radiation. This is evident in the curve for latent heat transfer due to vaporization of water. Evapotranspiration is primarily a daytime phenomenon; there is little or no vaporization of water at night. In a dry climatic region little soil moisture is available for evapotranspiration,

CHAPTER 3

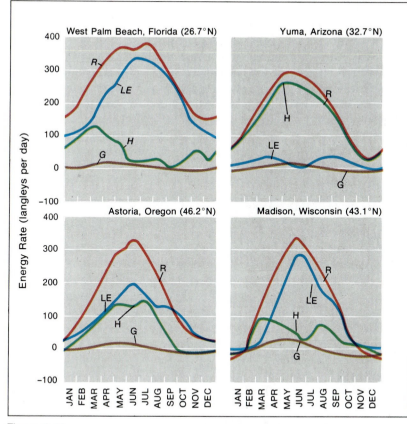

Figure 3.12

The graphs show the annual regime of the local energy balances at four places in the United States. Madison, Wisconsin, is representative of humid midlatitude regions with low solar radiation income during winter; West Palm Beach, Florida, of humid subtropical regions with moderate solar radiation income during winter; Astoria, Oregon, of humid and cloudy maritime climates with much winter cloudiness; and Yuma, Arizona, of hot desert regions with little cloudiness.

R is the radiation balance (solar energy received vs. longwave energy emitted). LE is the energy removed by evapotranspiration (latent heat transfer). H is the energy exchanged between the ground and the air by conduction and convection. G is energy conducted into the ground from the surface. (Doug Armstrong after William D. Sellers, *Physical Climatology,* © 1965, The University of Chicago Press)

and the net radiation gain goes almost entirely into sensible heat—the heating of the air and soil. On the average, equal energy inputs will cause dry regions to be warmer during daylight hours than regions with high rates of evapotranspiration.

Figure 3.12 presents annual energy balances for four contrasting locations within the continental United States. Significant differences may be observed among these locations on the basis of latitude, yet it is apparent that the balances also vary according to other environmental factors, such as moisture availability. The greatest contrast is between Madison, Wisconsin, and Yuma, Arizona. At Yuma, a desert location, very little of the net radiation gain can be used up by evapotranspiration, so nearly all is available for heating the air and soil. As a consequence, Yuma is well known as one of the hottest year-round locations in the United States. By contrast, at Madison and West Palm Beach, Florida, abundant moisture in the summer permits high evapotranspiration rates, so less energy is available to heat the soil and air. Astoria, Oregon, is intermediate between these extremes.

Differential Heating and Cooling of Land and Water

Land surfaces and water bodies respond differently to equal energy inputs. This is because the heat capacities of land and water are very different. As we saw in Chapter 2, water can absorb much heat without increasing its temperature greatly, whereas land heats rapidly as it receives energy. Whereas the heat capacity of water is about 1, the heat capacity of most land surfaces is from 0.2 to 0.4. For a given energy input an area of ground will increase in temperature three to five times more than an equal area of water.

On a summer day, the temperature of a soil surface exposed to the sun can easily reach 40°C (104°F). Dry beach sand and asphalt pavement may become too hot for bare feet at more than 50°C (122°F). Most of the energy gain is concentrated near the surface, since conduction to greater depths is relatively slow. At night, the soil surface cools rapidly by longwave radiation and, in turn, cools the atmosphere immediately above by conduction. If there is little wind to mix the lowest layers of air, the air near the ground will become several degrees cooler than the air a few meters higher. Vertical profiles of air and ground temperatures during a fair winter day in southern New Jersey are shown in Figure 3.13. Between 9:00 A.M. and 1:00 P.M., the warmest temperatures were within the first 5 cm above the surface, but during late afternoon, overnight, and early morning the coldest temperatures were immediately above the surface, which was losing heat by longwave radiational cooling.

Surface temperatures of large water bodies change much less during the year than land surface temperatures. In addition to the greater heat capacity of water, the energy gain of a water surface is distributed to much greater depths. Solar radiation penetrates water to a depth of tens of meters, and waves and currents also distribute heat energy downward.

Local Temperature Variations

Diurnal Regimes

The physical principles underlying the energy budget can help explain temperature variations on a local scale. Figure 3.14 shows air temperatures for a period of several days at Baton Rouge, Louisiana. Air temperature near the ground tends to reflect surface temperature. In the first three days clear weather permitted ground and air temperatures to rise markedly during the day and to fall sharply at night. The lowest temperatures occurred just after sunrise, and the highest followed in mid-afternoon. Although the solar radiation input peaks near noon and then declines, cooling does not begin until late afternoon, when radiation losses begin to exceed gains. During cloudy weather the temperature change throughout the day is much less. Clouds reduce incoming solar radiation during the day, and at night they trap outgoing longwave radiation and return much of it to the surface, smoothing out the daily temperature variation.

A large body of water also tends to reduce daily temperature ranges over adjacent land areas because the temperature of large water bodies changes little from day to day. Latent heat transfer between the water and the atmosphere helps maintain uniformly mild temperatures at coastal cities in comparison with cities in continental interiors.

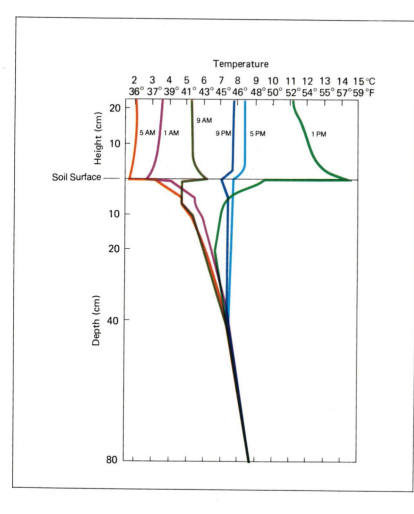

Figure 3.13
This figure illustrates steep temperature gradients immediately above and below the ground surface during fair days with little wind. The data were taken at an experimental site near Seabrook, New Jersey, during December 5, 1956. (After J. R. Mather, *Climatology: Fundamentals and Applications,* © 1974 by McGraw-Hill Book Company)

Annual Regimes

Monthly temperature variations reflect the same influences as daily variations. Figure 3.15 shows winter and summer temperatures at San Diego, California, a coastal city; Elko, Nevada, an interior city in an arid region; and Cleveland, Ohio, a midcontinent city with frequent cloud cover. The graphs show the distribution of temperatures for January and July, which is found by measuring the temperature every hour and plotting the number of hours each temperature occurred.

Because of the reduced energy input in winter, the January temperature distributions in all three cities fall well below the distributions for July. The January and July distributions for San Diego show the most overlap and the least separation between peaks because coastal temperatures tend to be uniform throughout the year. The monthly temperature distributions at Cleveland and Elko cover a greater range than those at San Diego because of the greater daily variation of temperature in the continental interior. The wide range of the

ENERGY AND TEMPERATURE

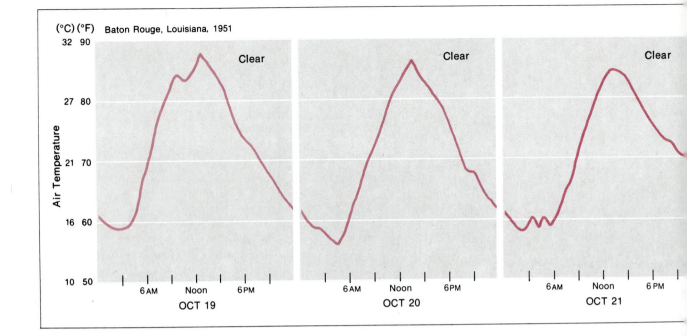

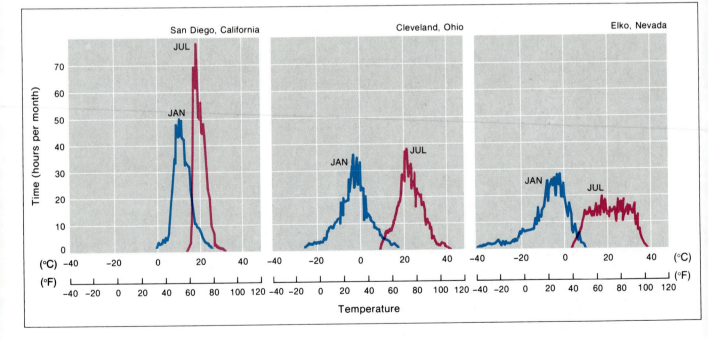

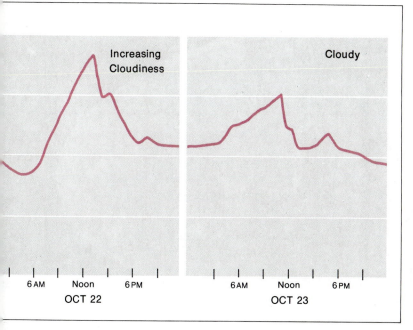

Increasing Cloudiness

Cloudy

| 6 AM | Noon | 6 PM |

OCT 22

| 6 AM | Noon | 6 PM |

OCT 23

Figure 3.14
The temperature records at Baton Rouge, Louisiana, for October 19 to 23, 1951, illustrate the daily variation in air temperature as the ground heats and cools under increasingly cloudy conditions.

Figure 3.15 (opposite)
The average temperature distributions during January and July are shown for three cities in different climatic regions over the five-year period 1935 through 1939. Each point on a temperature distribution tells the number of hours during the given month that the specified temperature was recorded; in San Diego in July, for example, a temperature of 20°C (68°F) was recorded in 57 separate hours. (Doug Armstrong after Arnold Court, *Journal of Meteorology, 8,* 1951, American Meteorological Society)

July temperatures for Elko reflects the city's high elevation (1,500 m; 5,000 ft) and desert location with minimal cloud cover, producing hot days and cool nights.

Average global temperatures for January and July are shown in Figure 3.16, and seasonal temperatures for the United States are depicted in Figure 3.17. In winter (January in the northern hemisphere, July in the southern hemisphere), temperatures generally depend on latitude and decrease regularly from the equator toward the poles. Note, however, the continental and maritime effects on temperature over North America and the North Atlantic during January. In summer, temperatures depend less strongly on latitude and more on land and water contrasts.

Modifications of Energy Balances

Changes in any of a number of factors can alter local energy balances. Cloudiness, atmospheric pollution, surface albedo, irrigation, vegetation—all of these influence the utilization of radiant energy. Results of human activities, both unintentional and planned, have modified local energy balances, giving rise to measurable changes in climate in both rural and urban areas.

Frost Protection

On the local scale, gardeners and orchard growers are concerned with the prevention of frost damage to plants. On clear nights the

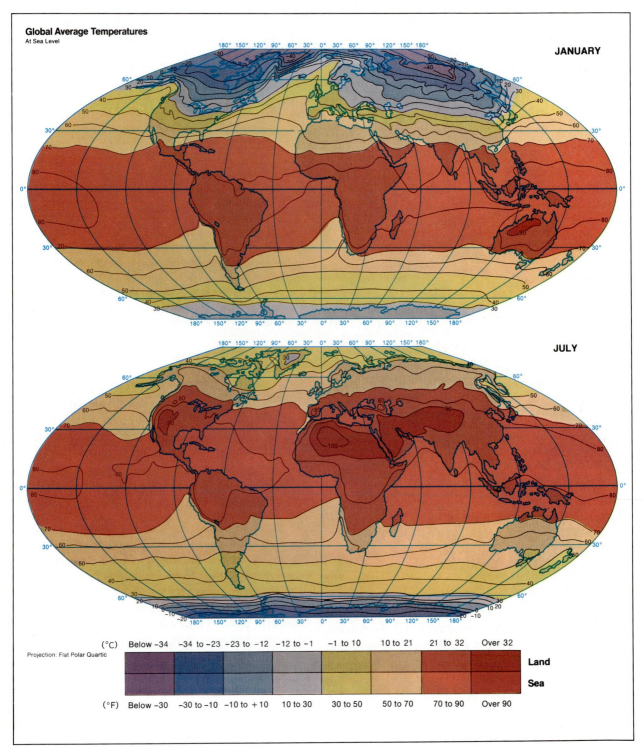

Global Average Temperatures
At Sea Level

JANUARY

JULY

Projection: Flat Polar Quartic

(°C)	Below −34	−34 to −23	−23 to −12	−12 to −1	−1 to 10	10 to 21	21 to 32	Over 32	
									Land
									Sea
(°F)	Below −30	−30 to −10	−10 to +10	10 to 30	30 to 50	50 to 70	70 to 90	Over 90	

ground rapidly loses heat to the atmosphere by longwave radiation, and return radiation from the atmosphere is minimal. During early spring and late fall, such conditions can cause temperatures near the ground to drop below the freezing point. Orchard growers often use large fans or even helicopters to disturb the cold air near the ground and mix it with warmer air above. These devices have largely replaced orchard heaters, or smudge pots, which produce a blanket of smoke to absorb and reradiate longwave radiation back to the ground. Radiant heaters are sometimes used to protect plants in courtyards and around patios. Often ornamental plants and small citrus trees are covered with plastic sheeting to reduce longwave radiation losses that lead to frost damage.

Urban Heat Island

Large urban areas produce their own distinctive climates by modifying local energy balances. Especially in winter, cities can be as much as 7° to 8°C (12° to 14°F) warmer than the surrounding countryside. Towering buildings and canyon-like streets trap above-average amounts of radiant energy, even when the sun is low in the sky. Asphalt, concrete, and bricks store sensible heat that would be used for evapotranspiration in a natural or agricultural environment. The burning of fossil fuels in homes, factories, and automobiles adds to the radiant heat input. Dust, smoke particles, and pollutants in the air trap outgoing longwave radiation and return it to the surface.

Figure 3.16 (opposite)
Global temperature distributions are shown for January and July. Temperatures have been reduced to approximate sea level values by applying a correction for the decrease in temperature with increased altitude. (Andy Lucas and Laurie Curran after Glenn Trewartha, *Introduction to Climate*, © 1968, McGraw-Hill Book Company and *Goode's World Atlas* © 1970, Rand McNally & Company)

These effects combine to produce the phenomenon of the *urban heat island* (Figure 3.18).

Other Human Effects on Energy Balances

Human modification of energy balances did not begin with the Industrial Revolution. Ancient agricultural practices, such as tropical slash-and-burn agriculture, midlatitude forest cutting, and overgrazing in subtropical latitudes, have altered surface albedo, the water-holding capacity of soils, and evapotranspiration rates over vast areas. Removal of natural vegetation for agricultural purposes has allowed the wind to lift soil particles high into the troposphere, causing dust storms that darken the sky over areas of thousands of square kilometers.

The modern industrial age has intensified the human impact on energy balances. Jet aircraft leave exhaust products and condensation trails high in the troposphere and even in the lower stratosphere. Jet condensation trails resemble some natural clouds and may significantly reduce shortwave radiation in regions of heavy air traffic. Perhaps more disturbing are findings that various synthetic chemicals, such as the chlorofluorocarbons in spray-can products and the methyl chloroforms in solvents, may have the potential to destroy the ozone layer in the stratosphere that prevents dangerous amounts of ultraviolet radiation from reaching the earth's surface.

At the surface artificial ponds and reservoirs, paved areas, forest removal, and the drainage of marshes all have altered albedo and evapotranspiration patterns. Data from satellite surveys of albedo, land use, and plant cover, and of the degree of cloudiness over the entire surface of the earth, are required to assess the effects of these changes. The launching of the Earth Resources Technology Satellite, ERTS-1, in the summer of 1972, was an important first step in this program, which has continued through data transmissions from a series of orbiting satellites, such as LANDSAT 1, 2, 3, and 4, and SEASAT.

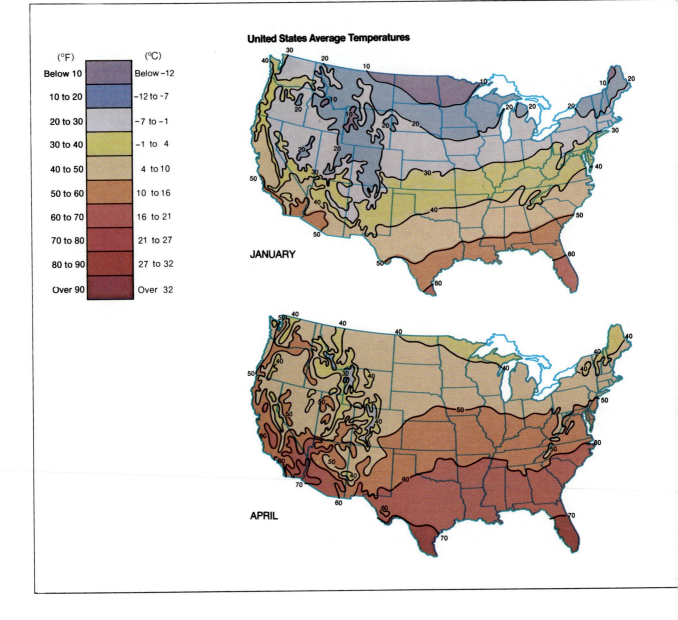

United States Average Temperatures

JANUARY

APRIL

(°F)

Below 10
10 to 20
20 to 30
30 to 40
40 to 50
50 to 60
60 to 70
70 to 80
80 to 90
Over 90

(°C)

Below −12
−12 to −7
−7 to −1
−1 to 4
4 to 10
10 to 16
16 to 21
21 to 27
27 to 32
Over 32

CHAPTER 3

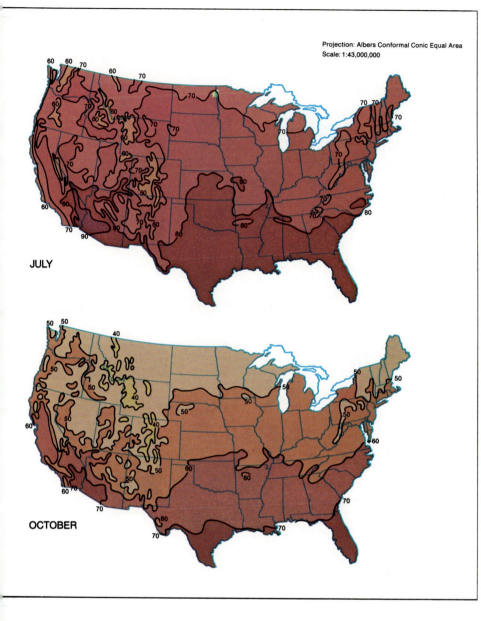

Projection: Albers Conformal Conic Equal Area
Scale: 1:43,000,000

JULY

OCTOBER

Figure 3.17
United States temperature
distributions (in °F) are shown
for selected months. (Andy
Lucas after *The National Atlas
of the United States of
America,* 1970)

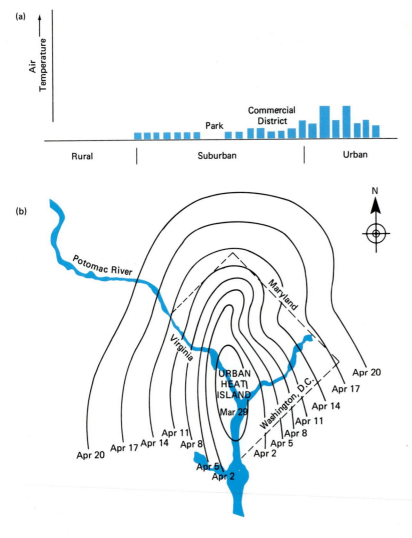

Figure 3.18
All metropolitan regions have urban heat islands where temperatures are significantly higher than over surrounding rural regions.

(a) An idealized temperature cross section over an urban region. (After T. R. Oke, *Boundary Layer Climates*, 1978, Methuen)

(b) Average dates of the latest freezing temperature in spring in the Washington, D.C., metropolitan region. (After Clarence A. Woollum, *Weatherwise*, Vol. 17, 1964)

KEY TERMS

energy flux
electromagnetic radiation
electromagnetic spectrum
shortwave radiation
longwave radiation
perihelion
aphelion
circle of illumination
parallels of latitude
meridians of longitude

prime meridian
tropic of Capricorn
tropic of Cancer
winter solstice
summer solstice
Arctic Circle
Antarctic Circle
vernal equinox
autumnal equinox
solar constant

langley
ozone
troposphere
stratosphere
albedo
insolation
greenhouse effect

conduction
convection
transpiration
evapotranspiration
net radiation
urban heat island

REVIEW QUESTIONS

1. Describe the electromagnetic spectrum. What is the relation of temperature to wavelength?
2. What factors determine the amount of solar radiation reaching the top of the earth's atmosphere?
3. Describe the variation of day length from pole to pole at the winter solstice.
4. Describe how the atmosphere intercepts and interacts with incoming solar radiation.
5. Discuss the various ways the surface of the earth loses energy.
6. What is the role of the atmosphere in heating the earth?
7. What local environmental conditions modify a region's energy budget?
8. In energy-budget terms, why does the maximum daily air temperature tend to occur between 2:00 and 6:00 P.M.?
9. Compare the utilization of net radiation in humid and in dry climates.
10. Explain the effects of landmasses and oceans on the seasonal regimes of air temperatures.
11. What are various ways in which the energy balance of the earth-atmosphere system might be modified by human action?

APPLICATIONS

1. Using Figure 3.4, plot the annual regime of solar radiation at the top of the atmosphere at the latitude of your campus. Compare this with the regimes at latitudes 15 degrees to the north of your location and 15 degrees to the south of your location. Explain the differences in the three regimes.
2. Estimate how the land use and land cover of your campus area have changed from presettlement days to the present. How would these landscape changes have affected the albedo of the campus area as a whole?
3. Over which regions of the earth would you expect the greatest seasonal changes in albedo?
4. Local daily temperature highs and lows are usually given on radio and TV and in newspapers: highs referring to late afternoon, and lows occurring in early morning hours. Keep track of the daily temperature range and the degree of cloud cover over a month-long period. How much do varying degrees of cloudiness affect the daily temperature range? Are other factors involved?
5. On a clear evening with no breeze, record the variations in temperature in different areas of your campus. You will need a thermometer that responds quickly to temperature changes (check with your instructor). Explain the pattern of temperature variations you measure.

6. On both a clear day and a cloudy day, make a record of hourly temperatures at an unshaded ground surface and compare these readings with hourly air temperatures 5 feet above the ground. Make the same experiment measuring temperatures on bare ground and grassy surfaces. Explain your results.

FURTHER READING

Gedzelman, Stanley David. *The Science and Wonders of the Atmosphere.* New York: John Wiley (1980), 535 pp. This beautifully illustrated book emphasizes introductory-level explanations of physical processes. Chapters 3, 4, and 8 include very helpful examples of earth-sun relationships, radiation laws, and atmospheric optics.

Geiger, Rudolf. *The Climate Near the Ground.* (Trans. of 4th German ed.) Cambridge, Mass.: Harvard University Press (1965), 611 pp. This classic stresses the results of field studies on each of the continents, with particular emphasis on radiation and temperature. Geiger is sometimes called the "father of microclimatology."

Mather, John R. *Climatology: Fundamentals and Applications.* New York: McGraw-Hill (1974), 412 pp. Chapter 2 includes basic discussions of radiation and temperature, instrumentation and data, as well as their applications and limitations.

Miller, David H. "A Survey Course: The Energy and Mass Budget at the Surface of the Earth." Publ. No. 7, Comm. on College Geog., Assoc. of American Geog. (1968), 142 pp. This very useful monograph is organized into study units that proceed from energy exchange processes through local energy budgets to regional synthesis. Emphasis is placed on professional papers from almost every region of the globe.

Oke, T. R. *Boundary Layer Climates.* London: Methuen (1978), 372 pp. This text is a more advanced analysis of surface energy exchanges. Emphasis is placed on local and geographical differences, and the text is an excellent primer for students who want to do special studies.

Sellers, William D. *Physical Climatology.* Chicago: University of Chicago Press (1965), 272 pp. This work focuses almost entirely on fundamental analyses, in both descriptive and mathematical forms, of the radiation and energy budgets at the earth's surface.

Trewartha, Glenn T., and **Lyle H. Horn.** *An Introduction to Climate,* 5th ed. New York: McGraw-Hill (1980), 416 pp. Focus on geographical distributions of solar radiation and temperature is found in Chapters 2 and 8–12.

A furious storm at sea emphasizes the complex
interactions of the atmospheric and oceanic systems.
Winds drive the surface ocean currents, and the
ocean temperatures influence the circulation patterns
of the atmosphere.

Snow Storm—Steam Boat off a Harbour's Mouth Making Signals in Shallow Water and Going by the Land by William Turner.
(The Tate Gallery, London)

4
General Atmospheric and Oceanic Circulations

A balloonist can determine only whether to rise higher or sink lower. But in 1979 and 1980 balloon trips were completed across the widths of both the Atlantic Ocean and North America. How can balloonists, with no control over their horizontal direction, decide that they are going to cross a continent, or an ocean? The answer is easy: the motion of the atmosphere does the deciding. In the middle latitudes all long-distance balloon trips go from west to east because the general motion of the atmosphere is in that direction. It is as impossible to go the other way as it is for a raft to drift up a river instead of down it. Similarly, a capped bottle that is thrown into the sea off the Florida coast will probably come ashore in Iceland, Norway, or Ireland, never in Brazil, or even nearby Cuba. To send a message to Cuba in a floating bottle, the best starting point would be Morocco; objects floating ashore in Brazil come from South Africa or Angola.

On a global scale, both the air and the seas move in paths that vary somewhat from season to season but are usually dependable from one year to another. The large-scale semipermanent pattern of atmospheric flow, which drives the oceanic circulation, is one of the most important facts of physical geography and is known as the *general circulation*.

The general circulation of the atmosphere and the oceanic circulation that results from it are the principal mechanisms by which energy is transferred from equatorial and tropical regions having net solar energy gains to high-latitude regions having net energy losses to space. These flows of air, carrying both latent and sensible heat, along with the ocean water carrying sensible heat, maintain the earth's thermal balance at all latitudes. The general circulation also delivers to the continents much of the water that evaporates from the oceans. It determines local weather and is the foundation for the entire pattern of global climates. This chapter explains the general circulation of the atmosphere and the resulting oceanic circulation.

FORCES CAUSING ATMOSPHERIC MOTION

The atmosphere is in ceaseless motion, whether churning with furious storms or drifting so slowly that hardly a breeze is felt. Although the details of atmospheric circulation are complex, the general features can be understood by looking at the forces involved.

Any force may be thought of as a push or a pull. According to the fundamental laws of motion developed in the seventeenth century by the English physicist Isaac Newton, a moving object's speed and direction of motion cannot change unless a force is made to act on the object. Horizontal motion is influenced only by forces pushing or pulling in the horizontal direction, and vertical motion is influenced only by forces acting in the vertical direction. Although the force of gravity pulls every parcel of air downward toward the earth, gravity has no direct effect on the horizontal motion of air. The forces that act on a parcel of air moving horizontally in the atmosphere are the pressure gradient force, the Coriolis force, and friction.

The Pressure Gradient Force

Pressure is force per unit area. If the pressure on one side of a parcel of air is greater than the pressure on the other side, the parcel will be pushed toward the area of lower pressure (Figure 4.1a). Recall from Chapter 2 that differential surface heating on the earth produces horizontal differences of atmospheric pressure. The greater the difference in pressure, the greater the net push and the more rapid the resulting motion. This "push," caused by the horizontal difference in pressure across a surface, is called the *pressure gradient force*.

On a weather map, meteorologists represent atmospheric pressure with lines called *isobars*. Isobars connect locations on the map that have equal atmospheric pressures (see Figure 4.1d). The pressures shown do not correspond to measured values: they have been corrected to sea level values to compensate for the lower pressures measured at higher elevation weather stations. The pressure gradient force is at right angles to the isobars, and is strongest where the isobars are most closely spaced—that is, where the pressure decreases most rapidly. Surface wind speeds and directions are determined by the interactions of the pressure gradient force, the Coriolis force, and the friction force, with wind speeds strongest where isobars are packed most closely together.

The Coriolis Force and the Geostrophic Wind

If it were not for the rotation of the earth, winds would simply follow the pressure gradients. But the earth's rotation complicates the motion of the atmosphere. Except over the equator, air moving down the pressure gradient is deflected, or turned, from straight paths when viewed from the surface of the earth.

Coriolis Force

Figure 4.2a shows the initial trajectories of rockets launched to the north and to the south

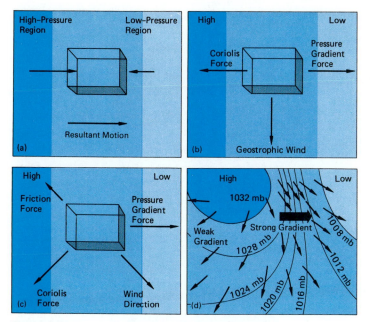

Figure 4.1

Effects of pressure differences on air movement.

(a) The pressure gradient force pushes a parcel of air away from high pressure toward low pressure.

(b) A parcel of air aloft moving horizontally on the rotating earth experiences a Coriolis force and a pressure gradient force. The Coriolis force acts at right angles to the direction of motion of the parcel, and will continue deflecting the parcel of air until it is moving parallel to the isobars. Air motion parallel to isobars is known as the *geostrophic wind.* This diagram is drawn for the northern hemisphere.

(c) Friction acts on air parcels moving near the ground, reducing wind speeds. Because the Coriolis force is proportional to the wind speed, the deflection is less than for the geostrophic winds above.

(d) Isobars are lines on weather maps connecting places with equal atmospheric pressure (adjusted to sea level). In this diagram pressures are shown in millibars (mb). Wind speeds are strongest where the pressure gradient is strong, as shown by isobars packed closely together.

from the central United States. As the rockets proceed toward the north and south as viewed from space, the surface of the earth rotates and turns underneath them; in other words, the east-west and north-south system of parallels and meridians turns underneath the rockets,

or *the moving air.* The right-hand diagram in the figure shows that from the earth's surface, the rockets, or the moving air, appear to be deflected to the right, as you look downwind. Figure 4.2b shows that there is a similar apparent deflection to the right when the moving air flows initially toward the east or west.

This deflection is known as the *Coriolis force* after Gaspard Coriolis, the nineteenth-century French engineer who first explained the phenomenon mathematically. Even though no real force is operating, the apparent deflection resulting from the earth's rotation affects all objects, such as air and ocean water, moving freely over the earth's surface. A correction for the Coriolis force must be added when aiming missiles and even long-range artillery.

The Coriolis force acts at right angles to the direction of the motion. In the northern hemisphere the apparent deflection is to the right and in the southern hemisphere to the left, so that the pattern of deflection in the two hemispheres is symmetrical. It is important

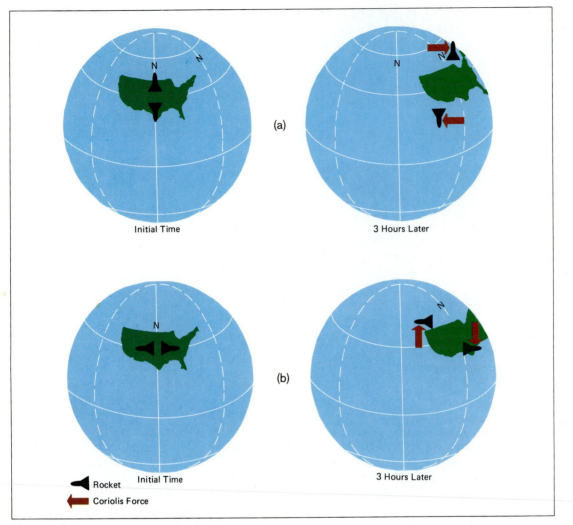

Initial Time

3 Hours Later

(a)

Initial Time

3 Hours Later

(b)

◄ Rocket

◄ Coriolis Force

Figure 4.2
The Coriolis force is associated with the rotation of the spherical grid or parallels and meridians as the earth rotates on its axis from west to east.

(a) Rockets are launched from the central United States to either the north or south. The diagram on the right, for some time later, shows how the grid of parallels and meridians rotates under the rockets, causing their paths to appear deflected to the right.

(b) These diagrams illustrate similar deflections to the right when the rockets are launched into east or west trajectories. (From Joe R. Eagleman, *Meteorology: The Atmosphere in Action.* New York: D. Van Nostrand Co., 1980)

to remember that deflection must always be thought of in terms of looking downwind in the direction toward which the air is moving. Air flowing toward the equator is turned to the west in both hemispheres, and air moving toward the poles is turned to the east. Likewise, air moving to the west is turned toward the poles, and air moving toward the east is turned toward the equator. The Coriolis force is zero at the equator. The deflection becomes significant in subtropical latitudes, and increases to a maximum at the poles. The Coriolis force is also proportional to the speed of the moving air, so that the deflection is greater with stronger winds.

Geostrophic Wind

In the upper troposphere, the Coriolis force is approximately equal to the pressure gradient force, causing air to flow at right angles to the pressure gradient force and parallel to isobars (Figure 4.1c). Meteorologists refer to airflow that is parallel to isobars as the *geostrophic wind.* The circulation of upper-level air, above the frictional effects of the earth's surface, approximates geostrophic flow. Technically, the geostrophic wind is limited to situations with straight isobars, but the airflow aloft does maintain a direction approximately parallel to curved isobars also.

Buys Ballot's Law

In the middle and higher latitudes of the northern hemisphere, the Coriolis deflection causes air to flow in a clockwise direction around high-pressure areas and in a counter-clockwise direction around low-pressure areas (Figure 4.1b and Figure 4.3). Using these facts, the Dutch meteorologist Buys Ballot in 1857 described the relationship between pressure systems and the direction of air flow aloft in simple terms. *Buys Ballot's law* states that if one's back is to the geostrophic wind in the northern hemisphere, low pressure is to the left and high pressure is to the right. In the southern hemisphere, low pressure is to the right, high pressure to the left.

The geostrophic winds of the upper atmosphere are especially important to meteorolo-

Figure 4.3
This figure shows the relationships between upper and surface winds for both northern and southern hemispheres showing the influence of the pressure gradient force, the Coriolis deflection, and the force of friction. (After Arthur N. Strahler, *Introduction to Physical Geography,* 1st ed., 1951, John Wiley & Sons)

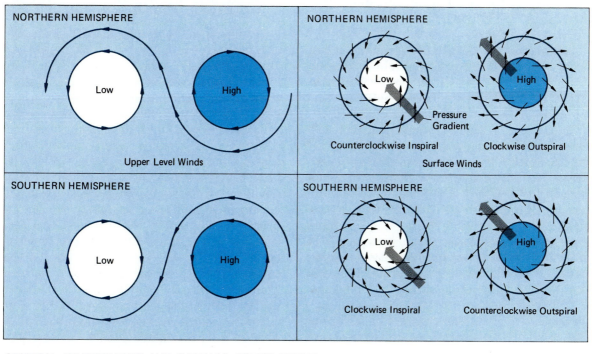

gists concerned with weather forecasting, for they influence surface air movements, as will be seen later in this chapter. Upper air measurements of pressure, wind speed, and wind direction, as well as temperature and humidity, are taken twice a day at more than 100 weather stations across the continental United States. From these measurements, meteorologists construct weather maps that estimate pressure and wind at various levels of the atmosphere over the United States. These projections are the basis for predicting weather as much as 72 hours in advance.

The Force of Friction

If the energy sources that drive the atmospheric circulation were suddenly to disappear, friction between the atmosphere and the earth's surface, and within the atmosphere itself, would cause all atmospheric motion to slow and eventually cease. Scientists regard friction as a "force" that is opposed to the motion of masses on the earth or in the atmosphere. For a parcel of air to maintain its movement, the driving force must overcome frictional resistance.

The effect of friction is at its maximum at the earth's surface and decreases upward. Recall that the Coriolis force is proportional to the speed of the object. It follows that near the ground, where surface winds are slowed by frictional forces, the Coriolis effect is reduced, while the pressure gradient force is not affected. Consequently, instead of flowing parallel to pressure isobars, air near the surface responds to the pressure gradient force by flowing from higher to lower pressure at an angle across the isobars (Figure 4.1c). This causes surface air to spiral *into* centers of low pressure and to spiral *out of* centers of high pressure. In the northern hemisphere, the inward spiral is counterclockwise, the outward spiral clockwise. In the southern hemisphere, the reverse is true (see Figure 4.3). Relating Buys Ballot's law to surface winds, if your back

is to the wind in the northern hemisphere, and you rotate 45° to the right, low pressure will be on your left, high pressure on your right. In the southern hemisphere, rotate 45° to the left, and low pressure will be on the right, high pressure on the left (Figure 4.3).

THE GENERAL CIRCULATION OF THE ATMOSPHERE

The pressure gradient force, the Coriolis force, and the force of friction are the basic factors that determine atmospheric motion. But other large-scale mechanisms are also involved in the atmospheric circulation. The following sections describe a generalized model of the global circulation of the atmosphere. This model represents a sort of grand average for one year; the actual circulation at any moment may be somewhat different.

Pressure Patterns and Winds on a Uniform Earth

To understand the general circulation of the atmosphere, it is helpful to think first of the atmospheric circulation that would exist if the earth's surface were absolutely uniform. Let us start by also assuming that there is no Coriolis force. Under such conditions, a low-pressure belt would develop over the equatorial region, where there is excessive radiational heating. A high-pressure belt would develop over the polar regions, where there is excessive radiational cooling. The air would rise over the low latitudes and subside over the poles. The atmospheric circulation would consist of a surface flow of air from the high-pressure regions at the poles to the low-pressure zone at the equator. To complete the circulation loop there would be a return flow of air toward the poles in the upper atmosphere.

Such an enormous vertical convective cell is sometimes called a *Hadley cell,* after the Eng-

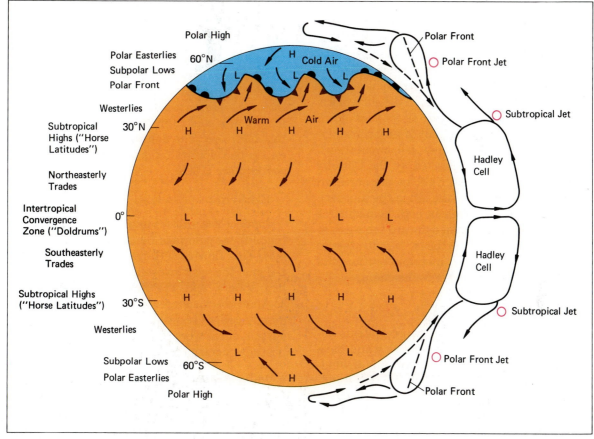

Figure 4.4
This sketch is a highly diagrammatic representation of the general circulation over a homogeneous earth with a water surface. The Coriolis effect is included. Semipermanent pressure and wind belts are shown on the surface of the sphere, and a much enlarged vertical cross section is shown on the right hemisphere only. The polar front is shown only for the northern hemisphere, and it is discussed in more detail in later sections of this chapter.

lish meteorologist who proposed this model of circulation in 1735.

Hadley Cells

The right side of Figure 4.4 shows a vertically exaggerated cross-section of the Hadley cells north and south of the equator. Note that modern interpretations of the general circulation restrict the Hadley cell regimes in each hemisphere to the lower latitudes between the equator and about 30° latitude. As the meridians of longitude converge poleward, so does poleward moving air. Thus cooling and convergence aloft force air downward in the subtropical latitudes.

If we add the Coriolis force to this circulation model, the circulation pattern is altered. Excessive radiational heating at the equator still produces rising air that flows out toward the poles aloft. Once away from the equator, however, the poleward-flowing upper air in the northern hemisphere is turned to the right by the Coriolis force, producing a westerly (west-to-east) subtropical "jet stream." Leftward deflection of poleward-moving upper-level air in the southern hemisphere creates a westerly subtropical jet stream there as well. The air aloft cools by longwave radiation to space and a portion sinks into the lower atmosphere at about 30° latitude in each hemisphere. This produces high-pressure zones in the subtropics. Much of this descending air flows back down the surface pressure gradient toward the equator, completing the Hadley cell circulations. This surface flow is deflected by the Coriolis force, again toward the right in the northern hemisphere and toward the left in the southern hemisphere (Figure 4.4). In both hemispheres the deflected flow is turned toward the west, so that this flow is from northeast to southwest in the northern hemisphere and from southeast to northwest in the southern hemisphere.

Subtropical Highs

The surface pressure and wind systems described above exist in a general way and have traditional names dating from the era of wind-dependent sailing vessels. The belts of descending warm dry air, fair weather, and weak surface wind in the subtropics, called the Calms of Cancer and the Calms of Capricorn, were also known as the *horse latitudes,* or in meteorological terms, the *subtropical highs.* Although other interpretations of this term have been made, it is generally believed that when sailing ships were becalmed here, and supplies of food and drinking water dwindled, horses were the first passengers to go overboard. Therefore, floating horse carcasses were sometimes sighted in this zone.

Trade Winds

Winds are traditionally named according to the direction from which they blow. Between the horse latitudes and the equator in the northern and southern hemispheres lie zones characterized respectively by northeasterly and southeasterly *trade winds.* The "easterly trades" are among the earth's steadiest and most persistent winds, providing sea traders in sailing vessels with reliable westward routes over the oceans.

Intertropical Convergence

Near the equator the trade winds converge into a low-pressure zone of generally light, variable winds and calms traditionally known as the *doldrums.* The doldrum belt is what present meteorologists refer to as the *intertropical convergence* (ITC).

Westerlies

Some of the subsiding air of the subtropical highs flows poleward near the surface (Figure 4.4). This flow is deflected toward the east to become the southwesterlies of the northern hemisphere and the northwesterlies of the southern hemisphere. Usually these winds are simply called *westerlies.* In the days of sailing vessels, these were the winds used for eastward passages across the oceans.

Polar Fronts

The surface air within the westerlies is normally of subtropical origin. When this warm "tropical" air meets colder "polar" air moving toward the equator, the tropical air flows up and over the denser polar air. The *polar front* is the boundary between the two types of air. Eastward-migrating low-pressure cells created within these zones of converging air cause them to be known as the *subpolar lows* (Figure 4.4).

As the warm air flows up over the wedge of colder air along the polar front, it is chilled by expansion and radiational cooling. This cooled air subsides over the polar regions, forming high-pressure centers known as the *polar highs.*

Cold surface air spreads from the polar highs in both hemispheres and undergoes westward Coriolis deflection to become the *polar easterlies.* The polar easterlies meet the westerlies in the zones of friction along the polar fronts.

From time to time, particularly during winter, the polar front bulges toward the equator, allowing polar air to penetrate to subtropical latitudes. These bulges, known as *polar outbreaks,* are shown by the broken arrows in the cross section in Figure 4.4.

Idealized Pressure and Wind Systems

Surface pressure and wind patterns on a uniform, rotating earth can be summarized as follows. There are three zones of low pressure and atmospheric convergence—the region of the polar front in each hemisphere and the intertropical convergence zone astride the equator. There are also four zones of high pressure and atmospheric divergence—the polar and subtropical highs in each hemisphere. In general, zones of low pressure and convergence are associated with clouds and precipitation, while fair weather prevails in zones of high pressure and divergence. Due to seasonal changes in the angle of the sun's rays (resulting from the tilt of the earth's axis), these zones shift poleward in summer and toward the equator in winter. The heat storage in oceans, however, delays poleward migration by one or two months, so that these belts do not reach their highest latitudes until late summer.

Observed Surface Pressure and Wind Systems

The actual patterns of pressure and wind at the earth's surface are more complicated than the simple belts associated with the uniform surface model. The presence of continents and oceans considerably influences the circulation. The specific heat of rock or soil is much less than that of water, so the continents become warmer than the oceans in summer and much cooler in winter. This differential heating and cooling of land and water masses affects surface pressure and winds, as thermal low-pressure cells develop over the land in the summer to be replaced by high-pressure cells in the winter. Mountain ranges also disturb the flow of the atmosphere, even at the atmosphere's upper levels.

Pressure and Circulation Cells

The observed pattern of pressure distribution shown in Figure 4.5 (see pp. 88–89) consists of separate cells rather than continuous zones of high and low pressure. Nevertheless, these cells tend to be distributed along the same bands of latitude as the idealized pressure zones on a uniform earth. The average direction of surface winds is closely related to the pressure distribution. Surface winds spiral outward from high-pressure regions and inward toward low-pressure regions, indicating geostrophic flow modified by friction with the earth's surface.

In the southern hemisphere, there is almost continuous ocean between latitudes 45° and 65°S, so that the observed circulation, particularly in winter (July), tends to be similar to the model for the homogeneous surface. In the northern hemisphere, the presence of large landmasses produces well-defined pressure and wind cells. These cells migrate north during the summer and south during the winter, reflecting changes in the receipt of solar radiation.

Seasonal Variations

In winter, the northern regions of North America and Eurasia become very cold in comparison with the adjacent oceans. This causes the polar high to be displaced in the direction of the equator, forming two high-pressure centers—the massive Siberian high over north-central Asia and the much smaller Yukon high over eastern Alaska and the Canadian Arctic. Two subpolar lows, the Icelandic and Aleutian lows, are situated over the oceans at similar latitudes.

During summer, when the continents become warmer than the oceans, the subtropical

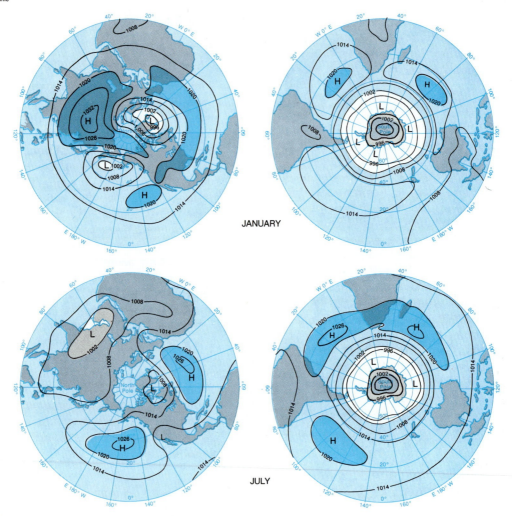

JANUARY

JULY

Figure 4.5
Global atmospheric pressure and average direction of surface winds in January and July. Pressure readings are in millibars.

highs strengthen and expand over the cooler oceans. The subtropical highs over the oceans of the northern hemisphere are known as the Pacific or Hawaiian high and the Azores or Bermuda high; these nearly stationary high-pressure cells dominate the weather over the subtropical oceans and adjacent continental areas all summer long. Meanwhile, thermal

low-pressure centers develop over the hot deserts of southwestern United States and southern Asia (Figure 4.5).

Monsoons

The seasonal shifts in pressure patterns generate seasonal shifts in winds. The inset in Figure 4.5 shows how the average position of the

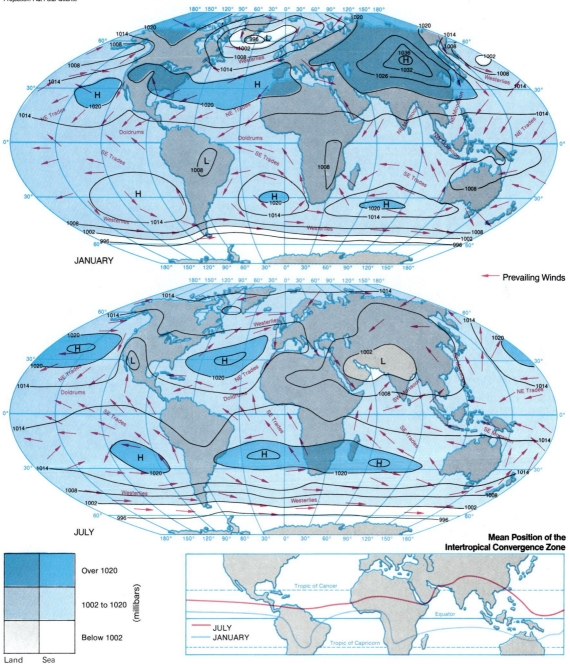

JANUARY

JULY

Prevailing Winds

Over 1020

1002 to 1020

Below 1002

(millibars)

Land   Sea

Mean Position of the Intertropical Convergence Zone

Tropic of Cancer

Equator

Tropic of Capricorn

JULY
JANUARY

GENERAL ATMOSPHERIC AND OCEANIC CIRCULATIONS

Electrical Energy from the Wind

Wind has been used as a source of energy throughout most of the history of civilizations. The majestic Yankee Clippers of the early nineteenth century come quickly to mind, and the variety of wind-powered vessels for marine transport is mind-boggling, to say the least. Some people try out wind surfing and sailboats, and many dream about a holiday to exotic islands on a sail-powered yacht.

The picturesque old windmills of Holland are important elements of the cultural heritage of many Europeans and Americans, and most older Americans can still remember the spartan windmills that dotted the American rural landscapes, especially throughout the Great Plains. The windmills were used to pump groundwater for homesteads and farms and even to provide water for cattle in the drier regions like the Great Plains. By filling sails or turning windmills, elements of the general circulation were made to provide energy directly for work. But inexpensive fossil fuels and electrical energy gradually replaced the sails that powered commercial vessels and windmills.

In the 1970s, rising energy costs renewed interest in windmills and wind-powered vessels. Experimental wind-energy "farms" have been developed in California. Here persistently strong winds associated with the pressure gradient from the Hawaiian high over the eastern Pacific and the thermal low over southeastern California and Arizona combine with vigorous sea breezes to offer especially favorable conditions for generating electricity directly from wind machines. Since the cooler airflow off the Pacific is channeled through gaps in the coastal mountains, certain locations have the greatest potential for economic development of wind power.

One such favorable location is San Gorgonio Pass, east of Los Angeles and just outside of the plush desert resort of Palm Springs. More than 1,000 wind machines have already been erected on 2,300 acres of federal land, and there are plans for another 1,000 or more. Each wind machine is more than 30 meters (100 feet) high, and the field looks like a strange forest of metal trees gyrating madly in the landscape. The power is fed into the southern California electrical grid, and a typical windmill can provide power for about 15 to 20 residential homes. A new and more efficient 50-meter windmill that looks like a giant eggbeater was erected in the fall of 1986 in the Texas Panhandle, another region plagued or "blessed" with strong winds.

Will the windmills remain cost-effective with the rapidly falling fossil fuel prices in the mid-1980s? Perhaps not in the short run, but in the long run they probably will.

intertropical convergence—the meeting place of the trade winds—shifts between January and July. Note that there is relatively little change in the position of the ITC over the Atlantic and eastern Pacific oceans. However, over land areas, especially those bordering the Indian Ocean, there is a large seasonal displacement. Associated with the extreme shift in the position of the ITC over eastern and southern Asia is a reversal in wind direction between winter and summer. This seasonal change in the direction of surface winds produces very strong seasonal changes in the prevailing weather. Half of the year the wind blows from the land to the sea; the other half it blows from the sea to the land. A wind system that reverses direction seasonally is known as a *monsoon.* Monsoons are discussed at greater length in Chapter 6 in connection with tropical weather types.

The Upper Atmosphere and Jet Streams

The circulation pattern of the upper atmosphere is much simpler than that near the surface. Poleward of the subtropical highs, the upper atmosphere flows from west to east in a vast circumpolar vortex. Since friction here is at a minimum, the Coriolis and pressure gradient forces are in balance. Therefore, airflow in the upper atmosphere approximates geostrophic conditions, with low pressure to the left and high pressure to the right in the northern hemisphere (Figure 4.6). The strongest flows are concentrated in the relatively narrow *jet streams*.

Wind directions and speeds in the upper atmosphere are usually determined by tracking freely drifting radio-equipped balloons that transmit meteorological data to the ground. In winter the upper-air winds are strong and form a well-defined pattern of circulation with several undulations. Many winter storms in the northern latitudes appear to be generated with the help of jet streams. The flow of upper-air winds tends to be weaker during summer than during winter.

High-velocity streams of air were first observed in the upper troposphere in the 1940s with the development of military aircraft that could fly at altitudes exceeding 10 km (6 miles). These so-called jet streams were found to occur at altitudes of 10 to 15 km. They are hundreds of kilometers wide and several kilometers thick. The wind speed along the core of a jet stream can exceed 300 km (200 miles) per hour for a distance of 1600 km (1,000 miles) or more. Figure 4.4 shows that there are two jet streams in each hemisphere: the subtropical jets associated with the subtropical highs and the polar front jets that are normally above the polar fronts. The jet streams are utilized by eastward-bound airliners and are avoided as far as possible on westward flights. Assistance from the jet stream causes east-bound flights across the full width of the United States to require about an hour less time than westbound flights.

Jet Stream Oscillations

The polar jet streams generally follow a sinuous path, as shown in Figure 4.7a. There are

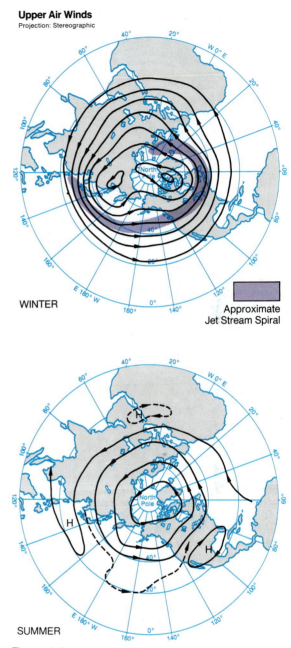

Upper Air Winds
Projection: Stereographic

WINTER

Approximate Jet Stream Spiral

SUMMER

Figure 4.6
Winter and summer wind patterns for the northern hemisphere in the upper atmosphere near an altitude of 5.5 km (3.4 miles), where the atmospheric pressure is approximately one-half the pressure at sea level. (Andy Lucas after Herbert Riehl, *Introduction to the Atmosphere,* © 1972 by McGraw-Hill Book Company)

surfaces and much greater than that of the air. The great heat capacity of the oceans makes them reservoirs of the energy that powers global weather. This energy is released into the atmosphere both by conduction and convection and by latent heat transfer during the evaporation process that moves enormous amounts of water vapor from warm sea surfaces to the atmosphere. While this water provides the earth's rainfall, it also helps power the atmospheric circulation by latent heat release during condensation and cloud formation.

Ocean temperatures are "conservative," changing very slowly through the seasons. But even small changes in ocean temperatures seem to be able to produce large changes in weather patterns. Oceanographers have recently discovered vast and persistent surface "pools" of ocean water that are 1° to 2°C warmer or cooler than the water around them. These ocean temperature abnormalities, or "anomalies," seem to produce changes in the average atmospheric circulation. For example, unusually warm water off the east coast of the United States during 1971 and 1972 was associated with very wet weather in the coastal states, and it probably helped provide energy for Hurricane Agnes, a tropical cyclone that was especially destructive between Virginia and Pennsylvania. Because of complex interactions with the upper air flow, the same anomaly may have played a role in freezes, droughts, and grain crop failures in the Soviet Union in 1972. A pool of colder than normal water in the central Pacific in 1976–1977 is believed to have caused changes in the general circulation, producing severe drought in the western United States along with the coldest winter in 60 years in the east, followed by a summer drought in the Midwest that equaled those of the Dust Bowl years of the 1930s.

Surface Currents

Wind blowing over the surface of the ocean exerts a push on the water, but the resulting motion of the water is only a fraction of the wind speed. This is due to the greater density and internal friction of water. Wind-driven *ocean currents* have speeds ranging from kilometers per hour to kilometers per day. Because friction causes current velocities to decrease rapidly with depth, strong wind-driven currents are confined to the upper hundred meters or so of the ocean. Deeper currents unrelated to wind stress have also been discovered, but their origin and pattern remain unclear.

A well-developed ocean current takes a long time to respond to changes in the wind, and even the waves produced by strong winds require many hours to reach their full development. Therefore, ocean currents reflect average wind conditions over periods of many months, and the circulation of the oceans is closely related to the general circulation of the atmosphere.

Gyres

We have seen that if the earth were uniform, without any land and water contrasts, the winds would form well-defined belts. Assuming complete cover by water, wind-driven ocean currents would flow around the earth in a similar pattern. But the actual distribution of land and water means that only the ocean encircling Antarctica can circulate freely in a continuous belt. The Atlantic, Pacific, and Indian oceans are bounded by continents or island archipelagos on the east and west, blocking free flow of ocean water around the earth at most latitudes.

As a consequence, the surface currents of the oceans consist of closed circulation loops, or *gyres*, which correspond to the global wind patterns. The Atlantic and Pacific oceans ideally would include three gyres on either side of the equator, as represented in Figure 4.9. The trade winds drive the low-latitude east-to-west currents of the subtropical gyres, and the mid-latitude westerly winds drive the higher latitude return flows from west to east. The actual oceanic circulation is shown in Figure 4.10.

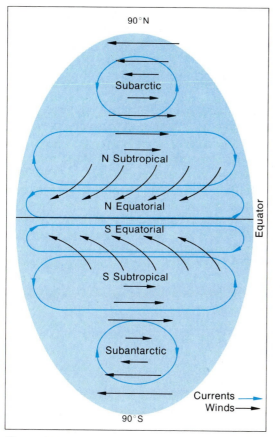

Figure 4.9
The idealized oceanic circulation in an ocean basin consists of loops, or gyres, which are driven by prevailing winds. (Doug Armstrong after P. Weyl, *Oceanography*, 1979, John Wiley & Sons)

The strongest currents are on the perimeters of the gyres (near the coasts), with much less movement near the centers of the oceans. Gyres are clearly developed in the Atlantic and Pacific oceans, but near Antarctica an eastward moving current flows around the entire earth; the idealized southernmost gyre in Figure 4.9 is replaced by the Antarctic continent.

Energy Fluxes

Ocean currents play key roles in redistributing heat around the earth. A current of warm tropical water flows poleward at the western edge of each ocean basin, and cool waters drift toward the tropics along the eastern margins of the seas. Along the east coast of North America, the Gulf Stream is the warm, north-setting current. Its counterpart in the Pacific is the Kuroshio current off Japan. Both are fast, narrow currents, moving several kilometers per hour. The volume of water transported annually by the Gulf Stream alone is more than 30 times greater than the total amount of stream-flow from all the continents.

The Gulf Stream merges into the slower-moving North Atlantic Drift near the Grand Banks off Newfoundland. The warm water of the North Atlantic Drift, which flows north-eastward to Europe, causes the average winter temperatures of western Europe to be significantly higher than those of eastern North America, even though Europe lies farther pole-ward. Thus, populous Denmark and Sweden are at the latitude of Hudson Bay in subarctic Canada.

Evaporation from such warm currents provides high-latitude maritime air masses with the latent heat and moisture supplies for the cyclonic storms that bring rain to Europe, eastern Asia, and western North America. Air moving over warm currents not only gains moisture, but is heated from below, producing steep lapse rates that cause air mass instability, favoring upward convection and heavy rainfall over the land.

Along the western coast of the United States the cold California Current is the source for the marine fog and cool air from the Olympic Peninsula in Washington all the way to southern California.

Upwelling

The ocean currents shown in Figure 4.10 are driven by the general circulation. But the ocean currents do not exactly parallel the prevailing winds. The Coriolis effect tends to turn the currents a bit to the right in the northern

Global Oceanic Circulation
Projection: Flat Polar Quartic

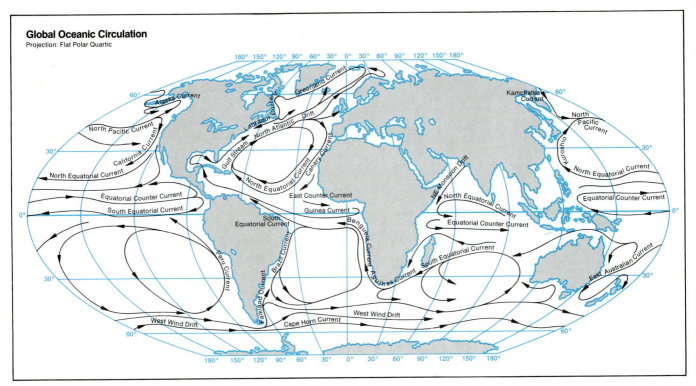

Figure 4.10

This map shows the principal currents in the surface layer of the oceans. The currents form loops of circulation, with strong currents on the perimeters and relatively little movement internally (compare with Figure 4.9). Ocean currents move warm water poleward and cold water toward the tropics, which helps to equalize the distribution of energy over the earth. (Andy Lucas after L. Don Leet and Sheldon Judson, *Physical Geology,* 3rd ed., © 1965, by permission of Prentice-Hall, Inc.)

hemisphere and to the left in the southern hemisphere. The surface water tends to drift off at an angle of 45° to the wind. Along the west coasts of continents, in particular, where winds blow equatorward around the eastern sides of the subtropical highs, the surface water moves away from the coasts. This is replaced by deeper water that rises to the surface along the coasts. This water is cold and produces several important effects.

Coastal *upwelling* of cool water occurs along the west coasts of all continents, but it is pronounced off California, Chile and Peru, Morocco, Namibia, and Somalia. The resulting cold water chills the air above it, producing fog. This chilled air is denser than the warmer air above it. Such a temperature inversion prevents the vertical motion of air needed to build rain clouds. Therefore, coasts where upwelling occurs are foggy but receive very little rainfall. All have desert climates, either seasonally or the year around.

El Niño

Upwelling is very important ecologically because it lifts nutrients up into sunlit surface waters, allowing the tiny marine plants and animals known as plankton to flourish and support a very rich marine food chain. This permits fishing industries to thrive along many desert coasts. The richest fishing grounds are off the Peruvian coast of South America. Here, however, disaster strikes every few years, when the waters become abnormally warm and upwelling seems to fail. Not being adapted to warm water, and lacking the nutrients supplied by upwelling, the plankton die off, and the entire food chain collapses. Fish die in great numbers, along with the birds that normally feed on them. Decomposing fish litter the beaches and the waters, giving off so much hydrogen sulfide that the white lead paint used on ships turn black. The phenomenon has been called the "Callao painter," after the port of Callao in Peru. Since this effect usually occurs around Christmas time, the southward drifting warm current is called *El Niño,* the Christ Child.

The cause of the disaster is periodic fluctuation in the strength of the west-blowing trade winds. The trade winds strengthen every few years, probably in response to sea surface temperature anomalies, and drive water into the western Pacific, actually raising the sea level there. When the trades slacken again, this warm water "sloshes back" eastward across the width of the Pacific, and a portion spreads southward along the Peruvian coast. This overruns the upwelling cold water and produces the disastrous El Niño phenomenon, which also appears to produce unusual weather in areas far from the coastal regions. The 1982–1983 El Niño seemed associated with weather disturbances bringing heavy rains to subtropical coasts and severe drought to inland areas as far away as Australia and the African Sahel region.

KEY TERMS

general circulation
pressure gradient force
isobars
Coriolis force
geostrophic wind
Buys Ballot's law
Hadley cell
horse latitudes
subtropical highs
trade winds
doldrums
intertropical convergence (ITC)
westerlies
polar fronts

subpolar lows
polar highs
polar easterlies
polar outbreaks
monsoons
jet streams
upper-air troughs
upper-air ridges
secondary circulations
ocean currents
gyres
upwelling
El Niño

REVIEW QUESTIONS

1. What is the relationship of the geostrophic wind to high and low pressure in the northern and southern hemispheres?
2. What does Buys Ballot's law tell us about the interaction of the pressure gradient force, the Coriolis force, and friction?
3. How does surface airflow in the vicinity of high- and low-pressure centers differ in the northern and southern hemispheres? Explain.

4. Compare and contrast the idealized patterns of atmospheric pressure systems and winds over the continents and ocean basins for winter and for summer.

5. How are the jet streams related to the energy balance of the earth-atmosphere system?

6. How do the continents and mountain barriers alter the idealized zonal patterns of pressure and wind systems over the earth?

7. Describe how ocean current patterns are related to atmospheric pressure and wind systems.

APPLICATIONS

1. Suppose your residence is in a North American urban area, with a bakery to your west, a chocolate factory to the east, an oil refinery to the north, and a stockyard to the south. Describe the sequence of odors you would sense when a strong low-pressure center passes from west to east along a line 100 km north of your location. Would the odors be the same if the low passed to the south of you?

2. By means of radio and TV weather reports and your own observations of cloud movements, keep a record of wind direction each day for the rest of the term. Do the wind conditions help you predict weather changes? How far in advance of weather changes do wind shifts occur?

3. Contrast your own data from the preceding exercise with the prevailing winds shown in Figure 4.5. What significant differences appear? How can these differences be explained?

4. Look at the existing pattern of ocean currents in Figure 4.10. Would it be possible to send a message in a floating bottle to every coastal location in the world? Could this always be done from another continent, or would some messages have to be sent from ships at sea?

5. The continents have not always occupied the positions they do today. About 250 million years ago all continents were united in one large elongated mass that formed a north-south strip extending from pole to pole. Draw a map of the earth's wind systems and ocean currents as they would have existed at that time. See the Appendix for information on map projections.

FURTHER READING

Ahrens, C. Donald. *Meteorology Today: An Introduction to Weather, Climate, and the Environment,* 2nd ed. St. Paul, Minn.: West Publishing (1985), 523 pp. Chapters 12 and 15 are especially helpful for study of pressure and wind relationships, and upper-air components of the general circulation.

Gedzelman, Stanley David. *The Science and Wonders of the Atmosphere.* New York: John Wiley (1980), 535 pp. Relationships between pressure and wind are developed in Chapters 15 and 16 by drawing upon commonplace examples and simple explanatory equations.

Hare, F. Kenneth. *The Restless Atmosphere,* rev. ed. London: Hutchinson (1956), 192 pp. This brief introductory classic contains outstanding regional and continental chapters that focus on the dynamics of the general circulation.

Lutgens, Frederick K., and **Edward J. Tarbuck.** *The Atmosphere: An Introduction to Meteorology,* 3rd ed. Englewood Cliffs, N.J.: Prentice-Hall (1986), 492 pp. An introductory text with a strong geographical perspective. Chapters 6 and 7 are particularly helpful for pressure, winds, and the general circulation.

Neiburger, Morris, James Edinger, and **William**

Bonner. *Understanding Our Atmospheric Environment,* 2nd ed. San Francisco: W. H. Freeman (1982), 453 pp. Fundamentals are presented in a nonmathematical framework.

Riehl, Herbert. *Introduction to the Atmosphere,* 3rd ed. New York: McGraw-Hill (1978), 410 pp. This nonmathematical text emphasizes the upper air circulation and its relationships to weather and climate.

Young, Louise B. *Earth's Aura.* New York: Avon Books (1979), 305 pp. A highly readable account of the various phenomena of the atmosphere, written for the general public. Full of anecdotes; informative and accurate.

Water enters the atmosphere through evaporation, is transported as water vapor, eventually condenses into clouds, and falls again to the earth as rain, snow, sleet, or hail. Condensation and precipitation are the results of a number of complex physical interactions that involve energy and moisture.

Seascape Study with Rain Clouds by John Constable, c. 1824–1828. (Royal Academy of Arts, London)

5
Moisture and Precipitation Processes

During the afternoon of June 9, 1972, clouds gathered over the Black Hills of South Dakota. Over large areas more than 30 cm (12 in.) of rain—equal to a full year's precipitation—poured down in a 6-hour period. Mountain streams became rampaging torrents, overflowing their banks and sweeping away cars and even houses. Automobiles and buildings blocked the spillway of a dam just upstream from Rapid City, causing water levels behind the dam to rise an additional 3 meters (10 ft). Late in the evening, the dam burst, and a flood wave swept through the city, causing enormous destruction and taking more than 200 lives. The total damage exceeded $120 million.

This type of disaster was repeated north of Denver, Colorado, on the night of July 31, 1976, when a similar deluge of rain fell in a 1- to 3-hour period, causing floodwaters to rush through the narrow canyons of the Big Thompson and Cache La Poudre rivers in the foothills of the Rocky Mountains. Within a few hours, the raging waters destroyed nearly all traces of human habitation in the canyons and drowned 144 persons.

These excessive rains can also occur over flat country where there is no extra component of rainfall associated with uplift of moist air along mountain fronts. Such an event occurred at Tulsa, Oklahoma, during the early Sunday morning hours of May 27, 1984, when as much as 38 cm (15 in.) of rain fell within 6 hours. About 7,000 buildings were inundated and 14 people drowned. Still, flash flooding in relatively level country is not as life-threatening as in the narrow mountain valleys where flood waters can pile up to incredible depths in a few minutes.

How is it possible for so much rainfall to be poured over small areas in so short a time? The water that fell was evaporated from the Gulf of Mexico, hundreds or even thousands of kilometers away. Moving masses of warm, moist air carried the water vapor northward across the Great Plains. Along the way, some of the molecules of water vapor condensed to form droplets of water only micrometers in diameter—much too small to fall to the ground as rain. Finally, however, the moisture-laden air was forced to rise into towering thunderclouds where water droplets coalesced to form raindrops that fell in a deluge.

The movement of water from the earth's surface to the atmosphere and then back again to the surface constitutes the atmospheric part of the hydrologic cycle that was described in Chapter 2 and illustrated in Figure 2.6 (page 35). The transfer of water between the surface and the atmosphere occurs by means of three related processes. *Evaporation* represents the transfer of liquid water at the surface to water vapor in the atmosphere. *Condensation* is the conversion of water vapor to water droplets or ice crystals in the form of fog or clouds, or dew at the surface. *Precipitation* involves the coalescence of water droplets or transformation of ice crystals into raindrops, snowflakes, or hailstones large enough to fall to the ground. During the condensation process the latent heat absorbed during evaporation is released as sensible heat in the atmosphere. Thus the hydrologic cycle can also be thought of as part of the energy balance of the earth-atmosphere system.

Under average conditions, nearly half of the earth's surface is covered by clouds. But only a fraction of the clouds produce significant rainfall. Precipitation is the result of complex atmospheric interactions, and the conditions that lead to it are so restricted that rain, snowfall, or hail can be considered an unusual rather than a commonplace event (Figure 5.1). This chapter describes how the atmosphere is continuously supplied with water vapor and how condensation, cloud formation, and precipitation occur.

TRANSFER OF WATER TO THE ATMOSPHERE

When water evaporates, water molecules are detached from the surface of the liquid and enter the air as water vapor, a dry gas. Molecules in liquids are in continuous motion but are kept from separating by attractive forces. The higher the temperature of the liquid, the greater the energy and motion of its molecules. Eventually, some molecules gain enough energy to break away from the liquid and enter the air. Because the escaping molecules carry kinetic energy with them, evaporation tends to cool the surface from which the molecules are removed. The evaporation of water molecules on your skin is what makes you feel a chill when you step out of a swimming pool, even when the air is warmer than the water. The sun supplies most of the energy needed to vaporize water—590 calories for every gram of water transformed from the liquid to the vapor phase under average conditions.

Evaporation from Water Surfaces

If a jar filled with water is left open in a room, all of the water will eventually evaporate. Heat stored in the room's air provides all the energy required for vaporization of the water. Even if the jar is sealed, some water will evaporate into the air above the liquid. Eventually, however, a condition of equilibrium is reached, and the water level in the jar remains constant. The air in the jar is then said to be saturated; it contains as much water vapor as it can hold at its particular temperature.

Figure 5.1
Because precipitation occurs only when certain conditions are present, the amount of moisture received by a given region may vary markedly from year to year. The coming of rain, or the lack of it, is a central concern each year.

(a) People in India rejoice in the coming of the summer monsoon rains. (Brian Brake/ Rapho/Photo Researchers, Inc.)

(b) This farm in Oklahoma was abandoned in the 1930s when a succession of drought years made farming impossible. (The Bettmann Archive)

MOISTURE AND PRECIPITATION PROCESSES

(c)

(d)

Figure 5.1 (continued)
(c) White Clay Creek at
Newark, Delaware, with near
average conditions.
(d) White Clay Creek during
the record-breaking flood
generated by Hurricane Agnes in
June 1972. (R. A. Muller)

CHAPTER 5

Saturation Vapor Pressure

Water vapor exerts pressure, just like any other gas. The amount of pressure exerted by molecules of water vapor in the air is called the *vapor pressure*. When the air is saturated, the vapor pressure is at its maximum, but this maximum varies with the temperature. In other words, *saturation vapor pressure* depends on the temperature of the air. As the air temperature rises, the saturation vapor pressure increases rapidly. Figure 5.2 shows that at the temperatures found near the earth's surface, the saturation vapor pressure nearly doubles for each 10°C increase in temperature. Hence, warm air has a much greater capacity to hold water vapor than does cold air.

Vapor Pressure Gradients

Although the rate of evaporation from water surfaces depends especially on the energy supply, it is also affected by the degree of saturation of the air above the water surface. Under normal atmospheric conditions there is variation in vapor pressure from the water surface through the air above. This is known as the *vapor pressure gradient*—the rate at which vapor pressure varies with distance. How much evaporation takes place depends on how "steep" the gradient is (how rapidly the vapor pressure decreases upward from the water surface).

The steepness of the vapor pressure gradient is determined mainly by the turbulence in the lower atmosphere. When there is no wind, a layer of nearly saturated air forms immediately above water surfaces. The vertical change in vapor pressure within this layer is small, so that the vapor pressure gradient is low, or "weak." Under such conditions the rate of evaporation is low. During windy conditions, however, water vapor is mixed through a much deeper layer of the atmosphere. This permits a steeper vapor pressure gradient, allowing evaporation rates to approach maximum values. To maintain high rates of evaporation,

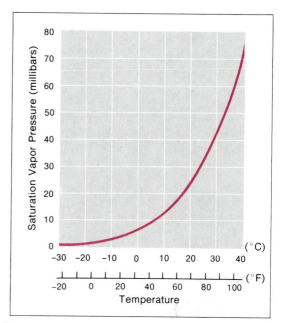

Figure 5.2
The curve shows that the saturation vapor pressure of air increases rapidly with increased temperature. The data extend to temperatures lower than 0°C (32°F), the normal freezing point of water, because small droplets of water can remain liquid and supercooled at temperatures as low as −40°C (−40°F). (Doug Armstrong after *Smithsonian Meteorological Tables*, 6th ed., 1971, edited by Robert J. List, by permission of the Smithsonian Institution)

energy for vaporization must be supplied by warm water, warm air immediately above the surface, or absorbed solar radiation. Evaporation, then, depends on both the energy supply and the vapor pressure gradient.

There are few regular measurements of evaporation from the world's oceans or large lakes. The global distribution of evaporation has to be estimated indirectly by means of energy-budget calculations. In general, average annual evaporation is related to latitude and the corresponding solar radiation gains and longwave radiation losses, as shown in Figure 3.9 (p. 62).

Evaporation rates also vary from place to place due to such factors as winds, cloudiness, water temperature, and water vapor content of the atmosphere. The highest annual evaporation rates occur over the subtropical oceans, where the water is warm, the air aloft is dry, and the weather is usually fair. Mean annual evaporation ranges from more than 100 cm (40 in.) over equatorial and subtropical oceans to less than 10 cm (4 in.) over polar oceans.

Evapotranspiration from Land Surfaces

Where plants cover the land, only a small part of the water vapor entering the atmosphere comes from direct evaporation of water in the soil. Instead, most is released through small openings in plant leaves called *stomata*

(Figure 5.3). After plants absorb soil water through their roots, water and dissolved nutrients are transported up the stem of the plant, and excess water is "transpired" through the stomata as vapor. During the day, when a plant is photosynthesizing, the stomata are open to allow the entry of carbon dioxide; at the same time, water vapor escapes by transpiration. At night, the stomata are closed and there is little transpiration. The energy requirement for transpiration is the same as that for evaporation: 590 calories per gram of water vaporized at normal temperatures. Hence transpiration is likewise a cooling process.

Water loss from land surfaces is usually regarded as the combination of transpiration from vegetation and evaporation from soil. As we saw in Chapter 3, the two processes together are known as evapotranspiration. In

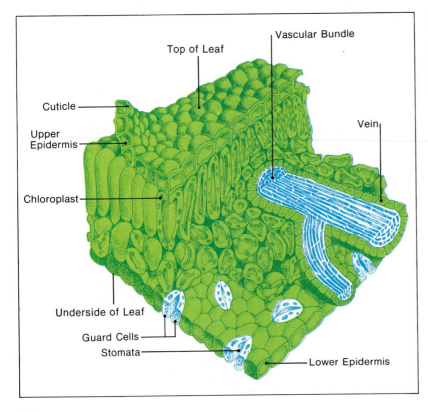

Figure 5.3
This vastly enlarged cross section shows the functional parts of a typical plant leaf. The leaves of green plants possess openings known as *stomata*. When the leaf is exposed to light, the stomata are open to allow the entry of carbon dioxide and the exit of oxygen and water vapor. (John Dawson)

Top of Leaf

Vascular Bundle

Cuticle

Vein

Upper Epidermis

Chloroplast

Underside of Leaf

Guard Cells

Stomata

Lower Epidermis

areas covered by vegetation, water loss by transpiration is usually three or more times greater than water loss by evaporation directly from the soil. The ratio between transpiration and direct evaporation can vary considerably, depending upon degree and type of vegetation cover, making it very difficult to evaluate patterns of evapotranspiration from the continents. Estimates of evaporation from the oceans are far less difficult. The effects of evapotranspiration on local water budgets and the hydrologic cycle are treated in detail in Chapter 7.

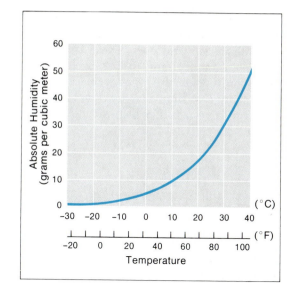

Figure 5.4
The curve shows the maximum amount of water vapor that can be contained in a cubic meter of air at a given temperature. The shape of the absolute humidity curve is similar to the saturation vapor pressure curve shown in Figure 5.2. (Doug Armstrong after *Smithsonian Meteorological Tables*, 6th ed., 1971, edited by Robert J. List, by permission of the Smithsonian Institution)

MOISTURE IN THE ATMOSPHERE

There is always some water vapor in the air. At a given temperature, the air may contain any amount of water vapor up to the maximum value at which saturation occurs. Figure 5.4 shows that air at 10°C (50°F) can contain up to 10 grams of water vapor per cubic meter. Ten grams, the weight of two American nickels, may not seem like much water, but rain clouds that contain 10 grams of water vapor per cubic meter have the potential to produce significant rainfall on the ground below. The water vapor in saturated air rises rapidly with increased temperature, so that warm air is able to contain more water vapor per unit of volume than cool air can contain.

Measures of Humidity

The water vapor content of the atmosphere is referred to as *humidity*. Humidity can be described by several different measures, each one useful for specific purposes.

The most direct measure of the air's moisture content is the *absolute humidity,* which is the weight of water vapor in a given volume of air (Figure 5.4). Absolute humidity is usually expressed in grams of water vapor per cubic meter of air. In the atmosphere, however, warming air expands and cooling air contracts. Even when there is no change in the water vapor content, the absolute humidity changes when the temperature changes. The use of absolute humidity, therefore, tends to be restricted to controlled scientific experiments.

In meteorology, other humidity measures are used to describe the water vapor content of air. The *specific humidity* is the weight of water vapor per kilogram of air, and the *mixing ratio* is the weight of the water vapor present per kilogram of dry air (air minus its water vapor).

Another important method of expressing the moisture content of air is by means of the *dew point* temperature—the temperature at which saturation occurs and moisture condensation begins. The higher the dew point temperature, the greater the moisture content of the air. The dew point is discussed more fully in the section on condensation processes. These three measures of humidity are useful because they change only when the amount of water vapor itself changes. They are used especially in studies of air masses, which are discussed in Chapter 6.

Relative Humidity

Finally, the humidity measure, *relative humidity,* is the ratio between the water vapor in the air and the amount the air could hold if it were saturated. The relative humidity is always expressed as a percentage and does not indicate the specific amount of water vapor present in the air. For example, air at 40°C (104°F) with a relative humidity of 50 percent contains about 25 grams of moisture per cubic meter, but cool air at 20°C (68°F) and 80 percent relative humidity contains only about 14 grams of moisture per cubic meter. The cooler air in this example has a lower water vapor content but a higher relative humidity than the hot air because the cooler air is closer to saturation. Figure 5.4 shows that air can become saturated merely by cooling without any change in water vapor content. As we shall see, cooling of air to its saturation point is the most important factor in moisture condensation and precipitation.

As a measure of atmospheric moisture, relative humidity has the same disadvantage as absolute humidity: whenever the temperature changes, the humidity measure also changes, even if there is no change in water vapor content. Figure 5.5 shows that temperature changes in a 24-hour period usually cause relative humidity to be highest at night and in the early morning and lowest in mid-afternoon.

Distribution of Water Vapor in the Atmosphere

Only a tiny fraction of the earth's water supply is stored in the atmosphere at any one time. If all the atmospheric water vapor fell to the earth as rain, it would produce a layer of water only about 2.5 cm (1 in.) deep. Heavy rains are possible only because there is a constant recycling of liquid water and water vapor within the hydrologic cycle. In the atmosphere, water that is lost by precipitation at one place is at the same time being replaced by evapotranspiration at another place.

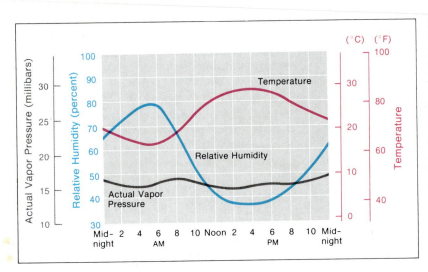

Figure 5.5
During fair weather, the vapor pressure of the air near the ground remains almost constant through the day. However, the relative humidity changes markedly through the day as the air temperature changes. (Doug Armstrong, adapted from *Hydrology Handbook,* 1949, American Society of Civil Engineers)

CHAPTER 5

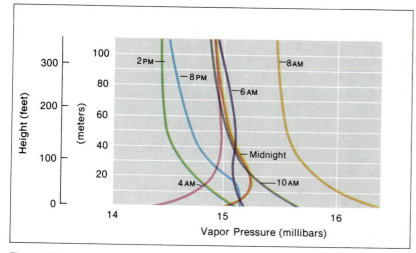

Figure 5.6

Each curve on this graph shows how the vapor pressure in the atmosphere near moist ground varies with height throughout a clear summer day. At 8:00 A.M. vapor pressure decreases rapidly with height, indicating a flow of vapor from the surface into the atmosphere. Vapor pressure decreases from 8:00 A.M. to 2:00 P.M. because thermal convection supports more efficient removal of moist air from near the surface. Water vapor continues to flow upward from the surface during the day while plants are transpiring. Later at night, the vapor pressure immediately above the surface increases with increased height, indicating a flow of water vapor from the atmosphere toward the ground. (Doug Armstrong after R. Geiger, *The Climate Near the Ground,* © 1965, Harvard University Press)

The transfer of moisture to the atmosphere by evapotranspiration proceeds most rapidly when the ground is warm and moist on a fair summer day in which there is a large net gain of radiation. Convection carries water vapor aloft quickly, so that a steep vapor pressure gradient (rapid pressure change with height) is maintained near the ground (see Figure 5.6). At night, however, the net radiation is negative and there is little or no energy available for evapotranspiration. At the same time, the air immediately above the surface usually is cooled to its dew point. Then water vapor condenses on the ground, roof tops, and automobiles as dew or frost.

The water vapor content of the atmosphere varies horizontally as well as vertically. Average atmospheric storage of water over North America ranges from about 5 cm (2 in.) near the Gulf of Mexico in the summer to less than 0.8 cm (0.3 in.) over central Canada during the winter. This pattern of water vapor distribution varies constantly, however. Daily rainfalls of as much as the 60 cm (24 in.) that fell near Houston, Texas, in tropical storm Claudette during July 1979 indicate that atmospheric circulations are capable of concentrating immense quantities of water vapor in a very short period of time.

THE CONDENSATION PROCESS

Condensation is the process by which water vapor in the atmosphere changes phase to tiny liquid droplets or ice crystals. At the earth's surface, condensation on cool or cold objects produces dew or frost. When water vapor condenses in the atmosphere, the result is a mass of water droplets or ice crystals known as a cloud. Fog is simply a cloud at the earth's surface.

For condensation to occur, the water vapor content of at least one layer of the atmosphere must reach saturation. Saturation can be produced by adding water vapor to the atmosphere, by cooling the atmosphere, or by a combination of added moisture and cooling. Most cloud formation and precipitation result from atmospheric cooling.

Condensation Nuclei

When moist air is cooled to saturation, condensation does not occur automatically. Fine particles, called *condensation nuclei,* must be present to act as centers for condensation and for the growth of water droplets. In moist air that is completely free of particles of any size, condensation takes place only when the vapor pressure is three or four times the saturation value. Air containing more than its normal saturation amount of water vapor is said to be *supersaturated.* However, since air is usually well supplied with microscopic particles, condensation generally begins almost as soon as the air is cooled to the saturation point.

Figure 5.7 shows that condensation nuclei are extremely small. Few exceed a radius of 1 micrometer (0.001 mm), and most have radii smaller than 0.1 micrometer (0.0001 mm). Giant condensation nuclei, with radii greater than 1 micrometer, number only about 1 per cubic cm. Principal sources of condensation nuclei are industrial pollutants and smoke, forest fires, salt crystals from sea spray, pollen, and dust particles. Natural processes and industrial and agricultural activities renew the supply of atmospheric particles as fast as they fall to the surface or are washed out by rain. Hence, nuclei for condensation are almost al-

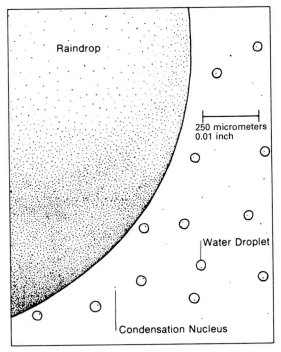

Figure 5.7
This drawing shows the relative sizes of condensation nuclei, water droplets, and a raindrop. Most condensation nuclei are less than 1 micrometer in radius; water droplets are usually less than 20 micrometers in radius; and raindrops have radii of 1,000 micrometers or more. (John Dawson)

ways available when the air becomes saturated.

The water droplets that form clouds usually grow to a radius of 10 or 20 micrometers before the addition of water molecules ceases. In a typical cloud there may be a million water droplets per cubic meter of volume, and the average water content of the cloud may be about 1 gram per cubic meter of air.

A cloud droplet with a radius of only 10 micrometers has a settling velocity (rate of fall) of less than 1 centimeter per second and would normally evaporate in the drier air below the cloud. Under average conditions, droplets do not grow larger than 10 or 20 micrometers in radius, nor can they easily combine in larger drops. Studies have shown that when such small droplets approach one another in a cloud, they are forced apart by the air between them. Thus most clouds do not contain water drops large enough to fall as rain. We shall see later in the chapter what processes are necessary to produce precipitation.

Dew and Frost

When a volume of air cools with no change in water vapor content, its relative humidity increases. The temperature at which the relative humidity becomes 100 percent is the dew point temperature. When the air cools to the dew point, water vapor begins to condense into small drops of liquid water. In humid summer weather the outside of a glass of ice water becomes covered with beads of moisture because the moist air near the glass has been chilled to the dew point temperature. The same process accounts for the formation of *dew* itself. Longwave radiation cools plants and the soil after sunset, and the air in contact with them is also cooled by conduction. If the air cools to the dew point, its water vapor condenses on the ground and exposed plant surfaces, forming dew. In urban areas, dew formation is often more apparent on automobiles parked outside during fair nights.

When the dew point is at or below 0°C (32°F), water vapor condenses as ice crystals, and *frost* appears on exposed surfaces. Because condensation liberates latent heat, dew and frost formation keeps plant and soil surfaces from cooling as much as they would without this release of heat.

Fog

Fog occurs when a thick layer of moist air near ground level is cooled to its dew point. This can occur in several ways, so that different types of fog are recognized.

Radiation Fog

Fog that occurs at night when the ground and the air immediately above lose heat by longwave radiation is called *radiation fog*. If the ground and air are moist, the lowest 10 meters or more of the moist air may be cooled to the dew point, resulting in condensation in the form of dense fog. Moist air and damp ground are particularly susceptible to radiation fog on clear nights, which promote longwave radiation losses to space. But even moderate breezes will prevent the formation of radiation fog by mixing the colder air near the ground with warmer air above.

Advection Fog

Advection refers to the horizontal movement of air across the earth's surface. When warm, moist air passes over snow or cold ocean water, the air may be cooled to its dew point, resulting in *advection fog*. Warm air moving over cold ocean water off the coasts of California and New England frequently produces advection fog. San Francisco is famed for its summer fog, which forms over cold coastal waters and is drawn inland in the afternoon by thermal convection over the land (Figure 5.8). As Figure 5.9 (see p. 115) shows, fog is most common along coastlines and in the Appalachian Mountains.

When advection fog occurs over snow, there may be a marked increase in snowmelt. Normally, snow melts slowly because it has a high albedo and reflects much incoming solar radiation back into space. However, the advection of warm, moist air over snow-covered areas causes condensation on the snow surface. Every gram of water vapor that condenses releases enough latent heat to melt more than 7 grams of snow. Thus when warm, moist air from the Gulf of Mexico moves northward over snow-covered landscapes in the American Midwest and Northeast, rapid melting of snow can occur even on a cool, overcast day. Quite reasonably, skiers hate fog!

Figure 5.8
This photograph shows San Francisco Bay and Oakland, California, shrouded in fog, with Mount Diablo in the distance to the northeast. Low-lying advection fog frequently occurs in the coastal region around San Francisco as moist air moving eastward over the Pacific Ocean cools to its dew point. (David Cavagnaro)

majority of ship accidents have occurred in fog. On one day in September 1923, seven destroyers of the U.S. fleet were wrecked by running aground in dense fog, one on the heels of another, near Point Conception on the southern California coast. Such accidents gain even more importance as an increasing number of the ships at sea are tankers laden with enormous quantities of oil that can produce disastrous pollution of coastal waters.

On land, massive chain accidents involving dozens to more than a hundred vehicles have occurred on fog-shrouded highways in California and New Jersey. Airports are usually located to avoid fog, but many continue to be affected by this hazard, causing frequent and inconvenient diversions of flights to alternate points.

Cooling of surface air by radiation or advection is restricted to a shallow layer of the atmosphere near the surface. The resulting fog and low clouds are simply not thick enough for the processes producing rainfall, snow, or hail to be effective. The air may be filled with a mist of small droplets, sometimes falling as light drizzle, but significant precipitation can occur only when a much thicker mass of air is cooled throughout its depth by being lifted to higher elevations.

Orographic Fog

Orographic fog is associated with mountain areas, such as the Appalachians. When moist air is forced up mountain slopes, it can be cooled to the dew point, resulting in the formation of fog and clouds. Orographic cooling is a major cause of condensation and cloud formation, and will be discussed in more detail later in this chapter.

Fog Hazards

Fog of any type can be very hazardous to human activities. Throughout history the vast

Clouds

Clouds are condensation forms that develop above the ground, usually by the lifting and chilling of moist air. Clouds can be organized into 10 basic types, which in turn can be grouped together into four classes according to height above the surface and vertical development. In Figure 5.10 the high *cirrus* clouds in the upper troposphere are composed of ice crystals that form thin filaments, wispy tufts, and translucent veils. The middle clouds are obviously intermediate in altitudes and composed of more dense masses of water droplets, ice crystals, or mixtures of both. The low clouds are similar to middle clouds but much closer to

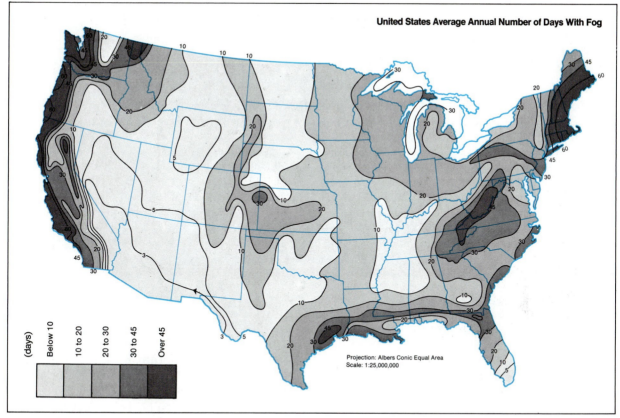

United States Average Annual Number of Days With Fog

(days)
Below 10 | 10 to 20 | 20 to 30 | 30 to 45 | Over 45

Projection: Albers Conic Equal Area
Scale: 1:25,000,000

Figure 5.9
This map shows the average annual number of days
of fog in the conterminous United States. (Adapted
from Arnold Court and Richard Gerston, *Geographical
Review, 56,* © 1966, American Geographical Society
of New York)

the surface. The fourth class are the vertical
clouds associated with strong updrafts. These
cumulus types usually appear as puffed-up
balls of cotton, sometimes forming awesome
white towers reaching all the way to the top of
the troposphere.

The 10 basic types reflect cloud height and
degrees of lifting and cooling. Figure 5.10 also
shows that cirrus clouds are subdivided into
sheets of *cirrostratus* that tend to cover the sky
and reflect gentle but widespread regional uplift, and *cirrocumulus* that represent localized

rising columns of air. The middle clouds are
similarly subdivided into *altostratus* (*alto* is a
prefix meaning middle) and *altocumulus.* The
low clouds include *stratus,* a low and often
thick continuous cover; *stratocumulus,* a
bumpy continuous cover; and *nimbostratus,*
simply low clouds producing precipitation. The
vertical clouds range from the *fair-weather
cumulus* to towering *cumulonimbus* clouds, or
thunderstorms, that can reach altitudes of
12,000 to 15,000 meters (40,000 to 50,000 ft),
with cirrus veils or anvils swept out ahead of
the storms by strong upper-air winds. Figure
5.11 is a mosaic of some of these cloud types,
including one example of "artificial" clouds
generated from heat and moisture above a petrochemical complex.

MOISTURE AND PRECIPITATION PROCESSES

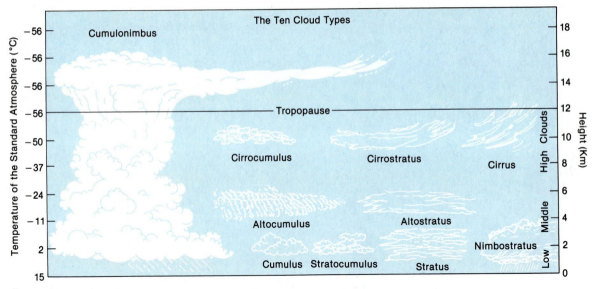

Figure 5.10
The ten most common cloud types shown here can be grouped into four classes representing height and vertical development. (Adapted from J. R. Eagleman, *Meteorology: The Atmosphere in Motion.* New York: D. Van Nostrand)

THE PRECIPITATION PROCESS

We mentioned earlier that most cloud droplets are only about 10 micrometers in diameter and that they will tend to remain aloft in clouds because of updrafts associated with most developing clouds. An average-sized raindrop, however, has a radius of 0.1 cm (1,000 micrometers), and geometric calculations show that about 1 million 10-micrometer cloud droplets are needed to form a single raindrop. The processes that cause water droplets to coalesce to drops large enough to fall from clouds have been studied in detail since the 1940s. Scientists have described two separate processes that are believed to be responsible for precipitation. These are known as the *coalescence model* and the *Bergeron ice crystal model.*

Formation of Rain and Snow

Coalescence Model

The generation of rain by the coalescence process depends on the occurrence of oversized water droplets that are larger than 20 or 30 micrometers in radius. A larger droplet falls just a bit faster than the small droplets, and it grows by colliding with and sweeping up smaller droplets in its path. Rising currents of air carry the drops upward faster than they can fall out of the cloud, allowing them more time to grow in size. A droplet requires about half an hour to grow to raindrop size by coalescence, and the rain clouds must be at least 1 km (0.6 mile) thick for the growing drops to remain in the cloud long enough to become raindrops. Thinner clouds limit the growth of drops by coalescence but may produce *drizzle,* a form of precipitation that consists of very tiny drops that "float" rather than fall to the surface. Wet pavements caused by drizzle can be very hazardous for motorists, but drizzle never produces significant quantities of precipitation.

Sea salt particles 1 micrometer or more in radius make particularly effective condensation nuclei for oversize droplets. That is be-

Acid Rain

Acid rain has become one of this decade's very important environmental issues. It is an international problem that cannot be effectively addressed by single states and countries acting alone. The present evidence suggests that acid rain is most serious over the highly industrialized regions of the middle latitudes, especially over the Great Lakes region of the United States and Canada and eastward to the Atlantic, and over western and central Europe, extending eastward well into the Soviet Union.

Pure water that is left in contact with clean air will soon become slightly acid, with a pH of about 5.6. The water absorbs some of the carbon dioxide in the air to become a weak carbonic acid. Precipitation is defined as "acid" only when its pH is less than 5.6. It should be understood that "acid rain" refers not only to rainfall but also to the gravitational settling or fallout of acidic compounds in particles and gases and to fog droplets intercepted by the trees of mountain forests.

The acidity is produced by sulphur and nitrogen compounds resulting in large part from the burning of fossil fuels for power generation, industrial processes, and transportation, but it is also formed by natural processes such as organic decay, lightning, and forest fires. In the atmosphere very complex photochemical and wet cloud reactions produce the sulphuric and nitric acids.

The impact of acid rain on the environment is of the greatest concern. Some lakes and streams in New York's Adirondacks and in Scandinavia have become so acidic that there is little fish life. Other lakes in these regions have been less seriously damaged because the waters are buffered by alkaline compounds that leach from bedrock of the surrounding drainage basins.

Mountain forests in these same regions have also been severely affected by acid rain, with especially sensitive plant species in rapid decline. The ecological interactions between the forests and the atmosphere are complex and not fully understood at this time. Acid rain is also gradually destroying historic buildings and monuments, especially those built from porous stones. Even modern building materials such as steel, aluminum, and paints are damaged by acid rain, which shortens the economic life cycles of these materials significantly.

Obviously, we need to learn much more about these complex interactions before still greater damage is done.

---

cause of the hygroscopic nature of salt—its natural tendency to absorb water vapor. Even far inland, salt particles blown off the ocean play an important role. Nor is it necessary for the salt particles, or other giant condensation nuclei, to be present in large quantities. Only about one giant condensation nucleus is needed for each cubic meter of cloud.

Ice Crystals Model

The coalescence process is most effective in thick vertical clouds, which are common in the tropics, but less so in higher latitudes, where some process other than coalescence is needed to produce significant quantities of precipitation. In the middle latitudes the tops of many rain clouds are in altitudes where the temperature is well below freezing. In 1933 Tor Bergeron, a Swedish meteorologist, proposed a process of droplet growth to explain rainfall from cold clouds that were not always thick.

Although puddles of water on the ground freeze at 0°C (32°F), tiny water droplets in the atmosphere can remain liquid at temperatures down to −40°C (−40°F). Because of this *super-*

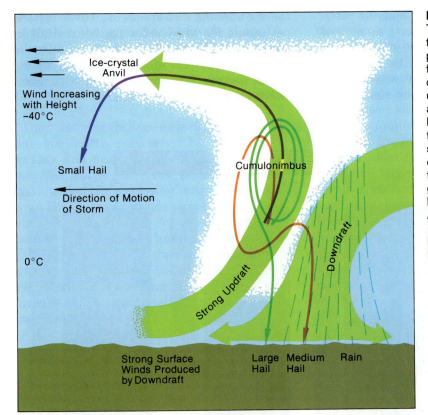

Figure 5.13
This cross section of an intense thunderstorm illustrates the cyclic processes involved in the formation of hailstones. Powerful currents of air in the central updraft carry falling raindrops and ice crystals upward again. Because of the forward motion of the storm, ice particles may be swept up in the updraft a number of times. On each passage through the cloud, the particle gains a new layer of ice, thus becoming a hailstone. (Doug Armstrong after Hermann Flohn, *Climate and Weather,* © 1969 by H. Flohn, used by permission of McGraw-Hill Book Company)

ATMOSPHERIC COOLING AND PRECIPITATION

To generate significant rainfall by either the coalescence or ice crystal model, large masses of air must be cooled throughout their depth. This can be accomplished only by lifting the air to higher elevations. Vertical displacement of large masses of air, which can produce heavy rainfall, snow, or hail from clouds, can be brought about in several ways. The remainder of this chapter discusses the phenomena that trigger precipitation from clouds.

Adiabatic Cooling

Thermals

Net radiation gains at the earth's surface heat the surface and cause evapotranspiration.

Largely because albedo and evapotranspiration vary from place to place, some locations heat up much more than others. Figure 5.14 shows how air above local "hot spots" becomes organized into rising bubbles of warm air that initially exchange little heat with the surrounding air. Soaring birds and glider pilots use these rising currents, called *thermals,* to gain or maintain altitude. Circling within the updraft, they are carried upward until the air cools by expansion and loses its buoyant tendency.

Dry Adiabatic Rate

The temperature decrease in rising air results from a process known as *adiabatic cooling.* The term "adiabatic" refers to the fact that the process occurs without energy gains or

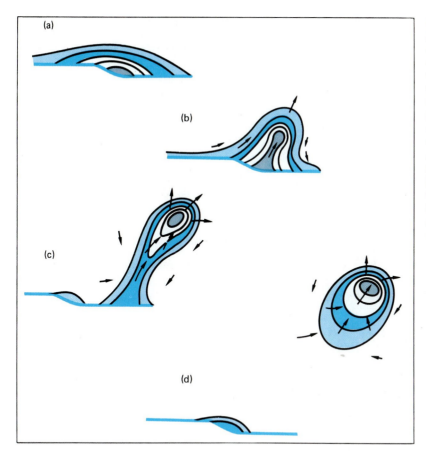

Figure 5.14
A thermal is a rising bubble of air emanating from a local "hot spot" on the earth's surface where the temperature exceeds that of the surrounding area. Each concentric circle, or isotherm, represents an increase of 0.1°C; thus the outer circle is 0.1°C warmer than the general atmosphere, the next circle is 0.2°C warmer, and so on. The highest temperature occurs at the center of the thermal. (After Betsy Woodward, *Cumulus Dynamics,* Pergamon Press, 1960)

losses by the air. As a parcel of air rises to higher levels, the surrounding air is increasingly less dense. The air in the rising parcel expands in response to the decrease in the density and pressure of the atmosphere. Because the molecules of the expanding air become more widely separated, they collide less often, so that the sensible temperature of the rising air decreases.

The rate of adiabatic cooling of rising air in which condensation is not occurring is known as the *dry adiabatic rate*. Its value is 10°C for every kilometer of increasing altitude (5.5°F per 1,000 ft), as illustrated in Figure 5.15.

The process can also be reversed. When a parcel of cool air descends, it becomes compressed. As a result, the molecules in the par-cel collide more frequently and the temperature increases. This is called *adiabatic heating*. A descending parcel of cool air, subject to adiabatic heating, increases in temperature at the dry adiabatic rate.

Moist Adiabatic Rate

Rising air cools at the dry adiabatic rate only until it reaches its dew point temperature and becomes saturated. After saturation, it may continue to rise, but it cools at a reduced rate called the *moist adiabatic rate*. The rate of cooling decreases at saturation because the water vapor in the parcel begins to condense, releasing latent heat. The added heat increases the buoyancy of the parcel and enables it to rise to

MOISTURE AND PRECIPITATION PROCESSES

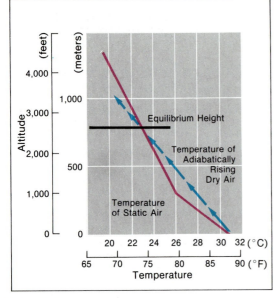

Figure 5.15
The figure compares the temperature of a typical parcel of rising dry air with the temperature of the air in the surrounding static atmosphere. The parcel experiences a net upward force, causing it to continue rising only as long as the temperature of the parcel exceeds the temperature of the static air. The equilibrium height of the parcel is near the altitude where the temperatures are equal. (Doug Armstrong)

greater altitudes. This mechanism permits towering clouds to develop with potential for high-intensity precipitation.

The moist adiabatic rate varies with the moisture content of the rising air. In warm, moist air it may be 5°C per km (2.8°F per 1,000 ft), but at very low temperatures it approaches the dry adiabatic rate of 10°C per km.

Not all the water vapor in the air parcel condenses at once. Condensation continues as the parcel rises, and at any given altitude the parcel contains just enough water vapor to maintain saturation. Figure 5.15 shows that when the rising air cools to the same temperature as the surrounding atmosphere, its upward motion by convection slows down and eventually ceases.

Thermal Convection

On a fair day, puffy clouds form within thermals at the altitude where saturation is reached and condensation begins. Figure 5.16a shows that the thermal continues to rise until the temperature of the rising air equals that of the surrounding atmosphere. This equilibrium point marks the greatest height to which the cloud can grow. The vertical movement of thermals responsible for fair-weather cumulus cloud formation is known as *thermal convection.*

Environmental Lapse Rate

Since the atmosphere is heated principally from below by longwave radiation from the earth's surface, air temperature normally decreases upward. A global average rate of temperature decrease for all weather conditions is about 6.5°C per km (3.6°F per 1,000 ft), which is known as the *average lapse rate.* Locally, however, the rate of change of temperature with altitude may deviate considerably from the average. The actual vertical temperature change measured at any location, known as the *environmental lapse rate,* has important effects on thermal convection and cloud formation.

Stability

The environmental lapse rate in Figure 5.16b shows warmer air aloft over cooler air at the surface. This *temperature inversion* is a common result of nighttime radiational cooling during fair weather. If dry air near the surface were forced to rise by flowing against the slopes of a mountain range, it would cool at the dry adiabatic rate and would soon become cooler and denser than the air above the inversion. Being cooler than the surrounding air, the lifted air would tend to sink downward rather than rise through the inversion. Therefore a temperature inversion restricts upward motion in the atmosphere. Meteorologists refer to environmental lapse rates that discourage

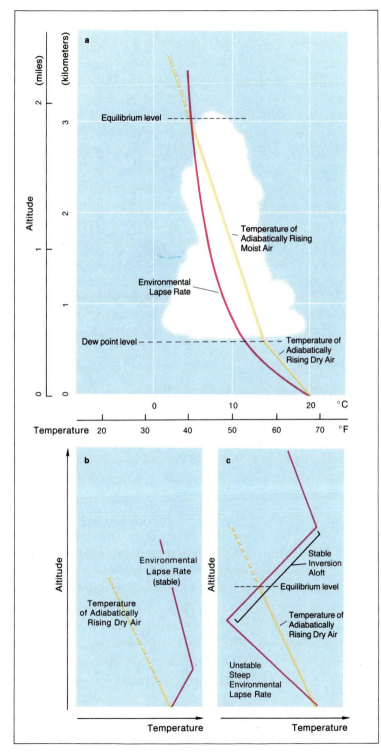

Figure 5.16

(a) When the land surface heats, it warms the lower atmosphere, causing the environmental lapse rate (red line) to bend to the right (higher temperature) at the surface. A rising thermal or parcel cools at the dry adiabatic rate (yellow line) until condensation begins. The base of the cloud at about 0.7 km marks the altitude at which the thermal cools to the dew point, initiating condensation. The parcel then cools at the moist adiabatic rate, which is less than the dry rate. The release of the latent heat of condensation enables the parcel to rise to much higher altitudes before it reaches equilibrium with the surrounding static air.

(b) This shallow inversion, with warmer air above cooler air at the surface, is characteristic of radiational cooling at night. If a parcel were forced to rise (yellow line), it would come quickly into equilibrium with the surrounding static air. Inversions are associated with atmospheric stability.

(c) With daytime heating, the inversion is much higher aloft, permitting thermals to reach much higher altitudes. The equilibrium level is the limit of thermal convection. (Doug Armstrong)

vertical movement of air as *stable lapse rates*. Temperature inversions are an extreme type of atmospheric stability, in which there is little opportunity for thermal convection or the development of thick clouds and much precipitation. By preventing the upward movement of air an inversion also acts as a lid that traps pollutants emitted by automobiles and industries and causes them to become concentrated in a layer of "smog."

Instability

On clear days the ground is warmed by solar radiation, and the lower atmosphere is heated by conduction and thermal convection. This usually eliminates any overnight temperature inversion at the surface. Air can rise, and thermals become active (Figure 5.16c). If the environmental lapse rate is steep (temperatures decreasing rapidly with height), rising thermals continue to be warmer and less dense than the surrounding air and can continue moving upward buoyantly. The rise of the air to great altitudes permits the development of towering cumulonimbus clouds that are associated with thunderstorms (see Figures 5.16a and 5.11d). Meteorologists refer to the steep environmental lapse rates leading to the development of thermal convection as *unstable lapse rates*.

Convective Showers

During summer, when rapid heating of the ground creates steep lapse rates and conditions of atmospheric instability, rain-producing cumulus-type clouds follow a regular pattern of development. Early in the day, the sky is usually clear. After a few hours scattered wisps of cloud begin to appear at an altitude of about 1 kilometer (3,000 ft). These are formed by condensation in convective updrafts. The wisps increase in size and eventually become large cumulus clouds with bases at the altitude at which the rising air has cooled to its dew point. Cumulus clouds indicate the tops of rising

thermals. Between the clouds air sinks to replace the air that rises in the thermals. For this reason, cumulus clouds never cover the sky completely, and precipitation from them takes the form of scattered showers and thunderstorms that typically break out in the afternoon and persist into early evening. Although an individual shower may produce more than 2.5 cm (1 in.) of rain locally, thermal convection rarely results in widespread rains.

Thunderstorms

Combinations of a deep layer of warm, moist, unstable air and intense surface heating can result in thunderstorms of convective origin. The lightning and thunder are associated with rising unstable cells that cool at the moist adiabatic rate to temperatures as low as $-15°C$ (5°F) with a mixed upper cloud of ice crystals and supercooled water droplets.

Figure 5.17 shows a schematic diagram of the three stages of a thunderstorm cell; most thunderstorms consist of several individual cells, with each cell progressing through a life cycle in less than 30 minutes. The mature stage is characterized by a strong downdraft of cooler air and heavy rain within the forward section, and an updraft in the rear section where water vapor is drawn into the cell and condensation occurs at rapid rates. A cell is rarely more than a few kilometers in diameter. Strong thunderstorms can produce damaging wind gusts and very heavy rains in the downdraft. Unusually strong downdrafts are called *downbursts,* and they have been shown to be related to airplane crashes during takeoffs and landings. Tornadoes, discussed in Chapter 6, are a product of the most severe thunderstorms.

Orographic Lifting

When moisture-laden air is forced to rise over a mountain barrier, the air cools, and condensation and precipitation often result. This

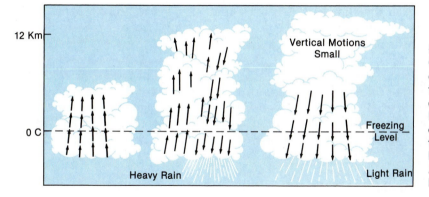

Figure 5.17
Cumulus, mature, and dissipating stages of a thunderstorm cell as described originally by Horace Byers. In the diagram the cell is moving from left to right. (After Battan, *Fundamentals of Meteorology,* 2nd ed., Prentice-Hall, 1979)

process, which exerts a strong local effect on precipitation, is known as *orographic lifting.*

To the frequent disappointment of vacationers, clouds, fog, and precipitation are characteristic of most mountainous regions. In western North America, the air normally flows from west to east, so the western, or windward, slopes of mountain ranges are often cloudy and wet. When the air descends the eastern slopes, however, the sinking air warms at the dry adiabatic rate. In the warmer descending air, water droplets evaporate and the clouds dissipate. The eastern slopes and adjacent lowlands are usually sunny and dry (Figure 5.18) and are said to lie in the *rainshadow* of the mountains. Other mountain areas are subject to different patterns of orographic lifting. In the eastern United States, moist air may advance against the Appalachian mountain system from either the east or the west, so that the mountains have no dry side.

A spectacular example of the effects of orographic lifting is found on the island of Kauai in the Hawaiian chain. The average annual rainfall on the windward side of Mount Waialeale is 1,170 cm (460 in.), but it is only 51 cm (20 in.) on the lee side, just 25 km (15 miles) away. This extraordinary rainfall record is produced by orographic clouds maintained by persistent moisture-laden winds from the northeast. It was orographic lifting of unusually moist unstable air, producing enormous

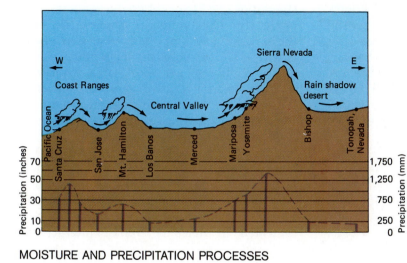

Figure 5.18
In mountainous regions average annual precipitation patterns are closely related to the terrain. This idealized transect across central California and western Nevada shows strong orographic effects on precipitation along the chains of mountains, and the rainshadow valleys between the mountain ranges east of the Sierra Nevada. The Central Valley is close to sea level; the Sierra Nevada rises to 4,200 m (14,000 ft); and Bishop and Tonapah are at about 1,200 m (4,000 ft) and 1,800 m (6,000 ft). (From C. Donald Ahrens, *Meteorology Today,* West, 1982.)

MOISTURE AND PRECIPITATION PROCESSES

cumulonimbus clouds, that caused the Rapid City and Big Thompson Canyon disasters noted at the beginning of the chapter. The truly unusual aspect of these storms was that the rain-producing clouds remained over the same areas for several hours, rather than drifting onward and spreading out their effects, as normally happens.

Frontal Lifting and Convergence

The lifting of large masses of moist air, causing widespread clouds and precipitation, does not require mountains. In the middle and higher latitudes, the atmospheric circulation often brings together large masses of warm and cold air. Largely because of density differences, the warm and cold air masses do not mix but remain separated by a boundary zone known as a *front*. Fronts, which are discussed more fully in Chapter 6, are zones of rapid transitions in the temperature and humidity characteristics of adjacent air masses. Where the air masses are converging, the warmer, less dense air either rides up and over the cooler air or is wedged upward by invading colder air. As it rises, the warm air cools to its dew point, clouds form, and precipitation is likely to occur.

Idealized cross sections of fronts are shown in Figure 5.19. When warm air is advancing horizontally over a surface previously occupied by colder air, a *warm front* exists (Figure 5.19a). The advancing warm air slides over the retreating cold air. The warm air rises only about 1 km in a horizontal distance of about 100 km; thus the ascent is gentle, unlike airflow over a mountain barrier. High cirrus clouds come first, followed by lower altostratus and nimbostratus clouds that produce widespread but gentle precipitation.

When cold air is advancing horizontally over a surface previously occupied by warmer air, a *cold front* is present. Figure 5.19b illustrates typical cloud distribution along a cold front, where advancing cold air pushes warm air aloft. The slope of a cold front is about twice as steep as that of a warm front, rising about 1 km over a horizontal distance of some 50 km. This more abrupt uplift causes cold front precipitation to be more intense but shorter in duration than warm front precipitation. In the eastern United States, cold fronts usually arrive in the form of a "squall line" of cumulonimbus thunderheads that produce violent rain. The thunderheads pass by rather quickly to be followed by clearing skies with scattered fair-weather cumulus clouds.

Surface air can also rise and cool by *convergence*. In areas of low atmospheric pressure, designated as "lows" on weather maps, surface air spirals toward the low-pressure center like the water moving toward a bathtub drain. As air converges into the low-pressure area, it is sucked upward and escapes outward aloft. This convergence and ascent frequently results in cloud formation and precipitation, even though no air mass fronts are present. The specific weather associated with atmospheric pressure systems, air masses, and fronts, and the resulting global precipitation patterns, are the subject of the next chapter.

GLOBAL PRECIPITATION DISTRIBUTION

A knowledge of the atmospheric and oceanic circulations and their interactions allows us to predict the distribution of precipitation over the earth—a most important element of global climates. The processes producing thick clouds and precipitation are most often associated with air mass convergence in low-pressure systems, causing the uplift of unstable moist maritime air. Where air is subsiding and diverging from surface high-pressure centers, cloud formation and precipitation are unlikely. It follows, therefore, that in equatorial and tropical latitudes precipitation should mirror the pattern of convergence of moist tropical air. In higher latitudes precipitation should be concentrated in zones of frontal interactions.

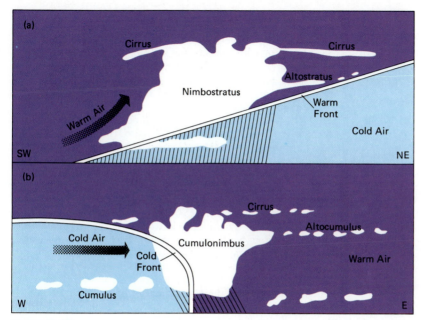

Figure 5.19

Warm and cold fronts.

(a) When a large mass of warm air moves into a region occupied by colder air, the less dense warm air flows up and over the surface cold air. This atmospheric cross section shows that the warm front slopes gently upward through the lower atmosphere. Characteristic clouds and precipitation from the warm air tend to be widespread, with precipitation usually at light to moderate intensities.

(b) When a cold air mass moves into an area of warmer air, the warm air is pushed upward more sharply. As a result, the slope of the cold front is steeper, and the precipitation associated with it tends to be more intense and localized. (Vantage Art, Inc., after Horace R. Byers, *General Meteorology,* 3rd ed., 1959, McGraw-Hill Book Company)

Coasts washed by warm ocean currents will be zones of heavy precipitation as will the windward sides of mountain ranges. Cold water coasts and the leeward sides of mountain ranges will be dry.

The map of average annual precipitation over the continents, shown in Figure 5.20a, bears out these expectations. Precipitation is especially high in three general locations: the equatorial region where the intertropical convergence zone is present most of the year; the monsoon regions of India and southeast Asia; and the region of the polar front in the mid-latitudes. Wet spots also appear wherever moist air is regularly forced up and over mountain ranges, as in the Pacific Northwest from northern California to southeastern Alaska, in the Andes of southern Chile, and on the south side of the Himalaya Mountains.

In contrast, precipitation is low over most subtropical regions, from the Sahara Desert through Arabia and Iran to Pakistan; over the interiors of Asia and North America, which are far from oceanic moisture sources; and over the polar regions where the capacity of the atmosphere for water vapor is low.

MOISTURE AND PRECIPITATION PROCESSES

Figure 5.20

Maps of average annual precipitation.

(a) This map of average annual precipitation illustrates the complex distributions associated with the positions of continents, ocean basins, and mountain barriers. In spite of the complexity, however, global patterns of high and low precipitation are evident, and these should be related to the global pressure and wind systems shown in Figure 4.5.

(b) The pattern of average annual precipitation over North America displays some broad regularities. The East tends to be moist because of atmospheric moisture from the Gulf of Mexico and the adjacent Atlantic. Precipitation decreases westward from the Mississippi River Valley. Moist air from the Gulf of Mexico seldom flows westward, and moist air from the Pacific releases most of its moisture as orographic rainfall on coastal mountain slopes facing west.

(c) The Great Plains are dry in winter because of the influx of cold dry air from the Canadian Arctic. The Far West receives most of its precipitation in winter, when midlatitude cyclones sweep the coast; a strong subtropical high over the eastern North Pacific inhibits precipitation during summer. The East receives precipitation during both winter and summer. (After *Goode's World Atlas*, 1970)

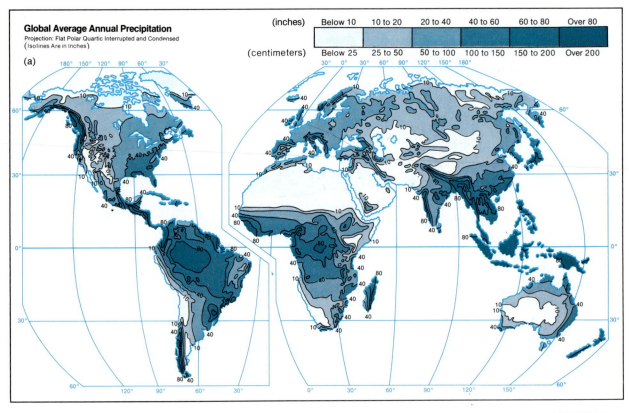

Global Average Annual Precipitation
Projection: Flat Polar Quartic Interrupted and Condensed
(Isolines Are in Inches)

(a)

(inches)	Below 10	10 to 20	20 to 40	40 to 60	60 to 80	Over 80
(centimeters)	Below 25	25 to 50	50 to 100	100 to 150	150 to 200	Over 200

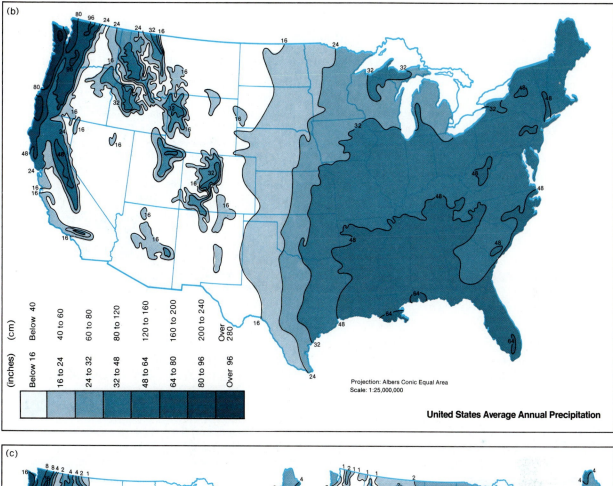

(b)

(inches)	(cm)
Below 16	Below 40
16 to 24	40 to 60
24 to 32	60 to 80
32 to 48	80 to 120
48 to 64	120 to 160
64 to 80	160 to 200
80 to 96	200 to 240
Over 96	Over 280

Projection: Albers Conic Equal Area
Scale: 1:25,000,000

United States Average Annual Precipitation

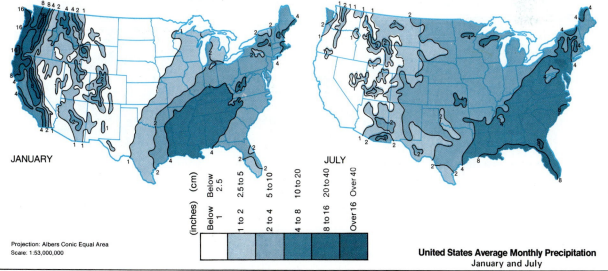

(c)

JANUARY

JULY

Projection: Albers Conic Equal Area
Scale: 1:53,000,000

(inches)	(cm)
Below 1	Below 2.5
1 to 2	2.5 to 5
2 to 4	5 to 10
4 to 8	10 to 20
8 to 16	20 to 40
Over 16	Over 40

United States Average Monthly Precipitation
January and July

MOISTURE AND PRECIPITATION PROCESSES

Next-Generation Weather Radar

Most television viewers are familiar with radar returns that show the distribution of precipitation around radar facilities of the National Weather Service. The WSR-57 radars emit short pulses of electromagnetic radiation, and the radar screen records the returns from clouds and precipitation particles. The returns, which reflect the intensity of the precipitation, are displayed on the screen in terms of brightness or on a scale of colors. Although the trained observer can estimate the arrival time of an approaching squall line, spotting tornadoes within the severe weather remains difficult. Many tornadoes are undetected on radar; even when they are, the lead time between formation and detection is often barely several minutes. But that could change.

NEXRAD is the acronym given by the Weather Service for its next generation of radars scheduled to be installed and operational in the early 1990s. NEXRAD takes advantage of the "Doppler effect," which enables the radar to see motion of precipitation particles and wind patterns within clouds. Research with experimental Doppler radars began in the late 1950s, and intensive research at the National Severe Storms Laboratory at Norman, Oklahoma,

showed that it was very possible to identify the "signature" patterns of wind flow in severe storms as much as 30 minutes ahead of tornado formation, then follow the tornadoes through their entire life cycles.

With the NEXRAD system, Weather Service personnel expect to be able to provide early warnings of severe storms and tornadoes, which could save many lives each year. The system should also provide early warnings of possible hail on the Great Plains and of excessive local rainfall that could generate treacherous flash flooding in urban areas. NEXRAD should also be extremely helpful at major metropolitan airports during stormy weather when commercial airplanes cluster like flys awaiting their turns to land.

Present NEXRAD plans call for a network of more than 110 radars and even more user processor subsystems to be shared by the National Weather Service, the Federal Aviation Administration (FAA), and the Air Weather Service, a branch of the U.S. Air Force. The first of a limited production of new units should be in place late in 1988, and research personnel of the Weather Service anticipate a near revolution in providing accurate weather information.

KEY TERMS

evaporation
condensation
precipitation
vapor pressure
saturation vapor pressure
vapor pressure gradient
stomata
transpiration
humidity
absolute humidity
specific humidity
mixing ratio
dew point

relative humidity
condensation nuclei
supersaturated air
dew
frost
radiation fog
advection
advection fog
orographic fog
clouds
   cirrus
   cumulus
   cirrostratus

cirrocumulus
stratus
stratocumulus
altostratus
altocumulus
nimbostratus
cumulonimbus
coalescence model
Bergeron ice crystal model
drizzle
supercooling
ice-forming nuclei
hailstones
adiabatic cooling
dry adiabatic rate
adiabatic heating

moist adiabatic rate
thermal convection
average lapse rate
environmental lapse rate
temperature inversion
stable lapse rate
unstable lapse rate
convective showers
thunderstorms
downburst
orographic lifting
rainshadow
warm front
cold front
convergence

REVIEW QUESTIONS

1. Compare the factors affecting the transfer of water to the atmosphere over oceans and continents.
2. What is the significance of the relationship between temperature and the atmosphere's capacity for holding water vapor?
3. What are the mechanisms by which water is returned to the atmosphere from the earth's surface?
4. What variables determine the vapor pressure gradient? What is its significance in terms of evaporation rates?
5. Compare the various measures of atmospheric humidity.
6. Describe the diurnal regime of vapor pressure and relative humidity during fair weather.
7. Discuss the process by which water vapor is transformed into drops or ice crystals large enough to fall from clouds to the ground as precipitation.
8. Discuss some situations in which atmospheric cooling might eventually result in clouds and precipitation.
9. Why is there very little chance of significant precipitation during radiation fogs?
10. Discuss the diurnal regime of stability and instability near the earth's surface during fair weather.
11. What are the physical bases for atmospheric weather modification?

APPLICATIONS

1. From Figure 5.20, identify the regions of the earth where the annual precipitation averages less than 25 cm (10 in.). Which of these regions appears to be associated with the subtropical highs, rainshadows, continental interiors, and cold ocean currents? Explain the reason for each of the following deserts: the Gobi, Atacama, Thar, Namib, Taklamakan, Kalahari, Sahara, Simpson, Sonoran, and Great Basin.

MIDLATITUDE SECONDARY CIRCULATIONS

Most midlatitude weather is dominated by the interaction of large air masses of unlike characteristics. In the northern hemisphere, especially during winter when the wave patterns of the upper-air westerlies have the greatest north to south amplitudes (see Figure 4.8) and the jet stream flows vigorously, warm moist air moving poleward out of the subtropical highs meets cool, dry air flowing equatorward out of the Siberian and Yukon highs. Where the air masses meet, a front is formed. The interacting airflows frequently become organized into vast, spiraling eddies that constitute "storms" of midlatitudes. These usually migrate from west to east and may be as much as 1,000 km (600 miles) in diameter. Such migrating secondary circulations and associated air masses and fronts produce the changeable weather of the middle latitudes.

Air Masses

An *air mass* is a large, nearly uniform segment of the atmosphere that moves as a unit in association with secondary high- or low-pressure systems. Air masses retain the temperature and moisture characteristics of their source regions even after traveling thousands of kilometers. Air mass types are therefore identified according to their source region: *polar (P)* or *tropical (T), continental (c)* or *maritime (m)*. The capital letter indicates air mass temperature characteristics; the lower-case letter suggests relative moistness. Thus, four general air mass types are commonly recognized: *cP, mP, cT,* and *mT*. Some classifications also include arctic *(A)* and equatorial *(E)* air masses. Air masses are also designated according to their stability at the surface and aloft: *K* and *W*, meaning colder and warmer, respectively, than the surface beneath, and *u* and *s*, meaning unstable or stable aloft. An *mTKu* air mass has maximum instability: it is colder than the surface, therefore subject to heating from below, which produces a steep lapse rate; and it is unstable aloft as well.

Continental air masses normally produce very little precipitation. Instead, they tend to gain moisture from evapotranspiration. The opposite is true of maritime air masses; once formed, they lose more moisture by precipitation than they receive from evapotranspiration. They are the sources of rain and heavy snowfall around the world.

Polar Air

Extremely cold winter weather in the United States is associated with *continental polar (cP)* or arctic *(A)* air masses that form over the snow-covered plains of northern Canada. Arctic air masses originate closer to the polar region and are colder in the summer than those air masses called polar. In this discussion of North American air masses we shall regard all cold air masses as being the polar type.

During the long subarctic winter, the sun often remains below the horizon and frequent clear skies promote intense radiational cooling of the snow-covered land surface. After a few days these conditions create a nearly homogeneous cold, dry cP air mass that extends horizontally for more than 1,000 km (600 miles).

Eventually, the cP air spreads out of the source region and moves toward lower latitudes (Figure 6.1). The air is modified only slightly as it sweeps southward, traveling more than 4,000 km (2,500 miles) across the Great Plains and down the Mississippi Valley to the Gulf of Mexico. During midwinter, even at New Orleans the mean temperature of the cP air is close to 0°C (32°F), with occasional temperatures lower than −5°C (23°F).

The mountains of western North America prevent most cP air from spreading to the Pacific coast. Instead, the west coast is usually under the influence of *maritime polar (mP)* air that has moved eastward from the seas off Japan and Alaska. This air is never as cold as

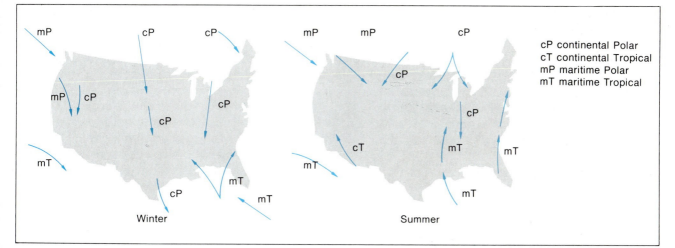

Figure 6.1
This figure depicts the average movements of air masses over the conterminous United States during winter and summer. During winter much of the northern half of the United States is invaded by cold polar air, whereas during summer the northward flow of tropical air dominates the weather in most regions. (Calvin Woo after Dieter H. Brunnschweiler, *Geographica Helvetica*, Vol. 12, 1957)

polar air that originates over the land, but it is much more moist than *cP* air. Thus, Pacific coast lowlands receive heavy rain in winter but seldom experience freezing temperatures. Similar *mP* air from the North Atlantic occasionally invades the northeastern United States, bringing freezing rain in the fall and winter.

Tropical Air

However, when the *cP* air moves out over the warm waters of the Gulf of Mexico, it gains moisture and is warmed rapidly. Modification is so quick that a "new" air mass is produced within 48 hours or so. This warm, moist air mass is designated *maritime tropical (mT)*, just like air masses that originate over the tropical oceans. Most heavy rains in the midlatitudes are produced by condensation of moisture within poleward-moving *mT* air masses.

Occasionally in the summer the Pacific coast

receives an outbreak of *continental tropical (cT)* air. This air is hot and extremely dry, coming from the deserts of Mexico and the southwestern United States. Globally, the most important source of *cT* air is the Sahara Desert of North Africa. Saharan *cT* air affects all the Mediterranean coastlands, occasionally crosses the Alps, and even penetrates as far as Great Britain and Scandinavia.

Great Lakes Snow Squalls

Air masses move over the earth's surface according to predictable routes, with movement determined by upper-air circulation patterns over North America. But the weather also depends on surface characteristics. For example, over the large land areas of the Great Plains and Mississippi Valley, *cP* air is generally associated with weather that is cold and windy but clear. However, over the Great Lakes the lower layers of *cP* air receive heat and moisture from large expanses of water. The results are spectacular snow squalls on the eastern and southeastern shores of the Great Lakes. The impact of these snow squalls on the pattern of average seasonal snowfall over the northeast and northcentral states is shown in Figure 6.2.

SECONDARY ATMOSPHERIC CIRCULATIONS

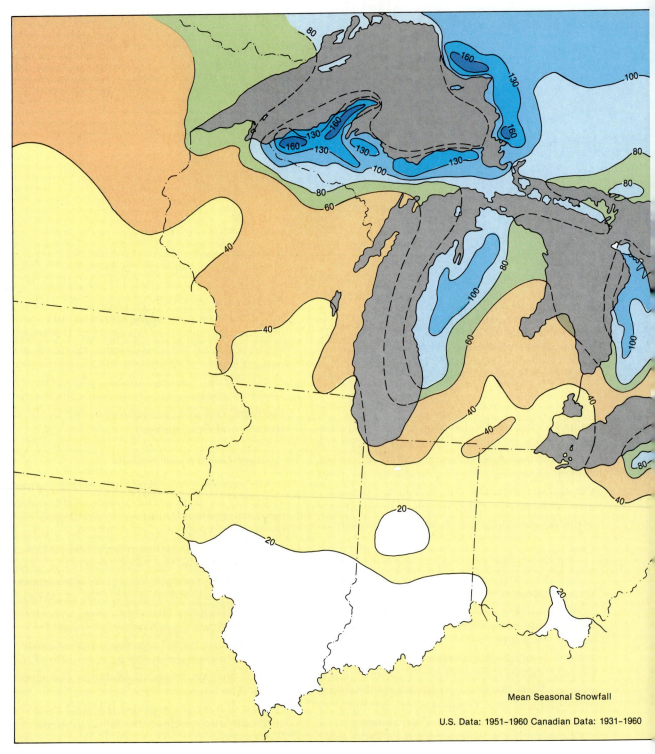

Mean Seasonal Snowfall

U.S. Data: 1951–1960 Canadian Data: 1931–1960

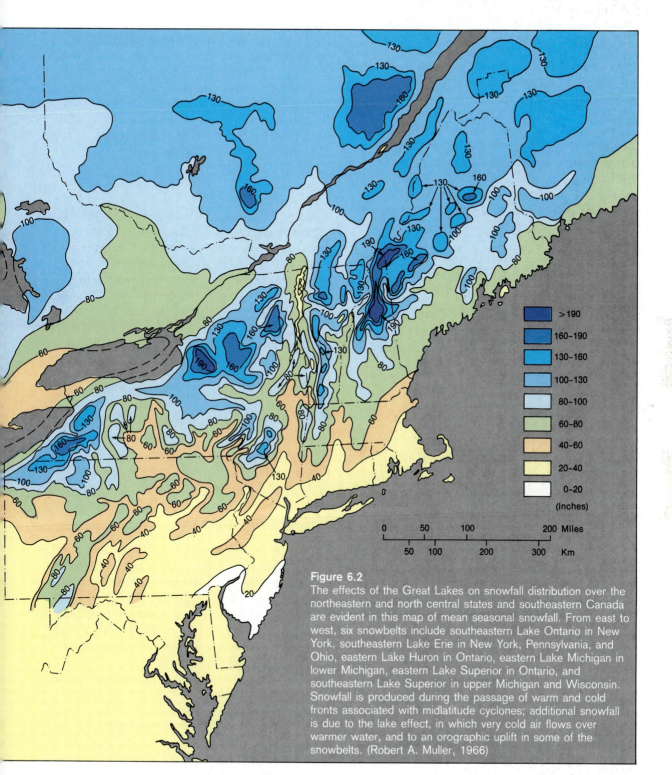

Figure 6.2
The effects of the Great Lakes on snowfall distribution over the northeastern and north central states and southeastern Canada are evident in this map of mean seasonal snowfall. From east to west, six snowbelts include southeastern Lake Ontario in New York, southeastern Lake Erie in New York, Pennsylvania, and Ohio, eastern Lake Huron in Ontario, eastern Lake Michigan in lower Michigan, eastern Lake Superior in Ontario, and southeastern Lake Superior in upper Michigan and Wisconsin. Snowfall is produced during the passage of warm and cold fronts associated with midlatitude cyclones; additional snowfall is due to the lake effect, in which very cold air flows over warmer water, and to an orographic uplift in some of the snowbelts. (Robert A. Muller, 1966)

SECONDARY ATMOSPHERIC CIRCULATIONS

Fronts

A *front* is an interface between two air masses that differ in temperature or humidity, or both. The most distinct fronts are those separating air masses whose properties contrast most sharply—in particular, the cold, dry *cP* and warm, moist *mT* types. Such fronts are common in eastern North America. On the Pacific coast most fronts merely separate successive *mP* air masses of varying temperature and humidity, although *mT* and *cT* air masses are occasionally involved, especially in southern California. On rare occasions *cP* air penetrates into central California and the Pacific Northwest, with disastrous consequences to ornamental vegetation and winter crops in these areas.

Warm Fronts

When warm air moves into a region previously occupied by a colder air mass, the forward edge of the warm air mass is designated as a *warm front*. Figures 5.19 (see page 129) and 6.3a show how a warm front slopes forward from the surface as the lighter warm air slides over the denser cold air it is replacing. The cool air is shown retreating toward the right, corresponding on a map to the east or northeast, the usual direction in which a warm front advances in the middle latitudes. At an altitude of 1 km, the warm air may be several hundred kilometers ahead of the front at the surface.

Weather conditions in the vicinity of a warm front depend on the properties of the air masses as well as the nature of the land surface. Nevertheless, there is a characteristic sequence of weather conditions associated with a warm front. Condensation and cloud formation begin where the warm, moist air rises over the cooler air and cools adiabatically to its dew point. Because the warm front slopes so greatly, it is usually detected through the appearance of high cirrus clouds a day or more before the surface front arrives (Figures 5.19, page 129, and 6.3a). As the surface front approaches, sheet-like stratus clouds become thicker and lower. Just ahead of the surface front is a broad band of precipitation. Warm fronts move slowly and often produce steady rains that may last a day or more.

Cold Fronts

A *cold front* develops when cold air advances into a region occupied by warm air. Figure 6.3b shows that the advancing cold air pushes under the warm air, forcing the warm air to rise abruptly, so that the frontal slope is much less gradual than that of a warm front. There is rapid development of towering cumulonimbus clouds, with precipitation occurring just ahead of the surface front. The passage of a cold front in the summer is usually associated with the sudden appearance of a line of thunderstorms and a rapid drop in temperature (Figure 6.4). The zone of precipitation along a cold front may pass by in only an hour or two. In those cases where the surface boundary between tropical and polar air masses remains at about the same location for a day or more, the front is designated as *stationary*.

Occluded Fronts

More complex *occluded fronts* consist of three or more air masses, with one front overtaking another so that the air mass between loses contact with the ground. This creates a broad band of heavy precipitation falling from the occluded (lifted) air mass (Figure 6.3c). Occluded fronts will be treated below in connection with cyclonic storms.

All fronts are segments of the polar front, which is a component of the general circulation. As such, they are associated with moving secondary circulations, including midlatitude cyclones and anticyclones.

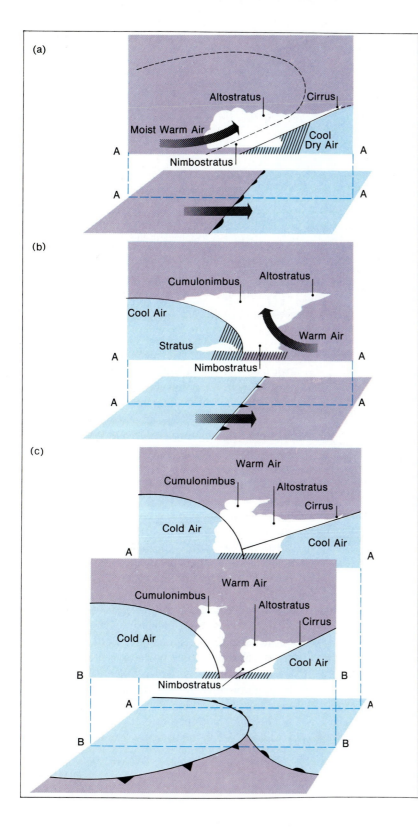

Figure 6.3
Three-dimensional representation of air-mass frontal zones.

(a) Vertical cross section of a warm front, together with the surface symbol as it would appear on a weather map. As the uplifted warm air becomes cooler, condensation and cloud formation may occur. The resulting precipitation arrives at a location ahead of the surface warm front itself.

(b) A cold front develops when a mass of cool air pushes into a region of warmer air. The warm air ahead of the front is forced upward over the incoming cool air. Cloud formation and precipitation may occur as the uplifted warm air cools. Precipitation normally arrives at a location just ahead of the surface front.

(c) An occluded front involves three air masses of different temperatures, and it combines some of the features of both warm and cold fronts. In the forward vertical section, the occlusion is not complete, and the warm air is in contact with the ground in close proximity to a cold front and a warm front. In the rear vertical section, the occlusion has formed, and the warm air is lifted completely above the ground. (Calvin Woo adapted from Hermann Flohn, *Climate and Weather,* © 1969 by H. Flohn, used with permission of McGraw-Hill Book Company)

Figure 6.4
Instability of the air in this squall line over the Texas Panhandle is indicated by the development of cumulus clouds overtopped by giant cumulonimbus thunderheads. (Robert A. Muller)

Cyclones and Anticyclones

Almost any daily weather map for a midlatitude region shows centers of high and low pressure. The spiraling flows of air into moving centers of low pressure, called "lows," are *cyclones*. The diverging flows around high-pressure regions, referred to simply as "highs," are *anticyclones*. The horizontal differences in pressure within large highs or lows usually amount to less than 10 millibars over a distance of 100 km (60 miles), only a small fraction of the normal sea level atmospheric pressure of 1,013 millibars.

Midlatitude Cyclones

As shown in Figure 6.5, a traveling cyclone consists of converging surface air that ascends and diverges in the upper atmosphere. As air rises in a cyclone, it cools adiabatically, resulting in cloud formation and perhaps rainfall. A traveling anticyclone, on the other hand, consists of air subsiding from aloft that diverges at the surface. The descending air in an anticyclone is heated adiabatically, reducing its humidity and producing clear skies.

Midlatitude cyclones develop along the polar front where warm and cold air come into contact. In North America, cyclones most often form just east of the Rocky Mountains and migrate toward the east and northeast, eventually passing out into the Atlantic Ocean. Most precipitation in the central and eastern United States and Canada is associated with the ascent of warm, moist mT air over the cold cP air within the cyclonic circulation. Along the Pacific coast most cyclonic disturbances involve the interaction of mP air masses of differing character. Having originated far away off the coast of Japan or Alaska, these cyclonic storms commonly arrive at the west coast in the form of mature occluded fronts.

The evolution of midlatitude cyclones usually follows a characteristic pattern. Cyclones begin along a stationary segment of the polar front. Figure 6.6a shows a stationary front with a northeast-southwest orientation. Such a front may remain over the same region for as much as 24 to 48 hours before a cyclone begins to develop, possibly due to an upper-air disturbance, such as a wave trough in the jet stream.

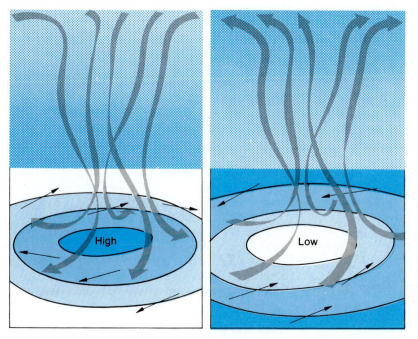

Figure 6.5
In a region of high pressure (left), air descends from the upper atmosphere and diverges outward along the surface of the ground. In a region of low pressure (right), air converges inward along the surface of the ground and ascends to the upper atmosphere. The directions of circulation are shown for the northern hemisphere. (Calvin Woo)

The isobars in Figure 6.6a show that pressure increases away from the front on both sides. On the polar side of the front, cold air flows from the northeast; on the tropical side, warm air flows from the southwest. As is normal where warm and cold air are in contact, there is often warm air aloft over the cold air at the surface. This produces a band of clouds on the polar side of the front, with some light precipitation.

The cyclone begins as a small wave disturbance caused by a local drop in pressure, as shown in Figure 6.6b. As pressure falls even more, air begins to converge toward the center of the low in a counterclockwise circulation. Ahead of the center, the warm air advances into the former domain of the cold air, forming a warm front. The cold air begins to sweep southeastward behind the center, forming a cold front. A broad apron of warm-front rain develops ahead of the center, and a narrow band of showers breaks out along the cold front.

Many small waves may move toward the northeast along the polar front, dissipating in 6 to 12 hours. A few waves grow into well-developed cyclonic circulations. One of the difficult tasks of the weather forecaster is to predict the development of a significant midlatitude cyclone from a small wave disturbance.

Figure 6.6c shows the cyclone at a more advanced stage. The central pressure of the cyclone may fall to less than 1,000 millibars (29.5 in.), and the converging counterclockwise circulation may expand to a diameter exceeding 1,000 km (600 miles). In eastern North America the cyclonic system normally moves in a northeasterly direction at a speed between 25 and 50 km (15 to 30 miles) per hour. Steady and substantial rains usually occur ahead of the surface position of the warm front, and occasionally severe thunderstorms and even tornadoes are associated with the cold front. The weather map in Figure 6.7 shows a small, intense midlatitude cyclone out of which several "killer" tornadoes were born.

Radioactive Fallout from Chernobyl

The dispersion of dangerous materials through the atmosphere is dependent mostly on the weather, especially wind patterns. Whether the accidental dispersion involves a cloud of toxic gas escaping from a disabled tank truck, radioactive particles emitted from a nuclear power plant, or even ash exploding from a volcano, the plume of materials will be swept along with the wind. Large volcanic bombs fall back to the slopes of the volcano adjacent to the crater, heavier ash may be carried for tens of kilometers, and very fine volcanic ash that is lifted to the stratosphere may stay aloft for a year or more and be swept around the hemisphere many times.

When Swedish nuclear engineers recorded sudden increases in radioactivity in the atmosphere on Sunday, April 27, 1986, they soon began to suspect a nuclear disaster within the Soviet Union because the low-level tropospheric winds were from the southeast. It was not until several days later that Soviet authorities issued the first official statements of the nuclear power plant disaster at Chernobyl, about 100 km (60 miles) north of Kiev in the Ukraine.

During the early phases of the fire and meltdown, a midlatitude cyclone was located to the northwest of Chernobyl over Scandinavia and the Baltic Sea, and the low-level counterclockwise converging air in the warm sector swept radioactive materials over Poland, the Baltic States, and Scandinavia. The storm system with its trailing cold front moved steadily toward the east, however, and high pressure settled over Eastern Europe within two days. The diverging clockwise airflow, and the frontal showers along the cold front, helped to cleanse the air of radioactive particles over northern and central Europe; instead, radioactive materials were swept to the southwest of Chernobyl over Romania and Bulgaria, and eventually over the Mediterranean Sea and adjacent countries by May 2. With winds first from the southeast and then from the northeast following the cold front, the nearby metropolitan population of Kiev apparently was spared from a heavy fallout of radioactive debris. A different wind pattern could have been catastrophic for the city of more than 2 million people.

Some of the radioactive materials were swept aloft into the jet streams of the upper troposphere, and by May 7 very low levels of radioactive materials had been measured more than half way around the northern hemisphere to the east in Washington and Oregon. Fortunately, the continued release of radioactive materials was greatly curtailed within a week, and except in the immediate vicinity of Chernobyl, the levels of radioactivity were not judged high enough to be significantly dangerous to health at that time. Even so, the effects on farmers all over Europe were severe, with animal products such as milk and lambs unmarketable because of real or suspected contamination. But different weather patterns could have produced a much greater disaster.

Figure 6.6 (opposite)
Cyclones in the midlatitudes usually follow characteristic patterns of evolution along the polar front. This figure should be studied in conjunction with Figure 6.3, which shows other properties of fronts.

(a) The evolution begins along stationary segments of the polar front where cold and warm air stream in opposite directions.

(b) The stationary front tends to be unstable, and a bulge, or wave, usually develops within one or two days. The waves move along the polar front toward the northeast, and most dissipate within 6 to 12 hours.

(c) A few waves, however, grow into well-developed circulations with diameters of 1,000 km (600 miles) or more.

(d) The cold front eventually overtakes the warm front, lifting the warm air away from the surface and forming an occlusion. The occlusion eliminates the surface air temperature differences that provide energy for the system, and the circulation then weakens and finally dissipates.

(e) This sequence of development can be traced on the three map sketches for March 5, 6, and 7, 1973. As a cyclone occludes, a new wave disturbance may form farther back along the trailing stationary front.

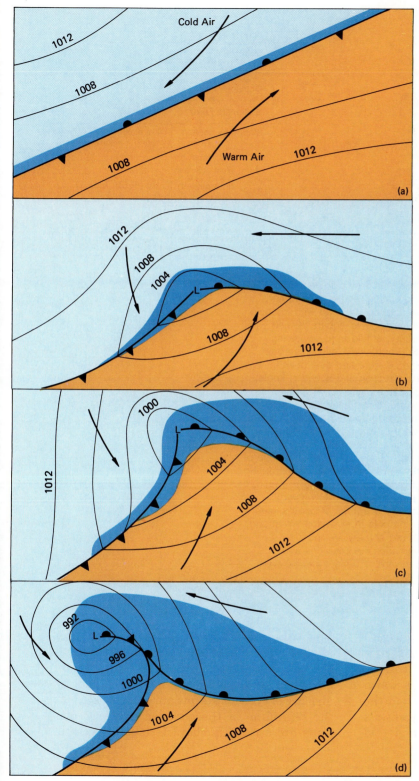

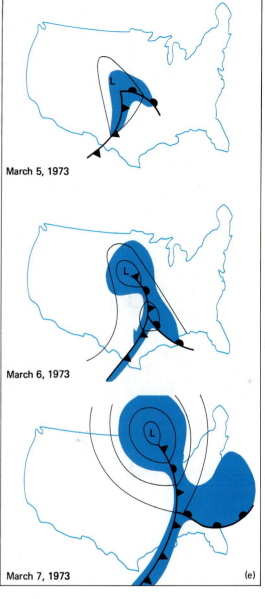

March 5, 1973

March 6, 1973

March 7, 1973

(a)

(b)

(c)

(d)

(e)

Cold Air

1012

1008

1008

Warm Air

1012

1012

1008

1004

L

1008

1012

1000

1012

L

1004

1008

1012

992

L

996

1000

1004

1008

1012

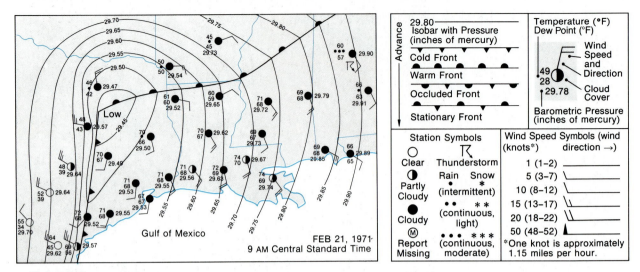

Figure 6.7
The weather map (left) portrays weather conditions in the south-central United States on February 21, 1971, based on ground station reports at 9:00 A.M. Central Standard Time. The information on the map can be read with the help of the accompanying symbol table (right). The map shows isobars labeled with barometric pressure readings in inches of mercury, a cold front advancing eastward through Texas, a warm front advancing northward, and numerous station reports. The station symbols indicate temperature and dew point in °F, barometric pressure in inches of mercury, wind direction and speed in knots, cloud coverage, and precipitation events. Note that a station in Alabama, near the eastern edge of the map, reports a thunderstorm and heavy rain.

The station symbols on the weather maps prepared by the National Weather Service carry more detailed information than do the station reports on the simplified weather map shown here. Among other additional station data, the Weather Service maps present cloud type, height of the cloud base, change of barometric pressure in the previous three hours, and weather conditions during the previous six hours.

This weather map shows an intense low-pressure system over eastern Texas. Note that most of the surface winds cross the isobars at small angles and converge toward the low center. West of the cold front, winds are strong, with speeds of 25 knots. During the afternoon of February 21, the low-pressure area generated thunderstorms and tornadoes in Louisiana and Mississippi. Damage from the tornadoes was severe, and many people were injured or killed. (After Robert A. Muller, 1973)

Figure 6.6d shows the cyclone in its mature phase. The cold front moves more rapidly toward the east than the cyclonic system does, so the cold front eventually overtakes the warm front. This lifts the warm air away from the surface, forming an occlusion. Occluded fronts have the properties of both warm and cold fronts (Figure 6.8). Heavy precipitation usually falls from the moist warm air that is lifted during the occlusion. Thus precipitation occurs on both sides of the surface front. The intensity and size of cyclonic storms are often greatest at the beginning of the occluded phase. Occlusions are especially characteristic of cyclonic

Figure 6.8
This weather satellite image of the eastern Pacific shows the typical comma-shaped cloud pattern produced by an occluded front that has moved into British Columbia and is approaching Washington and Oregon. The low pressure center is marked by the cloud spiral south of Alaska. The cloud band at the bottom of the view is the result of convection in the intertropical convergence zone. (National Oceanic and Atmospheric Administration)

systems over northern California and the Pacific Northwest, where the Coast Range, Cascade Mountains, and Sierra Nevada are often inundated by heavy rains (and snows at higher elevations) during winter and spring.

In its final stage, the energy supply of the cyclone undergoes a significant alteration. The developing occlusion eventually eliminates the surface air temperature differences that power the circulation. Warm, moist air from the warm front no longer feeds into the center of the system. It is not long before the low-pressure center begins to fill with cooler, drier air, and the cyclone itself loses its energy. The life cycle of a midlatitude cyclone ranges from 24 hours to as much as five days, so that it is possible for a particularly long-lived cyclone to travel across most of the United States.

The passage of midlatitude cyclones and associated fronts accounts for most of the late fall, winter, and spring precipitation over North America. Cyclones that develop along the eastern slopes of the Rocky Mountains from Alberta to Colorado and along the Gulf and Atlantic coasts migrate along the western, southern, and eastern margins of polar outbreaks, following paths that lead them toward the northeast and eventually to the region of the Icelandic low. Cyclones that reach the Pacific coast of North America in an occluded stage have formed along the eastern coast of Asia.

Jet streams play a part in the formation of midlatitude cyclones. The presence of waves in a jet stream aids in the ascent of surface air in the low-pressure center of a cyclone. Cyclones intensify when the surface flow of air coincides with favorable conditions aloft; when airflow aloft is unfavorable, development of the cyclone seems to be suppressed. That is why weather forecasters constantly monitor the upper air.

SECONDARY ATMOSPHERIC CIRCULATIONS

Tornadoes

A *tornado* is a narrow vortex of rapidly whirling air. It is almost always associated with severe thunderstorm activity. The rotating vortex extends downward from a cumulonimbus cloud, and it becomes visible when water vapor condenses, producing the familiar *funnel cloud*. Figure 6.9a shows dust and debris swept up from the ground, creating a much darker and more ominous-looking funnel. Occasionally, several funnels may dangle from the same cloud, and many funnel clouds aloft never reach the ground. As the mother cloud moves on, the funnel is often retarded at the surface by friction, so that it becomes tilted or crooked. Tornado funnels over coastal waters and seas are called *waterspouts*.

Tornadoes usually advance at a speed of 32 to 48 km (20 to 30 miles) per hour. Thus a tornado will normally pass a given point in a matter of seconds. The tornado and its track along the surface are usually only a few meters wide, although occasionally extending up to several hundred meters. Some tornadoes skip across the landscape, leaving a broken track of destruction. Most tornado tracks are 5 to 10 km (3 to 6 miles) in length, but a few are much longer; on May 26, 1917, a single tornado tracked more than 400 km (250 miles) across Illinois and Indiana in a little less than eight hours.

Although tornadoes are short-lived, they are extremely violent. The winds of the tornado vortex have been estimated to reach speeds of up to 600 km (370 miles) per hour. Within the funnel cloud, atmospheric pressure is as much as 50 millibars lower than the adjacent air, and pressure drops of up to 100 millibars have been estimated from damage patterns. As a tornado passes over a building, the strong winds of the vortex rip at the exterior, and the abrupt pressure drop causes the building literally to explode, with the roof and walls blown out. Cellars, interior halls, and bathrooms offer greater safety than rooms with outside walls.

House trailers are especially vulnerable to tornado winds.

Tornadoes break out most commonly along thunderstorm or squall lines ahead of cold fronts trailing from midlatitude cyclones, where temperature contrasts are large and instability great. In the United States instability is usually associated with a deep layer of warm, dry cT air from the Southwest above warm, moist mT air from the Gulf of Mexico. Tornadoes usually occur in the warm sector, and track from the southwest toward the northeast (Figure 6.9b).

The seasonal distribution of tornado outbreaks tends to follow the geographical distribution of fronts separating mT, cP, and cT air masses, midlatitude cyclones, and maximum instability through a deep layer of the troposphere. In the United States, therefore, tornadoes are most frequent in the South during late winter and early spring, with the hazard migrating northward into Kansas and Missouri, and finally to Iowa and Nebraska by June, when there are few tornadoes near the Gulf Coast. Many tornadoes occur during late afternoon, when instability tends to be greatest, but unfortunately occurrences remain relatively common at night, when visual detection is difficult.

In the United States, weather situations favorable for tornado development occur most frequently over the Great Plains, the Mississippi Valley, and the Southeast. The region extending from the Texas Panhandle northeastward across Oklahoma and eastern Kansas is sometimes called "tornado alley" (Figure 6.9c). Tornadoes are very infrequent west of the Rocky Mountains and across northern New England and the upper Great Lakes region. The mean annual number of tornadoes reported has increased significantly in recent decades. Since 1970 the annual average has been about 900. The recent increase is attributed mainly to improved observations and detection.

Microbursts Around Airports

On a sultry afternoon in July 1982 at New Orleans, Pan American Flight 759 took off into a broken sky with scattered thundershowers—not an unusual summer weather pattern along the Gulf Coast. Less than half a minute after takeoff, the plane crashed into a cluster of houses, killing all 145 people on board and 8 people on the ground. Not only was it one of the most deadly American aviation accidents up to that time, but careful study showed that takeoff and landing accidents in similar weather situations have occurred from time to time.

The National Transportation Safety Board, after a careful investigation, concluded that the crash was associated with a wind condition known as a "microburst." Microbursts were first identified by Dr. Theodore Fujita of the University of Chicago, who has continued to study severe local storms and tornadoes for almost 40 years.

Microbursts are strong downdrafts no more than 3 km (1.9 miles) in diameter that occur with thunderstorms. The descending air in the downdraft approaches the surface below and fans out quickly in all directions. As a plane landing or taking off flies through the microburst, in less than a minute it must pass through headwinds, giving the plane extra lift, then tailwinds, forcing the plane toward the ground. The microburst only lasts for 5 to 10 minutes as it is dragged along by the parent thunderstorm overhead, and it may be hidden within a rain shaft or invisible in rain-free air below the clouds.

The microbursts are most treacherous to aircraft that are taking off or landing. In this mode, the airspace between the plane and the ground is very limited, which makes it difficult for the pilot to compensate for unexpected headwinds, downdrafts, and tailwinds within a few moments. Pilots have always been trained to avoid thunderstorms, and airborne radar usually enables them to navigate around the most dangerous downdrafts and updrafts. Fortunately, microburst incidents are rare events.

Nevertheless, strong efforts are being made to help pilots cope with these most difficult situations. Training programs have been developed to help pilots instantaneously recognize microburst situations, and cope with them if caught. And most commercial airports are installing new almost instantaneous alert systems that measure wind speeds and directions at six locations around the airports. The data are monitored and processed by a computer programmed to search for sudden patterns that could be related to microbursts.

Anticyclones

A traveling anticyclone consists of air subsiding from aloft and diverging at the surface. The descending air in an anticyclone is warmed adiabatically, thus reducing its relative humidity and resulting in clear skies.

In the United States many of the anticyclones spill southeastward from the Yukon high with clear skies, dry air, strong winds, and cold temperatures. During fall, winter, and spring, meteorologists often refer to these fair-weather systems as polar outbreaks; the most severe *Arctic outbreaks* are capable of bringing near-zero temperatures (0°F or −18°C) almost to the Gulf Coast, from eastern Texas to the Florida panhandle. In a few hours these systems can wreak great damage on citrus orchards, winter vegetables, and suburban landscapes that feature exotic tropical plants.

In North America during winter, cyclonic storms and anticyclones that bring clear skies often cross the continent in rapid succession,

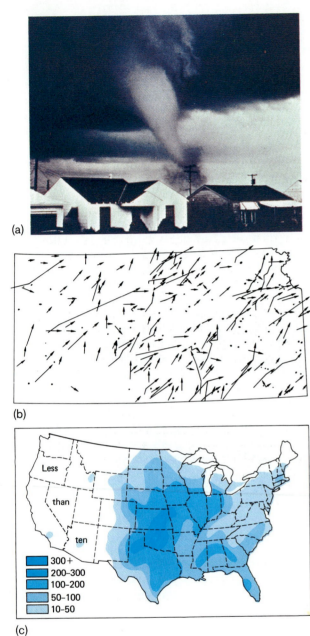

(a)

(b)

(c)

resulting in very changeable weather. In some winter weeks as many as three pairs of cyclones and anticyclones move down the St. Lawrence River valley toward the Atlantic Ocean, or invade the Pacific coast from the west.

During summer and fall, on the other hand, high pressure is often associated with the western extension of the "Bermuda" subtropical high, with warm and moist maritime tropical air settling over the eastern third of the United States. These anticyclones are notorious for their light winds and calms, for subsidence inversions aloft, and for the deteriorating air quality that results when a stagnant high stalls over the same region for days.

Local Circulations

The configuration of coastlines and land surfaces significantly affects local circulations and weather.

Land and Sea Breezes

Under weak anticyclonic conditions, *land and sea breezes* dominate the weather along coasts. Land and sea breezes result from the fact that the land warms during the day and cools at night, whereas the temperature of the adjacent ocean surface remains nearly constant. Heating of the land during the day generates localized low-pressure centers that draw air onshore in the form of sea breezes from the cooler ocean regions. Such breezes may be chilly in the midlatitudes but are very welcome along hot tropical coasts. Where cold water is present offshore, the sea breeze may bring fog inland daily in the later afternoon; this occurs in the San Francisco area in summer. Cooling of the land relative to the sea surface at night

Figure 6.9
(a) A close view of the tornado which swept across sections of Dallas, Texas, on April 2, 1957. (ESSA Weather Bureau)
(b) The direction and path lengths of tornadoes that occurred in Kansas between 1950 and 1970. (Joe R. Eagleman, Vincent U. Muirhead, and Nicholas Willems, *Thunderstorms, Tornadoes, and Building Damage,*

1975, Lexington Books)
(c) The number of tornadoes that were reported between 1955 and 1967 in the United States. The main features of the map include "tornado alley" over the Great Plains and the very small number of tornadoes westward from the Rocky Mountains.

Arctic Outbreaks

During late fall, winter, and spring, large pools of very cold polar air evolve over the frozen landscapes of northern Canada and interior Alaska. The cold air "sits" for several days over the even colder snow, and an icy cold, shallow inversion develops as longwave radiation cools the snow surface. Surface pressures rise in the dense cold air, and ultimately a strong anticyclone surges behind a cold front that moves rapidly toward the United States border. These cold-air surges are often termed *polar outbreaks,* but when temperatures near the Canadian border are extremely low—for example, −40°C (−40°F)—they are usually termed Arctic outbreaks.

Over the Northern Plains the early stages of Arctic outbreaks are often experienced as blinding blizzards. The air is filled with falling snow mixed with old snow swept aloft by the gale-force winds. Temperatures fall many degrees each hour, and persons have lost their lives when caught on foot without warm clothing in open country or stranded by impassible drifts in an automobile in the rural countryside.

In the vicinity of the Great Lakes, the Arctic outbreaks bring very deep snow squalls that plague the big cities on the southern and eastern shores of the lakes such as South Bend, Cleveland, Erie, Buffalo, Rochester, and Syracuse.

But in the Deep South, Arctic outbreaks simply bring temperatures well below freezing to regions that are not well prepared for their occurrence. The more extreme Arctic outbreaks produce dangerously low water pressure in urban regions because many residents leave their faucets open a bit—a trickle flow reduces chances of the freezing of exposed water pipes. Nevertheless, broken pipes are common, and plumbers and home-repair staffs are busy for weeks. Even more costly are the losses sustained by subtropical and tropical landscaping in the urban areas, by citrus crops in northern and central Florida and southern Louisiana, and by sugar cane in Louisiana and Texas.

Three excessively cold Arctic outbreaks—an unprecedented number—swept across the Deep South between 1981 and 1986, and citrus growers in many areas have given up. Climate specialists are unable to predict when there will be an extended run of years without extreme events. They point out that the extreme events recur in an almost random pattern through the years. But none of the recent Arctic outbreaks was as cold as the February 15, 1899, outbreak, when temperatures fell to close to −18°C (0°F) along the Gulf Coast from Houston east to Tallahassee!

causes an offshore flow, or land breeze. Land and sea breezes involve closed circulation loops, as shown in Figure 6.10a and b, with currents of air aloft that are the reverse of those at the surface. The upper (reverse) flow occurs at a height of 1.5 to 2 km (5,000 to 7,500 ft).

Mountain and Valley Winds

A similar diurnal reversal of wind direction is experienced in mountain regions, where the upper mountain slopes heat more strongly than the valleys during the day, causing air to be drawn upward through the valleys. These weak flows are called *valley winds.* The ascent of air over mountain summits often produces clouds in the late afternoon. At night the summit areas cool by radiation, and cold dense air drains away from them, moving down and through the valleys as chilly *mountain winds* (Figure 6.10c and d). This cold air drainage often produces temperature inversions and frigid nights in mountain valleys.

SECONDARY ATMOSPHERIC CIRCULATIONS

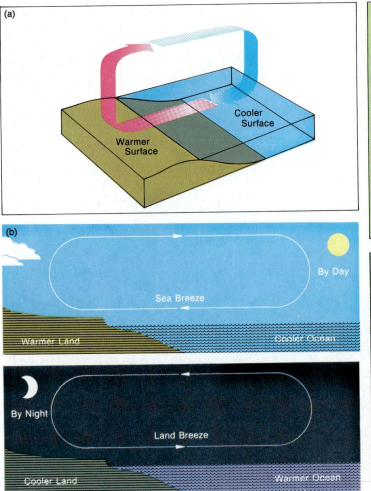

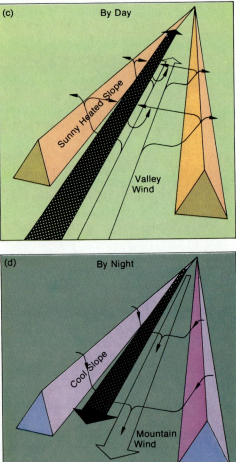

Figure 6.10

Small-scale circulations of air can significantly modify local weather conditions.

(a) This diagram shows a small convective cell generated near the boundary between warm and cool regions, such as land and water areas.

(b) Land and sea breezes develop because of the difference in temperature between the ocean and the land. During the day, the land heats more rapidly than the ocean, and a surface *sea breeze* develops from the ocean toward the land. At night, the land cools rapidly, and the surface flow of air is a weaker *land breeze* from the land to the ocean.

(c) Similar local winds develop because of the temperature differences between valleys and mountain slopes. During the day, as air rises up the warm slopes, *valley wind* flows up the valley to replace the ascending air.

(d) At night, cool air descends the slopes as a *mountain wind* and flows down the valley. (Calvin Woo)

(e) This view shows cloud formation resulting from convective rise of air above heated mountain slopes adjacent to Lake Brienz, Switzerland. (T. M. Oberlander)

(f) Orographic fog and cloud shrouding Neuschwanstein castle, near Füssen in Bavaria. (Robert A. Muller)

CHAPTER 6

(e)

(f)

TROPICAL SECONDARY CIRCULATIONS

Much less is known about tropical weather patterns than about the weather of the middle latitudes. There are far fewer observing stations in the tropics than in the middle latitudes, and weather data from tropical ocean areas are especially sparse. The Coriolis force is small near the equator, so equatorial winds are not geostrophic. Rotating cyclones and anticyclones do not form near the equator, and air mass contrasts and frontal activity are absent.

In the continental tropics, the principal influence on the weather is the daily cycle of heating and cooling of comparatively homogeneous humid air. The temperature variation from day to night in the tropics often exceeds the variation in average monthly temperatures through the year.

The tropics may lack variety in day-to-day weather, but many areas within the tropics have strong seasonal weather contrasts and exhibit weather phenomena found nowhere else.

Easterly Waves

Poleward of the band of persistent clouds and precipitation that marks the intertropical convergence, weak troughs of low pressure occasionally form in the trade wind zone and drift slowly westward (Figure 6.11). These *easterly waves* extend roughly north and south for a distance of several hundred kilometers. They form most often over the seas during the high-sun season. In the tropics, a warm layer of subsiding air usually overlies the surface layer, which produces a weak temperature inversion and prevents the surface air from rising to higher altitudes. This subsidence inversion is temporarily destroyed by an easterly wave, resulting in weather disturbances that vary from mild to quite violent.

Ahead of a wave, to the west of the trough, the surface winds diverge. In this area warm, dry air descends from the upper atmosphere, and the weather is fair. Behind the wave, to the east of the trough, the winds converge. Here, the inversion is broken, moist air can ascend to great heights, and severe thunderstorms may be generated. Each year a few easterly waves increase in intensity and develop into tropical cyclonic storms.

Tropical Cyclones

A dangerous interruption of the monotony of tropical weather is the late summer appear-

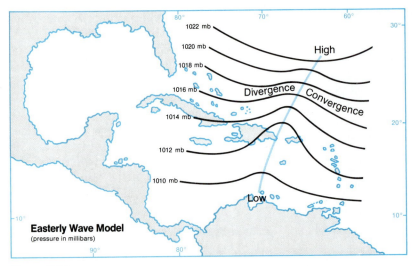

Figure 6.11
The kink in the isobars denotes the presence of an easterly wave centered between Puerto Rico and the Dominican Republic. Easterly waves move toward the west with the trade winds. Each year a few easterly waves increase in intensity and develop into the severe tropical cyclonic storms known as hurricanes. (Andy Lucas after Hermann Flohn, *Climate and Weather*, © 1969 by H. Flohn, used with permission of McGraw-Hill Book Company)

ance of intense cyclonic disturbances. If their wind speeds exceed 120 km (75 miles) per hour, such disturbances are classified as *tropical cyclones*. In the Caribbean area and on North American coasts these are called *hurricanes;* in the western Pacific they are known as *typhoons;* in the Indian Ocean and Australia they are simply called *cyclones.*

A tropical cyclone is an unusually compact and intense low-pressure center, much smaller in diameter than midlatitude cyclones, but having a pressure gradient that can exceed 30 millibars per 100 km. The winds that whirl around the center of a tropical cyclone commonly have velocities greater than 200 km (120 miles) per hour, accompanied by driving rain and severe thunderstorms.

Hurricanes begin as a rotating tropical storm generated from a strongly developed easterly wave. Why some storms die out and others continue to build to hurricane strength is not known, although upper-air flows may play a role. The characteristic structure of a fully formed tropical cyclone consists of a relatively cloudless central *eye* surrounded by a rapidly rotating wall of towering clouds, as illustrated in Figure 6.12. The height of the wall clouds is typically 10 to 15 km (6 to 9 miles), while the eye may be several tens of kilometers in diameter. Dry air from the upper atmosphere descends in the eye and is heated adiabatically, keeping the eye relatively free of clouds. Moist air spirals upward around the eye, and massive condensation produces the cloud walls and releases enormous amounts of latent heat.

Because of the weakness of the Coriolis force at low latitudes, tropical cyclones rarely form within 5° of the equator. Nor do they usually form at latitudes higher than 30° (Figure 6.13). The warm ocean surfaces between these two latitudes provide the necessary conditions for the formation of tropical cyclones. Unlike midlatitude cyclones, which are powered primarily

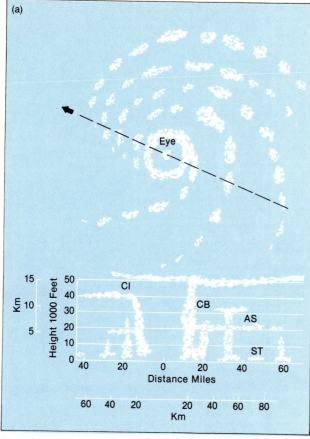

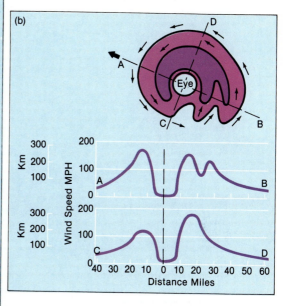

Figure 6.12

(a) Idealized vertical cross section of a severe hurricane showing clouds and map of spiral bands of heavy rainfall. Tropical storms and hurricanes are easily identified on the satellite images shown on TV; the bright circular pattern represents cirrus outflow aloft that normally covers the entire storm system.

(b) Graphs of idealized surface winds along the two transects on the map, which also shows wind direction (small arrows). The darker shading shows winds greater than 218 km (136 miles) per hour, with the lighter shading showing winds of hurricane strength. (Adapted from Walter Henry, Dennis Driscoll, and Patrick McCormack, *Hurricanes on the Texas Coast,* 1975, Center for Applied Geosciences, Texas A&M University.)

by the temperature differences between dissimilar air masses, the energy source for tropical cyclones is the latent heat released by massive condensation of water vapor. Water vapor is most abundant in warm tropical air over oceans with surface temperatures greater than 27°C (81°F). The oceans reach their maximum temperature in the late summer or fall, making this the hurricane season. Once a tropical cyclone begins to travel over land or cold water, it is cut off from the water vapor that is its source of energy, and its strength diminishes.

On a densely populated low-lying coastline, a direct hit by a strong hurricane is devastating. When gale-force hurricane winds drive coastal waters up onto low shores, usually called the *storm surge,* the sea may rise several meters above the normal high tide, surging tens of kilometers inland. The resulting destruction and loss of life can be enormous.

SECONDARY ATMOSPHERIC CIRCULATIONS

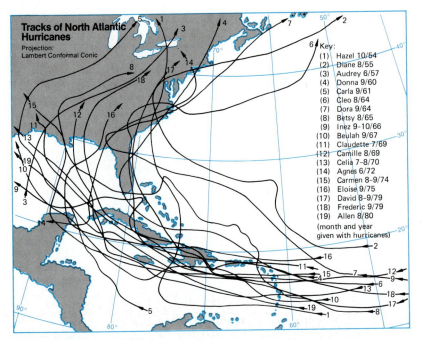

Figure 6.13
This map shows the tracks of some devastating North Atlantic hurricanes for the years 1954 through 1980. The path of an individual hurricane tends to be erratic, but as the map indicates, many of the hurricanes generated in the western Atlantic follow the same general course westward and northward. Hurricanes lose strength over land because of friction and lack of water vapor, but some hurricanes, such as Hazel, 1954, and Agnes, 1972, have traveled long distances across the United States, bringing heavy rains that caused rivers to overflow. (T. M. Oberlander and Andy Lucas after *The National Atlas of the United States of America,* 1970)

Some 250,000 people were drowned by such an event in Bangladesh in 1970. A rainfall of more than 25 cm (10 in.) is not unusual as a hurricane passes by, and, if the hurricane moves over land, heavy rains soon bring rivers to the flood stage. The number of hurricanes generated in the Atlantic and the Caribbean varies from three or four to a dozen per year, largely depending on the sea surface temperature and the behavior of the subtropical jet stream.

While it is difficult to predict the course of a hurricane, its distinctive cloud formations are easy to see on weather satellite images (Figure 6.14). Therefore, the moving storm can be tracked in time to give warning to threatened areas. Improved warning and communications systems have steadily reduced the loss of life resulting from hurricanes in the United States, even though those in peril often underestimate the danger and do not always take the warnings seriously. In less developed areas of the world, lack of adequate communications leaves great numbers of people unaware of the hazard, even after it is detected. When the cyclone in the Bay of Bengal struck Bangladesh in 1970, vast numbers received no warning, which accounts for the catastrophic loss of life.

Figure 6.14
(a) This photograph of Hurricane Gladys, 1968, was taken from the Apollo 7 spacecraft. The spiral clouds of the counterclockwise circulation of the hurricane are visible. (NASA)

(b) This satellite image of weather conditions on August 30, 1975, shows three hurricanes in various stages of development. All are moving westward in the zone of easterly winds. Hurricane Caroline, which originated in the western Atlantic, is moving onshore from the Gulf of Mexico, and caused destruction on the mainland. Hurricane Katrina is seen near its point of origin off the west coast of Mexico, and Hurricane Jewell is dissipating farther to the northwest. An occluded midlatitude cyclonic storm can be seen decaying over British Columbia and Washington. (National Oceanic and Atmospheric Administration)

SECONDARY ATMOSPHERIC CIRCULATIONS

Although every hurricane appears to be a potential disaster, these powerful storms are vital to the earth's heat balance. They are a means of transporting energy from areas of excess to areas of deficiency. If there were fewer hurricanes each year, those that occurred would have to be even larger and more violent to perform their vital function of spilling energy poleward.

Tropical Monsoons

We have already seen that monsoons are secondary circulations that involve a reversal in the direction of surface winds. This wind regime results from seasonal pressure changes on a continental scale. The most extreme case is in Asia. During the winter, the Asian interior is cold, and an almost steady outflow of dry *cP* air from the Siberian high brings clear weather to most of the continent. The dry northwest monsoon affects China and Japan, and the somewhat warmer northeast monsoon prevails in India and Pakistan. But during the summer, the heating of southern Asia produces a thermal low, easily seen in Figure 4.5 (pages 88–89). This causes the wind direction to reverse, as warm, humid *mT* air sweeps over the continent from the Indian Ocean and the southwestern Pacific. The southwest winds of the Indian summer monsoon soak up moisture as they cross the warm Indian Ocean. Summer precipitation from this air accounts for 70 percent of India's annual rainfall. The mountains of India trigger very heavy orographic precipitation, producing annual totals of more than 1,000 cm (400 in.) in some locations—85 to 90 percent of which falls between May and September.

The Indian monsoon also seems to be related to seasonal changes in the circulation of the upper air in the tropics. At the start of the summer monsoon, the westerly subtropical jet stream located above the south slope of the Himalaya Mountains shifts to the north of Tibet. At the same time, another jet stream blowing from the east appears over northern India. This easterly jet accentuates the ascent of air in the low-pressure center seen in Figure 4.5, and draws the ITC into India, far north of its usual position.

Weaker monsoon circulations also occur in China and Japan, southeast Asia, northern Australia, and the Guinea coast of Africa. There is a slight monsoon effect over the Mississippi Valley, with frequent outbreaks of polar air (called "northers") in winter, and waves of humid, sultry air invading from the Gulf of Mexico during summer.

KEY TERMS

air mass
    continental polar (cP)
    maritime polar (mP)
    maritime tropical (mT)
    continental tropical (cT)
front
warm front
cold front
stationary fronts
occluded fronts
cyclones
anticyclones

midlatitude cyclones
tornadoes
funnel cloud
waterspout
Arctic outbreaks
land and sea breezes
valley winds
mountain winds
easterly waves
tropical cyclones
hurricanes (typhoons)
hurricane eye

storm surge
monsoon

easterly jet
ITC

REVIEW QUESTIONS

1. Compare and contrast source regions and properties of *cP, mP, cT,* and *mT* air masses over North America.
2. Why is *mT* air frequently lifted over *cP* air?
3. Why might air-mass contrasts be greater over the southeastern United States than over eastern China?
4. Outline the stages of a midlatitude cyclone and associated weather over the eastern United States during winter.
5. Compare and contrast vertical and horizontal airflow in cyclones and anticyclones in the northern and southern hemispheres.
6. Contrast the form, processes, and properties of midlatitude and tropical cyclones.
7. Which oceanic areas and coastal regions are normally visited by tropical storms and hurricanes?
8. How are monsoon circulations related to the geographic distributions of land and water?

APPLICATIONS

1. Keep a daily log of the air mass types present in your area, using your own judgment and recollection of past extreme conditions to establish the actual air mass types. How abrupt are changes in air masses? Are some air masses transitional between the basic types discussed in the text? What is the nature of the weather during periods when air mass types change in your area?
2. On the basis of the midlatitude cyclone model described in Figure 6.7, determine the most likely progression of cloud types and weather events when a cyclonic storm approaches your area from the west in winter and passes to the north and northeast. What would change if the cyclonic disturbance passed to the south?
3. How far toward the equator do midlatitude cyclones penetrate? Does Hawaii experience such storms? Are there places that experience both midlatitude cyclones and tropical cyclones? If so, would both be possible at the same time of year? If the NOAA periodical publication *Environmental Satellite Imagery* is available on your campus, check the

images to answer the above question.
4. What was the heaviest rainfall ever received in your area? What unusual meteorological conditions occurred or combined to produce it?
5. Study the hurricane history of a segment of the North American coastline. Annual summaries of hurricane tracks are published in the periodical, *Weatherwise* (February issue), as well as by NOAA, and newspaper files are excellent sources for day-by-day accounts of weather events. As an alternative, do a similar study of tornadoes or of memorable midlatitude cyclones in some region—such as those that paralyze urban areas for days because they create snow removal problems.
6. What would be the long-range effect of a workable program to stop the growth of every storm that threatened to be destructive or costly?
7. What have been the greatest floods in your region's history? Were they produced by similar meteorological events?

area. In this region irrigation is necessary to produce crops. The contrasts on either side of the Cascades and many similar mountain systems illustrate how different and complex the hydrologic regime can be on the land, and how human activities are adjusted to accommodate these differences.

If we add up all the ways in which water is used, we find that the average daily consumption in the United States is equal to 6 cu meters (1,500 gal) per person—not including water used for the generation of hydroelectricity. The public uses only about 10 percent directly for such things as cooking, sanitation, and lawns and gardens. Industry uses another 10 percent. The remaining 80 percent is divided about equally between agriculture and thermal power plants that use water for cooling.

If all this water were truly consumed and permanently removed from the hydrosphere, disastrous water shortages would develop in short order. But water is a renewable resource. Nearly all the water used on farms, in homes, and by industry takes some path back into the hydrologic cycle. Most of the water diverted from rivers or pumped from wells to be used in irrigation is returned to the atmosphere through evapotranspiration in the fields. The water used in cities and in suburban homes is usually channeled into sewage systems to return to surface streams and finally to the ocean.

Throughout history people have labored to ensure reliable water supplies. As early as 5,000 years ago, the floodwaters of the Nile River in Egypt were channeled through long canals to vast basins surrounded by clay dikes. Aqueducts carried to ancient Rome a water supply that was ample even by modern standards. Today similar aqueducts carry water hundreds of kilometers to the Los Angeles basin and agricultural areas in the California and Arizona deserts. Water projects of amazing scale, involving gigantic reservoirs and transfers over thousands of kilometers, have been proposed, but costs and environmental issues have so far prevented their development.

The need for fresh water grows ever more pressing as populations increase and expand into areas where natural supplies are scarce. Because of this growing need to manage available water, there is increasing emphasis on the study and analysis of water on and below the earth's surface—a field known as *hydrology*.

In this chapter we shall first review the global water budget before examining in detail the pathways of water on the continents. A key concept in this chapter is use of the local water budget as a tool for evaluating the availability of water at a particular place.

THE GLOBAL WATER BUDGET

As we saw in Chapter 2, the water of the earth's hydrosphere is stored in several different conditions. Figure 7.1 shows that the oceans, which cover about 70 percent of the earth's surface, contain nearly 98 percent of the total water supply. Since ocean water contains dissolved minerals, primarily salt, it is unfit for consumption by humans, land animals, and even land plants. But when water evaporates from the oceans, the dissolved salts are left behind. Hence, water vapor evaporated from the oceans is a source of liquid water that is fresh and essentially "pure."

About three-fourths of the earth's nonsaline fresh water is stored as glacial ice, mainly in the ice sheets covering Antarctica and Greenland. These ice sheets receive an annual snowfall equivalent to only about 10 cm (4 in.) of liquid water. Glacially stored water returns to the oceans hundreds or even thousands of years later, when ice margins melt on the land or icebergs break away from coastal glaciers and melt in the sea. The ice sheets, then, represent long-term storage of fresh water.

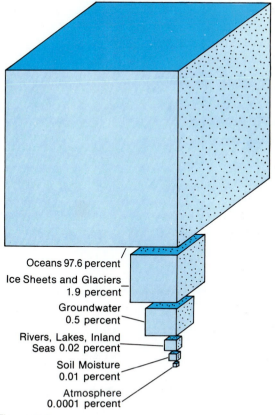

Figure 7.1
The volumes of the cubes show the relative amounts of free water in storage on the earth. (After R. L. Nace, *Water, Earth, and Man,* edited by R. J. Chorley, 1969, Methuen & Co., Ltd., Publishers)

Oceans 97.6 percent

Ice Sheets and Glaciers 1.9 percent

Groundwater 0.5 percent

Rivers, Lakes, Inland Seas 0.02 percent

Soil Moisture 0.01 percent

Atmosphere 0.0001 percent

The next largest reservoir, accounting for only one-half of 1 percent of the total water supply, is groundwater. Much of this is unusable because of mineral contamination or problems of extraction. Surface water supplies are an even smaller part of the total, as is moisture stored in the soil.

The amount of water stored in the atmosphere at any one time is even less. As noted previously, all the water in the atmosphere at any moment would form a layer only 2.5 cm (1 in.) deep over the earth. Over an entire year,

however, the atmosphere transports and recycles enough water to cover the earth with a layer of precipitation about 95 cm (37 in.) deep.

For the global water budget to remain in balance, the amount of water that leaves the atmosphere as precipitation must return to the atmosphere through evapotranspiration. Figure 7.2 (see pp. 168–169) shows how the global water budget is kept in balance. If we consider the oceans alone, we find that precipitation into them is less than evaporation from them. Over the continents precipitation exceeds evapotranspiration. But ocean levels are not falling, and the continents are not becoming flooded. These imbalances are offset because water continues to be exchanged between the oceans and continents. The land sheds its excess precipitation by contributing moisture to continental air masses that move out over the oceans, and by the flow of rivers to the sea.

WATER ON THE LAND

What happens to precipitation that falls on the continents? Depending on the characteristics of the land surface, this water can take several different pathways back to the atmosphere or to the seas to complete the hydrologic cycle.

Interception, Throughfall, and Stemflow

Not all the precipitation that falls reaches the soil. In urban environments, rain strikes roofs, building walls, and areas paved with concrete and asphalt. Most of this water runs off quickly into gutters and subsurface storm drains that empty into streams, lakes, or the ocean. The little that remains in puddles eventually evaporates, returning directly to the atmosphere.

Beyond urban areas, except where it is arid, plant growth covers the surface much of the year and prevents some rain from reaching the

Transport of Water
Vapor from the Oceans
(94)

(12)

Storage as Ice and Snow

Precipitation over the
Continents (106)

Evapotranspiration
from the Continents
(69)

Interception by Plants

Temporary Surface Storage

Surface Runoff

Infiltration

Soil Moisture Storage

Percolation

Groundwater Storage

Storage in Rivers and Lakes

Groundwater Runoff to Streams

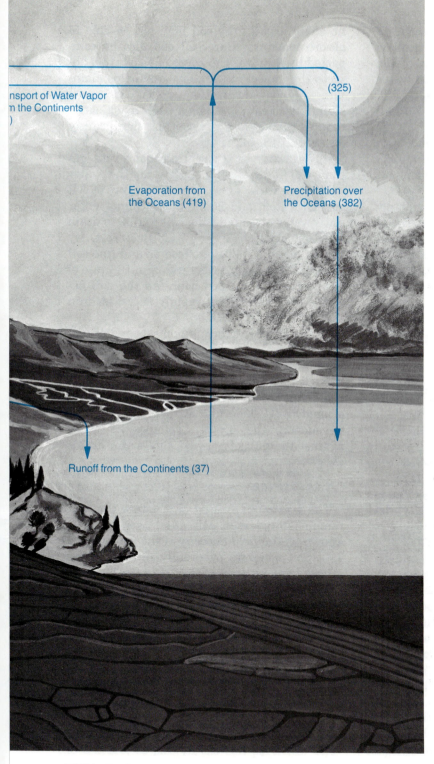

nsport of Water Vapor
m the Continents
)

(325)

Evaporation from
the Oceans (419)

Precipitation over
the Oceans (382)

Runoff from the Continents (37)

Figure 7.2
The movement of water through the hydrologic cycle involves numerous interactions and storage processes. Each year about 419,000 cu km of water evaporate from the oceans into maritime air masses. Evapotranspiration from the continents into continental air masses amounts to an additional 69,000 cu km (values in parentheses in the figure are water volumes in 1,000 cu km). This total volume of 488,000 cu km is equivalent to a mean annual precipitation over the globe of about 95 cm (37 in.).

Precipitation back to the oceans amounts to only 382,000 cu km, with 325,000 cu km originating from maritime air masses and 57,000 cu km supplied by continental air masses that move over ocean areas. Over the continents, precipitation (106,000 cu km) is greater than evapotranspiration (69,000 cu km); the difference (37,000 cu km), represents stream runoff that eventually returns to the oceans. Note that precipitation over the continents is supplied largely by water vapor from the oceans.

When precipitation falls on the land, a portion of the moisture is intercepted by vegetation and evaporates from temporary storage on leaves. The moisture that reaches the ground either infiltrates the soil, runs off across the surface, or evaporates from temporary storage in pockets and depressions. Some of the water that infiltrates the soil is stored as soil moisture, and a portion percolates deeper into the ground and enters groundwater storage. The flow of streams is maintained both by direct surface runoff and by groundwater contributions. (John Dawson)

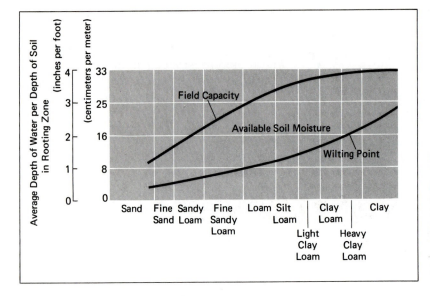

Figure 7.4
The field capacity, the wilting point, and the available soil moisture all vary according to the texture, or size, of the soil particles. (After Guy D. Smith and Robert V. Ruhe, 1955. *The Yearbook of Agriculture*, U.S. Government Printing Office, Washington, D.C.)

and groundwater flow. It represents the minimum depth to which a well must be drilled for a reliable supply of water.

Aquifers

Deposits from which groundwater may be obtained either naturally (in springs) or artificially (in wells) are *aquifers*. Subsurface aquifers contain about 30 times the amount of fresh water in the streams, lakes, and swamps of the earth.

Under natural conditions the quality of groundwater is usually good. The major exceptions are in arid and coastal regions, where aquifers may be contaminated by dissolved salts. Generally, the porous rock of an aquifer filters the water and removes suspended particles and harmful bacteria. But it is possible for urban and industrial hazardous wastes and agricultural pesticides to seep into aquifers. This is a particular problem where geological material is very permeable, as in areas underlain by sand or gravel deposited by streams or past glaciers, or by limestone that contains interconnecting cavities caused by the dissolving

action of groundwater. From New York's Long Island to California's Central Valley, wells have been declared unsafe because of recent contamination by pesticides and industrial wastes.

The problem of maintaining groundwater quality is a growing one. This is important because it is more practical for cities to utilize comparatively pure groundwater than to purify badly polluted river water. In the midlatitudes the temperature of groundwater from depths of 10 to 20 m (30 to 60 ft) is usually only 1° to 2°C higher than the average annual temperature. The relatively constant cool temperature makes groundwater very attractive for urban and industrial users.

A variety of subsurface materials are porous and permeable enough to act as useful aquifers. Most of the aquifers used in North America are beds of sand and gravel. Some were deposited as outwash from glaciers during the Ice Ages, and others are the result of much earlier deposition. Individual sand and gravel aquifers are commonly 50 m (160 ft) thick and often cover several thousand square kilometers. Among solid rocks, sandstones, limestones,

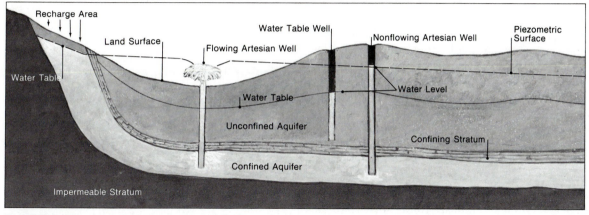

Figure 7.5
This diagram shows the principal features of aquifers in schematic form. (John Dawson after Raphael G. Kazmann, *Modern Hydrology,* 2nd ed., © 1972 by R. G. Kazmann, used by permission of Harper & Row)

and lava beds form the best aquifers because they have interconnected openings that collect and transmit water.

All the water stored in subsurface aquifers comes originally from precipitation. After field capacity is reached, additional rainfall that does not flow directly over the surface percolates through the ground to the water table. In arid regions, water may also seep downward from stream beds and lake bottoms. In humid areas, by contrast, groundwater seeps *out* into streams and lakes.

The surface region from which water drains into an aquifer is called the *recharge area.* Groundwater usually moves laterally in an aquifer at rates varying from meters per year to kilometers per day. When an extensive aquifer slopes gently for a long distance, the principal recharge area may be hundreds of kilometers distant from the wells where the water is extracted (Figure 7.5).

Aquifers may be either *unconfined* or *confined.* In an unconfined aquifer the water is not under pressure and will not rise above the level of the water table unless it is pumped to the surface. In confined aquifers the water-bearing layer is covered by a layer of impermeable material. Confined aquifers may lie far below the level of the water table.

Piezometric Surfaces

If the confining layer slopes downward, the difference in elevation between the upper and lower portions of an aquifer can result in a considerable difference in water pressure. If wells are drilled into a lower section of the confined aquifer, where the water pressure is high, the water in the wells will rise considerably above the level of the aquifer, as is shown in Figure 7.5. The elevation to which water will rise in such wells is known as the *piezometric surface.* The piezometric surface occasionally lies above the land surface. This creates flowing *artesian wells,* from which water gushes out at the surface with no necessity for pumping. Natural artesian springs occur where fractures in rocks permit water to escape to the surface from a confined subsurface aquifer.

Water can also collect on top of impermeable layers that lie above the normal water table. This creates *perched water tables.* These sometimes feed springs along canyon walls. Perched water tables are common in areas of layered sedimentary rocks (Chapter 12) and in thick masses of sediment deposited by streams and glaciers.

THE HYDROLOGIC CYCLE AND THE LOCAL WATER BUDGET

Groundwater Extraction

Approximately 50 percent of all the groundwater extracted in the United States is used for irrigation in Texas, Arizona, and California. Groundwater also supports extensive agricultural development and livestock industries in the Great Plains east of the Rocky Mountains and in eastern Australia, North Africa, Arabia, Iran, and other arid regions. Groundwater must be carefully managed so that the amount of water pumped out does not exceed the net flow into the aquifer. Some aquifers were recharged under climatic conditions that no longer prevail and are receiving little or no input of water today. This is especially true in desert areas. But in many areas the groundwater is being "mined"—the rate of pumping greatly exceeds the natural recharge so that water tables and piezometric surfaces are falling. In such areas pumped wells have to be deepened, and former artesian wells have to be pumped. In some areas groundwater withdrawal has caused the land surface to subside by as much as 8 meters (25 ft). This disrupts irrigation canals and has caused damage to streets and buildings in such scattered locations as Venice (Italy), Mexico City, Shanghai (China), Tokyo (Japan), and Phoenix (Arizona).

In some areas groundwater is being recharged artificially (Figure 7.6). This is common in southern California, using water brought some 600 km (400 miles) by aqueduct. Artificial recharge is also used in coastal regions where excessive pumping of groundwater has lowered the water table, allowing saline sea water to seep under the land, contaminating aquifers. This problem has appeared on Long Island, in southern California, in Israel, and in many other coastal areas.

Runoff and Streamflow

If rain falls at a rate greater than the infiltration capacity of the soil, water begins to collect on the surface. Surface irregularities and vegetation store some of the water for evaporation or later infiltration, but water also begins

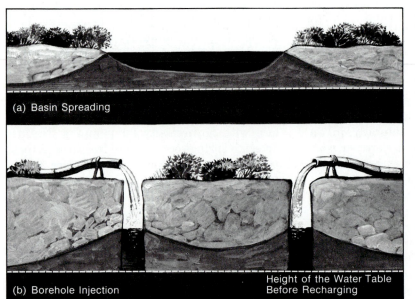

(a) Basin Spreading

(b) Borehole Injection

Height of the Water Table Before Recharging

Figure 7.6
Two methods that have been used to recharge aquifers and raise the level of the water table in regions where supplies of groundwater have been depleted by excessive withdrawal.

(a) Water pumped into shallow surface depressions in the natural recharge area of an aquifer seeps through the permeable rock into the aquifer.

(b) Water pumped into boreholes sunk into an aquifer seeps into the permeable rock to recharge the aquifer. (John Dawson)

As we have seen in this chapter, all groundwater is initially derived from precipitation that infiltrates the surface and ultimately percolates to the groundwater table. It follows, then, that any surface pollutants that can be transported downward have the capacity to reduce the quality of groundwater and even transform safe groundwater into deadly concoctions.

Hazardous waste sites that can contribute to groundwater pollution have received much media attention. EPA has identified at least 19,000 such sites across the country where toxic chemicals and other hazardous wastes have been dumped. From a climatic perspective, the toxic materials travel farther from the initial dumping sites in regions where heavy precipitation percolates through the subsoils. In the dry climatic regions, opportunities for translocations are limited, but even in desert regions runoff from occasional rainstorms will move toxic materials over or beneath the surface.

Entering the 1960s, we understood little about the long-term dangers of chemical wastes. Unwanted chemicals were indiscriminately dumped in out-of-the-way places around the margins of most metropolitan regions. Some of these sites are now on the EPA list of the most dangerous situations, and tens of millions of public and industry dollars are being allocated for very complex and difficult clean-up operations. Large, technologically sophisticated hazardous waste disposal centers, including pits, injection wells, and high temperature furnaces, are now being developed, but the disposal procedures at these centers continue to be controversial, and the centers are unwelcome wherever they attempt to locate. Modern societies covet the products of today's technology, but they are largely unwilling to cope effectively with the hazardous wastes that are byproducts of manufacturing processes.

There are many other examples of perhaps more innocent hazardous waste pollution. For example, there are more than two million underground storage tanks for oil and gas products, and an estimated quarter of a million of them will be leaking by 1990. And in agricultural areas, streams draining farm fields and groundwater are becoming overburdened with nitrate fertilizers and manure from feedlots where cattle and hogs are concentrated and from poultry complexes. Control of pollution and hazardous waste should be a growth industry in coming years.

to trickle across the ground or move in sheets under the influence of gravity.

Surface runoff increases from small rivulets at the beginning of a heavy rainstorm to a steady torrent when no more storage opportunities are available. Toward the end of a prolonged rain, nearly all the rainfall may become surface runoff. Virtually all land surfaces not covered by ice or loose sand are laced with networks of small erosional channels created and maintained by episodes of surface runoff. The initial channels carry water to a smaller number of larger channels that, in turn, feed major rivers (Chapter 14). The runoff carried to the sea by rivers represents precipitation that did not go into subsurface storage or return to the atmosphere by evapotranspiration.

Runoff

Hydrologists are concerned with the variable flow of rivers over time, especially the effect of precipitation on runoff. *Runoff*, used technically, is a measure of the average depth of water that flows from a drainage basin during a specified amount of time. Let us imagine that an intense rainstorm lasting an hour dropped 2.5 cm (1 in.) of rain on an impervious parking lot equipped with storm drains. The storm drains are efficient, and 15 minutes after the storm an average depth of only 0.05 cm of water, concentrated in a few depressions and cracks, is left on the parking lot. Later this water will return to the atmosphere by evaporation. The runoff from the parking lot amounts to 2.45 cm, or 98 percent of the rainfall. At the same time, we know that the runoff from an equal area of nearby parkland would be far smaller, if there were any runoff at all, for much of the rainfall there would be stored in the soil, eventually returning to the atmosphere by evapotranspiration.

Streamflow

Streamflow, on the other hand, represents the volume of water flowing down a stream channel during a short time period. It is usually expressed as *stream discharge* in cubic feet per second (cfs) or cubic meters per second (Chapter 14). A plot of stream discharge fluctuations is called a *stream hydrograph;* Figure 7.7 shows a hypothetical example for a stream of intermediate size in a humid region. The stream continues to flow between rains due to an almost steady input of groundwater that seeps into the channel. This minimum flow between storms is known as the *base flow.* During or shortly after each rainstorm, the water table rises and surface runoff reaches the stream. Figure 7.7 shows the characteristic rapid rise and slower recession associated with storm runoff, groundwater outflow, and recharged groundwater supplies. Daily fluctuations in discharge for three small streams in different climatic regions are shown and explained in Figure 7.8. The peaks rising from the base flow are caused by surface runoff during storms.

THE LOCAL WATER BUDGET: AN ACCOUNTING SCHEME FOR WATER

Water is not always available in the desired amounts just when and where we need it. The

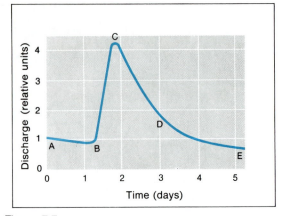

Figure 7.7
Graph showing the effect of an upstream rainstorm on the amount of water carried by a stream, measured from the time of the storm. From *A* to *B*, water from the rain has not reached the stream, and the discharge represents the base flow supplied by groundwater. From *B* to *C*, the discharge rises rapidly as direct surface runoff from the upstream drainage basin reaches the stream. From *C* to *D*, the discharge falls slowly as the last of the surface runoff makes its contribution. From *D* to *E*, the discharge again primarily measures groundwater inflow. (After R. C. Ward, *Principles of Hydrology,* © 1967, McGraw-Hill Book Co. (U.K.) Ltd., used with permission)

supply of moisture is so variable that a special technique is needed to estimate moisture availability. This is why climatologists and hydrologists make use of the *water budget* concept. The water budget is the local version of the hydrologic cycle. It takes into account four principal components of water distribution: precipitation, soil moisture, evapotranspiration, and runoff. When appropriate, snow accumulation and snowmelt may also be included. Of all these components, evapotranspiration is the most difficult to estimate accurately because of its dependence on complex meteorological and biological factors.

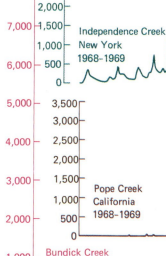

Figure 7.8

Hydrographs for three small streams in different climatic regions, each draining an area between 194 and 310 sq km (75 and 120 sq miles). The hydrographs are organized by water years, which begin October 1 and end September 30. The water year is a useful calendar for water-resource management because streamflow tends to be lowest in late September, with minimum groundwater outflow to support base flow. Typically, storm runoff is superimposed as spikes, or peaks, on the base flow contributed by groundwater. Storm runoff tends to rise quickly and to recede more slowly.

Bundick Creek is located in the warm and humid climate of southwestern Louisiana. Groundwater provides for some base flow all year, but streamflow is largest on the average during winter and spring. A considerable proportion of the annual flow is produced in just a few days by the very large spring flood flows,

and the maximum daily discharge of 7,980 cfs on March 25, 1973, was the highest daily flow in 17 years of measurements.

Independence Creek is located on the western flanks of the Adirondack Mountains in northern New York. This drainage basin is representative of climates where persistent low temperatures during winter allow most of the precipitation to accumulate as snowpack. Much of the annual flow occurs during spring as the snow melts, but there is usually a secondary discharge peak in autumn before the winter snows begin to accumulate.

Pope Creek drains a low mountainous region of the coastal ranges of northern California west of Sacramento. The climate is hot and dry during summer with no precipitation. Groundwater contributes to streamflow only during winter and spring, and normally there is no water in the creek from June through October.

Potential and Actual Evapotranspiration

Measurements of evapotranspiration are limited to detailed studies of small field plots using expensive instrumentation.

Potential Evapotranspiration

To overcome the difficulties of actual measurement, and to allow water budget components to be studied over large regions, the American climatologist C. Warren Thorn-

---

THE HYDROLOGIC CYCLE AND THE LOCAL WATER BUDGET

thwaite introduced the concept of *potential evapotranspiration* (*PE*) and developed formulas for estimating *PE* under varying conditions. Potential evapotranspiration, the key to the water budget, is worth examining in detail.

Potential evapotranspiration is the rate at which water would be lost to the atmosphere from a land surface completely covered by growing vegetation that has been supplied with all the soil moisture it can use. *PE* is normally expressed as the depth of liquid water that is converted to vapor in a given time. A typical value for *PE* during a summer month in the eastern United States is about 15 cm (6 in.). During a winter month in the same region, *PE* is usually less than 2.5 cm (1 in.).

Actual Evapotranspiration

Evapotranspiration normally proceeds at the potential rate as long as moisture is readily available in the soil. After a number of days without rain, however, soil moisture becomes partly exhausted, and it is increasingly difficult for plants to extract the remaining moisture. The actual rate of evapotranspiration will then fall below the potential rate. For this reason Thornthwaite distinguished between *PE* and *actual evapotranspiration* (*AE*). During wet seasons, the *AE* will be the same as the *PE*, but in prolonged dry periods, the *AE* is less than the *PE*. The term "actual evapotranspiration" is somewhat misleading; in Thornthwaite's water budget analysis, the *AE* is an estimation, rather than an "actual" measurement. Many environmental scientists simply use the term evapotranspiration (*ET*), rather than Thornthwaite's term, *AE*.

Solar Energy Inputs

Solar radiation is the principal factor that determines *PE*. In fact, one way to think about potential evapotranspiration is in terms of solar energy input and utilization. Both evaporation and transpiration require about 590 calories per gram of water. At average temperatures near the earth's surface, one gram of

water has a volume of one cubic centimeter. Therefore, we can calculate the energy needed to evaporate a pan of water 1 cm deep if we know the surface area of the pan in square centimeters. For example, if the pan absorbs 1,500 calories of solar energy for each square centimeter of water surface, we calculate that the "potential" evaporation amounts to about 2.5 cm (1 in.) of water.

PE, therefore can be thought of as both an index of energy supply to an area and, at the same time, an index of the climatic demand for water in the landscape. Rigorous estimates of *PE* from atmospheric data require the measurement and analysis of the terms of the radiation budget and energy balance, especially solar and net radiation, as well as temperature, humidity, and wind at two or more levels above the surface.

Estimation of Potential Evapotranspiration

Potential evapotranspiration from a field, or even from an entire landscape, is almost independent of the type of plant cover. Imagine looking down from an airplane on a forest or field during the middle of the growing season. The plants usually present a nearly uniform cover of overlapping green leaves. The albedo of almost all green fields or forests is between 10 and 25 percent. Therefore, an acre of forest and an acre of soybeans are not very different as regards solar radiation absorption on a clear day.

Thornthwaite's estimation of monthly *PE* depends mostly on two factors: (1) average monthly air temperature, and (2) latitude, which controls the length of the daylight period from month to month. Data on temperature and duration of daylight are available for most places on earth. Together, these two factors give an indication of solar energy input, which is actually measured in only a few places.

Estimated average annual *PE* variations over the United States, calculated from

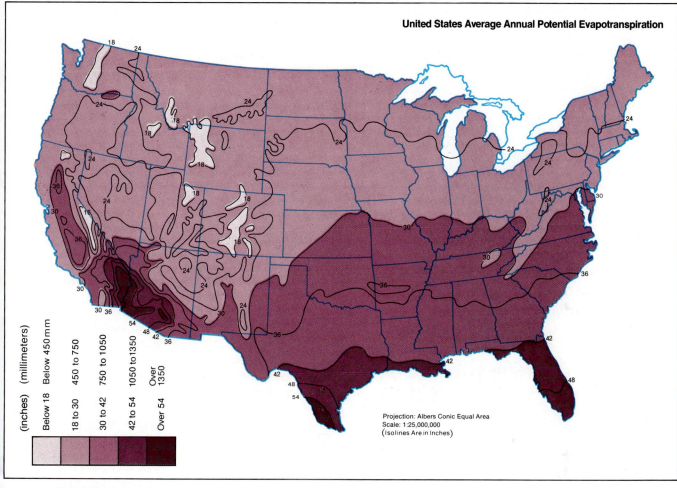

United States Average Annual Potential Evapotranspiration

Projection: Albers Conic Equal Area
Scale: 1:25,000,000
(Isolines Are in Inches)

(inches) | (millimeters)

Below 18 | Below 450 mm
18 to 30 | 450 to 750
30 to 42 | 750 to 1050
42 to 54 | 1050 to 1350
Over 54 | Over 1350

Figure 7.9
Map showing the average annual potential
evapotranspiration, or *PE*, for the conterminous United
States, calculated from Thornthwaite's formula.
(Adapted from *Geographical Review,* Vol. 38, © 1948,
American Geographical Society of New York)

Thornthwaite's formula, are shown in Figure
7.9. The Southwest, Texas, and Florida have
high values of *PE* because of greater solar radiation income and very warm weather. *PE* is
much lower farther north because of decreased
solar radiation income in winter and much cool
and cloudy weather. Across the high mountain

areas of the West, where temperatures are low,
the *PE* is also low. There is also a strong seasonal regime of *PE* throughout the United
States: *PE* is low in winter and high in summer, although this seasonality is least marked
along the West Coast and across the South.
Thornthwaite's formula is widely used for regional climatic analyses, but much more complex formulas that require more data have
been devised for detailed local studies.
   Potential evapotranspiration at a particular
location can be measured in a device called an
evapotranspirometer, which is an open tank

THE HYDROLOGIC CYCLE AND THE LOCAL WATER BUDGET

Figure 7.10
This is a weighing lysimeter, or evapotranspirometer, as it was being installed at Lompoc, California. When ready for measurement, plants growing in the lysimeter should be exactly the same as those around the lysimeter, so that the sun will not "see" the lysimeter. (Robert A. Muller)

about 60 cm (2 ft) in diameter and 90 cm (3 ft) deep. The tank is sunk into the ground so its top is flush with the surface (Figure 7.10). The tank is filled with soil and usually planted with a cover of grass. A weight increase represents water added to the tank by precipitation or irrigation; a weight loss, on the other hand, represents evapotranspiration, and downward percolation that is measured as it collects in a false bottom. The various measurements con-

stitute a water budget of the tank. As long as the grass in the tank does not experience water shortage, AE is equal to PE; the measured PE, averaged by months, is assumed to represent the climatic energy available in the surrounding area.

Regular operation of an evapotranspirometer is expensive and requires skilled personnel. Therefore such measurements have been limited to experimental studies, mostly in dry regions where it is important to know how much water is needed for particular types of vegetation or crops.

Calculating the Water Budget

Thornthwaite and John R. Mather developed the local water budget to represent a systematic accounting of the input, output, and storage of water at a location (Figure 7.11). The computation is essentially a comparison between PE and AE. If PE exceeds AE, there is a water deficit—there is no soil moisture recharge and no runoff; there is not enough water available to satisfy the climatic demand for it. If AE equals PE, there may be additional water that can result in moisture storage in the soil or surface runoff. In the computation, the incoming precipitation (P) during a given time period is allocated to evapotranspiration (AE), *soil moisture recharge* (Δ ST), and surplus (S). Surplus represents surface water runoff and groundwater recharge, and eventually finds its way to streams and rivers.

A Local Water Budget: Baton Rouge

To illustrate the local water budget, consider the example of Baton Rouge, Louisiana, in 1962. In an average year, Baton Rouge receives more than 125 cm (50 in.) of rainfall. Although Baton Rouge experiences one of the highest average annual rainfall totals for cities within the 50 states, a water budget analysis shows that water deficits existed and that

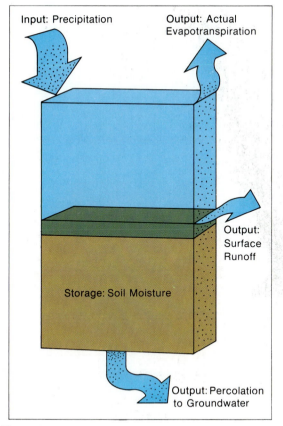

Figure 7.11
Schematic diagram illustrating the principal components of the local water budget.

crops in the Baton Rouge area needed irrigation water during several months of 1962.

The Water Budget for 1962

To illustrate the 1962 water budget at Baton Rouge, water income, output, and storage are considered month by month in Figure 7.12. Weekly or daily periods could also be used. The first row in Figure 7.12 shows monthly precipitation in inches, the units still used by the National Weather Service and for most applications in agriculture and engineering. The second row shows monthly PE estimated from Thornthwaite's formula. Since PE tends to fol-

low the seasonal regime of temperature, it is much smaller in winter than summer. The third row shows the difference between precipitation and PE. When $P - PE$ is positive, we have a "wet" month, and when $P - PE$ is negative, we have a "dry" month. For 1962, six months were wet, and six were dry.

Soil moisture is the storage component within the local water budget. Because 1961 had been a wet year, 1962 began with the soil moisture storage in Baton Rouge at its full capacity of 6.0 in. This storage capacity figure is representative of the Baton Rouge area; on a global basis, soil moisture storage capacity ranges from about 2 to 12 in. Our simple water budget will assume that plants use all the water they need from the soil until there is no more moisture in storage. Rows 4 and 5 show the change in soil moisture storage during each month and the amount of soil moisture storage at the end of each month.

During wet months, AE is always equal to PE. During dry months, plants will draw on soil moisture as needed (note the storage change in July). AE represents water passing through the plant system, and the amount of AE is related in a general way to the building of plant tissue. According to the table, AE was less than PE during August and September. This indicates a *deficit* (D), shown in row 7. The deficit represents the amount of additional water that the plants could have used. Therefore, it is an index of irrigation needed to maintain crops at their full growth potential.

Any excess water that remains after soil moisture is brought to capacity is *surplus* (S). During 1962, surpluses occurred during only four months: January, March, April, and June. Surplus water is available for surface runoff and groundwater recharge. It is this water that enters streams and changes the land surface by erosion and sediment deposition.

The Average Water Budget

Figure 7.13 shows both tabular and graphic representations of the average annual water

	JAN	FEB	MAR	APR	MAY	JUN	JUL	AUG	SEP	OCT	NOV	DEC	Total
1. Precipitation (P)	6.4	0.7	3.3	9.7	1.6	11.4	2.0	4.5	4.3	5.2	0.9	2.9	52.9
2. Potential Evapotranspiration (PE)	0.4	1.7	1.3	2.6	5.3	6.1	7.4	6.8	5.2	3.4	1.1	0.6	41.9
3. Precipitation minus Potential Evaporation (P-PE)	6.0	−1.0	2.0	7.1	−3.7	5.3	−5.4	−2.3	−0.9	1.8	−0.2	2.3	11.0
4. Change in Stored Soil Moisture (ΔST)	0	−1.0	1.0	0	−3.7	3.7	−5.4	−0.6	0	1.8	−0.2	2.3	−2.1
5. Total Available Soil Moisture (ST)	6.0	5.0	6.0	6.0	2.3	6.0	0.6	0	0	1.8	1.6	3.9	——
6. Actual Evapotranspiration (AE)	0.4	1.7	1.3	2.6	5.3	6.1	7.4	5.1	4.3	3.4	1.1	0.6	39.3
7. Deficit (D)	0	0	0	0	0	0	0	1.7	0.9	0	0	0	2.6
8. Surplus (S)	6.0	0	1.0	7.1	0	1.6	0	0	0	0	0	0	15.7

Figure 7.12
The local water budget for Baton Rouge, Louisiana, during 1962. The water budget equation is $P = AE + S \pm \Delta ST$. For 1962 the equation works out: $52.9 = 39.3 + 15.7 - 2.1$. Similarly, the energy budget equation is $PE = AE + D$, and for 1962 this is $41.9 = 39.3 + 2.6$. Each equation balances, so we can be quite confident that we have not made any calculation errors.

budget over a 30-year period for Baton Rouge. The long-term average monthly precipitation is much less variable than the actual monthly precipitation in individual years. This is because a wet July in one year compensates for a dry July in another year, and so on. Long-term average monthly precipitation in Baton Rouge, therefore, is never low, and in the average water budget for Baton Rouge there are no deficits.

The Continuous Monthly Water Budget

The seasonal and annual variations in energy and moisture regimes revealed by water budgets are extremely important to the activities of plants, wildlife, and humans. A more complex continuous monthly water budget model was used in Figure 7.14 to illustrate this variability. In the water budgets for Figures 7.12 and 7.13, it was assumed that soil moisture was "equally available" to meet the climatic demand of PE during months when P − PE was negative. In the model used for this figure, it was assumed that the soil moisture would become "decreasingly available" as it was depleted; in other words, the vegetation would not be able to withdraw all the needed soil moisture even though the soil still con-

Figure 7.13
Average water budget for Baton
Rouge, Louisiana, based on standard
climatological data for 1941–1970.
  (a) The distinction between the
components of an average water
budget and the budget of an individual
year (Figure 7.12) should be kept
clear.
  (b) This graph of the average water
budget for Baton Rouge is an example
of standardized graphs that have
appeared in research publications. In
graphical form it shows the seasonal
regimes of *P*, *PE*, *AE*, and *S*, as well
as soil moisture withdrawal and
recharge.

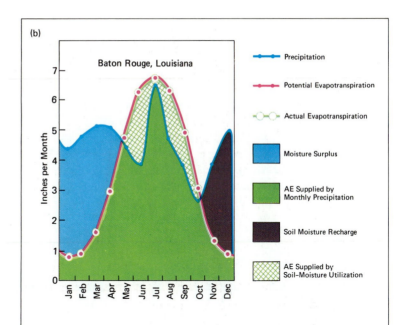

(a)	Jan	Feb	Mar	Apr	May	Jun	Jul	Aug	Sep	Oct	Nov	Dec	Year
1. Precipitation (*P*)	4.4	4.8	5.1	5.1	4.4	3.8	6.5	4.7	3.8	2.6	3.8	5.0	54.0
2. Potential Evapotranspiration (*PE*)	0.7	0.9	1.6	3.0	4.7	6.2	6.7	6.3	4.9	2.8	1.2	0.8	39.8
3. Precipitation minus Potential Evapotranspiration	3.7	3.9	3.5	2.1	−0.3	−2.4	−0.2	−1.6	−1.1	−0.2	2.6	4.2	14.2
4. Change in Stored Soil Moisture (Δ*ST*)	0	0	0	0	−0.3	−2.4	−0.2	−1.6	−1.1	−0.2	2.6	3.2	0
5. Total Available Soil Moisture (*ST*)	6.0	6.0	6.0	6.0	5.7	3.3	3.1	1.5	0.4	0.2	2.8	6.0	—
6. Actual Evapotranspiration (*AE*)	0.7	0.9	1.6	3.0	4.7	6.2	6.7	6.3	4.9	2.8	1.2	0.8	39.8
7. Deficit (*D*)	0	0	0	0	0	0	0	0	0	0	0	0	0
8. Surplus (*S*)	3.7	3.9	3.5	2.1	0	0	0	0	0	0	0	1.0	14.2

THE HYDROLOGIC CYCLE AND THE LOCAL WATER BUDGET

age year, and extensive irrigation is required if crops are to be grown. The idea in this case was to change the climate by building a gigantic dam across the far-off Bering Strait between Siberia and Alaska. The dam would have to be 100 kilometers (60 miles) long, far surpassing in size any dam existing on the earth. How could a dam at the Arctic Circle affect the climate in a desert 5,000 kilometers (3,000 miles) away?

If such a dam could be built, it would disrupt the oceanic circulation by preventing the flow of cold arctic waters into the Pacific Ocean. This would cause the Pacific Ocean to become warmer. We have seen (Chapter 4) that unusual cooling in the Pacific Ocean seems to trigger drought conditions in the western United States. Accordingly, advocates of the Bering Strait dam anticipated that *warming* the Pacific Ocean would *increase* rainfall in the American Southwest. Russian proponents of the plan hoped that it would indirectly improve the cold, dry climates over much of Siberia as well. But warming the Pacific would almost certainly induce a longer hurricane season in the tropics, and perhaps would bring hurricanes into higher latitudes. Furthermore, the subsequent cooling of the polar region could produce stronger outbursts of polar air and more violent cyclonic storms along the polar front. Changing the temperature of the Pacific Ocean would change the temperature of the air over the ocean and would affect the general circulation of the atmosphere. Because the general circulation controls the climates of the earth, the effects of the dam would be far-reaching. Beneficial effects in the American Southwest would quite likely be outweighed by harmful effects elsewhere. It is impossible to predict all the consequences of such an enormous scheme to modify climate.

The point of this rather far-fetched example is that the earth's climates are not independent phenomena—they are determined by the general atmospheric circulation. To change climates significantly, the circulation system as a whole must be modified, and this would cause climates everywhere to be affected. It is much more sensible to learn how existing climates are produced and how to take advantage of what they have to offer.

This chapter focuses on an understanding of global climatic systems. Two different but related methods for organizing or classifying climates are developed. The first focuses on atmospheric dynamics—the climates associated with the general atmospheric circulation and the "delivery" of energy and moisture *to* the surface of the earth. The second focuses on the interactions of energy and moisture *at* the surface of the earth that make possible plant and animal communities and the earth's agricultural systems.

MEASURING CLIMATE

The relative stability of the general circulation gives rise to characteristic regional weather patterns that make each region distinctive. Winter vacationers in Florida, for example, have every reason to expect a succession of warm, sunny days during their stay. Because the distribution of climatic energy and moisture has implications for other systems in addition to weather, it is useful to classify the types of climate experienced in the various regions of the earth. Climate classification helps reveal the general patterns in other environmental systems and serves to organize a wealth of information about the earth's surface.

Climatic Indexes

Climatic indexes are numerical measures used to distinguish one climatic type from another. The choice of an appropriate index of climate depends on its intended use. Someone interested in constructing tall buildings might want to consider wind speed and direction to assess the possibilities of wind damage. If the concern is transportation, interest would center on the frequency of fog and icing conditions. An irrigation engineer would be concerned with the amount of evapotranspiration from cropland. To assess a region's potential for air pollution, wind regimes and the timing and duration of temperature inversions must be known.

Classifications by physical geographers of regional climates on a global basis generally employ temperature and precipitation as indexes of climate. These are easily measured, data about them are generally available for most countries, and they clearly influence the distribution of vegetation and water supplies and the potential for various types of agriculture.

Time and Space Scales of Climate

The general factors that control climates are those we have considered in previous chapters. They include latitude, which determines intensity of solar radiation and day length and influences temperature; proximity to warm and cold ocean currents, which affect temperature and air mass stability; proximity to moisture sources and rain-producing mechanisms, including the polar front and the intertropical convergence zone; topography, which causes ascent of air and high rainfall or descent of air and rainshadows; and the general circulation, which determines the sources of air masses and the general directions in which they move.

The movement of air masses can change local weather conditions significantly within a few hours. Weather conditions can also vary markedly over short distances. Even if the indexes of climate are accepted as being temperature and precipitation, geographers must still decide when and how often to analyze these factors. The following example shows the importance of the time scale—the "when"—of measurement.

The average annual temperatures at Aberdeen, Scotland, and at Chicago, Illinois, differ by less than 2°C (3°F). The average annual precipitation at both locations is the same (84 cm or 33 in). Despite similarities in such annual averages, the climates at the two locations are quite different. The monthly averages give a much more accurate indication of the climate in each city. During January, Chicago averages 8°C (14°F) colder than Aberdeen. During July, Chicago averages 9°C (16°F) warmer than Aberdeen. The continental climate at Chicago is therefore considerably more extreme than the marine climate at Aberdeen. It should be evident that the seasonal distributions of temperature and precipitation, as well as the average annual values, are important in characterizing the climate of a place.

The "where" of measurement—the location and size of the climatic region—is also important. Again, the best spatial scale depends on the intended use of the information. For example, on a world map most of the west coast of the United States is considered to be a single climatic region. But someone concerned with water resources on the Pacific coast would need to see a much more detailed picture, showing the gradation from the abundant rainfall in the Pacific Northwest to the general dryness of southern California. The complex patterns of orographic precipitation and rainshadows, illustrated for California in Figure 5.18 (p. 127), need to be known and understood in detail.

Even a single residential lot has several *microclimates,* with a wide range of temperatures over a small area. Figure 8.1 shows temperatures that were read at 11 P.M. at various points in a backyard located in Baton Rouge, Louisiana. The microclimates of the yard are too small to appear on even a city-wide climate survey, but climatic conditions in different parts of the yard are important to someone trying to protect subtropical plants from freeze damage.

When examining global climatic types, excessive detail is not desirable. The Köppen sys-

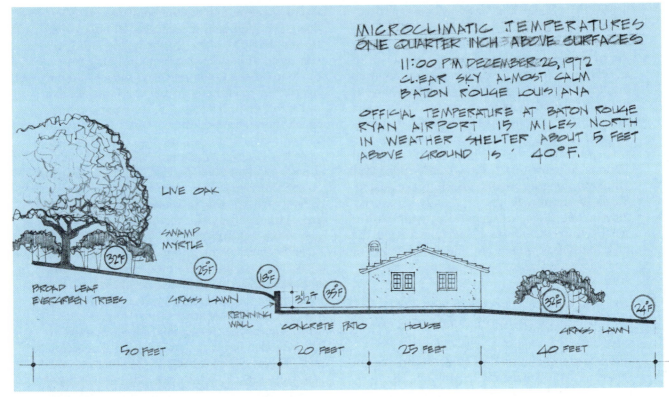

Figure 8.1
The nighttime temperatures in this Louisiana backyard show a large degree of variation from place to place because of such factors as local movements of air, differences in the amount of heat stored in the ground and buildings, and differences in cooling rates. Climate measured on such a small scale is called the *microclimate* of the area. Such details are lost when the climates of large regions are classified into broad categories. (Ron Wiseman after Muller, 1973)

tem of global climatic classification (discussed later in this chapter) assigns New York City and Nashville, Tennessee, to the same climatic region. On a nationwide scale, the climates of the two cities are obviously different, but on a global scale their climates are more like each other than they are like climates of tropical rainforests or Asian steppes, for example, which is the scale of distinction appropriate for world maps.

ONE VIEW OF GLOBAL CLIMATES: ATMOSPHERIC DELIVERY OF ENERGY AND MOISTURE

The availability of energy and moisture in different regions of the earth is a result of the global distribution of solar radiant energy and the general circulation of the atmosphere. To understand how solar energy and the general circulation control the gross features of the earth's climates, we need to analyze the climatic regions of an idealized hypothetical continent, on which no surface features impose distorting effects.

Distribution of Climatic Regions on a Hypothetical Continent

The hypothetical continent shown in Figure 8.2 is flat and featureless, with no mountains, seas, or gulfs. Nevertheless, it embodies some of the features of the actual continents: it is surrounded by oceans; it is broad at the north and extends to high latitudes, like North America and Eurasia; and it tapers toward the south and ends near latitude 60°S, like South America.

The distribution of average temperature on the hypothetical continent is directly related to the patterns of incoming solar radiation. Tropical and subtropical regions receive the greatest annual amounts of solar energy and are warmest year-around (see Figures 3.8 and 3.16, pp. 59 and 70). Regions at higher latitudes receive less solar energy and have lower average temperatures.

The annual temperature range varies according to latitude and distance from the oceans. Because the tropical regions receive a nearly uniform seasonal input of solar energy, they maintain comparatively constant temperatures all year. But because of the tilt of the earth's axis, the higher latitudes receive a large amount of solar energy in the summer and a small amount in the winter. As a result, the difference between average summer and winter temperatures on the hypothetical continent is greater the farther a region is from the equator. The annual temperature range also increases with distance from the moderating effect of the sea. This phenomenon is known as *continentality.* The continental interiors heat strongly in the summer because of the low heat capacity of land areas and become cold in the winter as a consequence of rapid heat loss by longwave radiation.

The distributions of atmospheric moisture and precipitation are controlled by air masses, wind patterns, and pressure systems. Air masses reaching the hypothetical continent after crossing the ocean would be moisture-laden as a result of evaporation of water from the ocean. This humid, maritime air would become a source of precipitation if it were cooled sufficiently to cause condensation. Most cooling takes place when the air is forced to rise by convergence in a low-pressure center or by convection due to heating at the continental surface.

Tropical Wet Climate

On the hypothetical continent the equatorial trough of low pressure (or intertropical convergence zone) occupies the region near the equator, as shown in Figure 4.4 (p. 85). The easterly trade winds that converge into this trough are usually moist due to evaporation of ocean water. As the converging air rises into the circulation of the Hadley cells, cloud formation and rainfall occur. Rainfall is frequent year-around near the equator; thus the climate of the equatorial region is classified as *tropical wet.* Because the trade winds sweep onshore and across the eastern side of the hypothetical continent between about 15°N and 15°S, the eastern side receives more moisture than the western side in equatorial latitudes. This is the climate of the equatorial portion of the Amazon Basin in Brazil.

Tropical Wet and Dry Climate

The tropical wet zone astride the equator is succeeded both north and south by *tropical wet and dry* climatic regions. These zones receive rain in the summer high-sun season as a result of the presence of the intertropical convergence zone with its rising moist air. Drought occurs in the winter low-sun season when the ITC moves into the opposite hemisphere and is replaced by the subsiding air and temperature inversions of the subtropical high-pressure cells, which are below the polar margins of the Hadley cells. Here winter is a time of drought rather than cold. The tropical wet and dry climate characterizes the African Sahel, the broad region south of the Sahara plagued by drought in recent years.

when midlatitude cyclones occur along the polar front. The southeastern United States has such a humid subtropical climate.

Humid Continental Climate

Like the actual northern hemisphere continents, the hypothetical continent is broad in the middle latitudes of the northern hemisphere. Here we find two regions of *humid continental* climate, separated by the interior dry climate, as shown in Figure 8.2. Because of the minimal moderating effect of the far-off oceans, these interior climatic regions are characterized by very wide annual temperature ranges, or continentality. In the southern hemisphere the continent is narrow and the moderating effect of the oceans is greater, so there is no development of continental climate. In the northern hemisphere, radiational heating of the land causes hot summers, and radiational cooling leads to cold winters. The prevailing westerly winds carry the continentality effect from the interior to the east coast. During winter and summer, precipitation is associated mainly with midlatitude cyclones, but precipitation over most areas is greater during summer, when the warmer air masses are supplied with abundant moisture. Midlatitude cyclones tend to become more intense during winter and spring, but the stormy periods occur between stretches of cold, fair weather. This humid continental climate is found in the North American Plains east of the Rocky Mountains.

Subarctic Climate

Poleward of the humid continental climates in the northern hemisphere is the *subarctic* climatic region, which extends from coast to coast. In winter the subarctic climatic region is dominated by the polar high-pressure cell displaced southward over the colder continent. The winter air is always cold and snowfall is light except near the coasts. Continentality again produces a large annual temperature range. Precipitation falls mainly during the mild summer in association with midlatitude cyclones. Most of northern Canada and the major part of the Soviet Union have such a subarctic climate.

Polar Climate

Beyond the subarctic climate region in the northern hemisphere is the region of *polar* climate. The region poleward of latitude 60°N is dominated by the polar high-pressure cell much of the year and is also influenced by the nearby cold and often ice-covered ocean. Winters are very cold with meager snowfall. Most precipitation occurs during the short summer when temperatures are above freezing over all but extensive ice-covered areas. Only Antarctica and the fringes of the Arctic Ocean experience a true polar climate.

Distribution of Climatic Regions on the Earth

The more complex patterns of climatic regions of the real continents shown in Figure 8.3 differ from those of the hypothetical continent mainly as a result of variations in the sizes, shapes, and topographic features of the land areas. For example, the mountain ranges in western North and South America prevent the west coast marine climates from extending as far inland as they do in Europe. The regions to the east of the Rocky Mountains are dry because moist air from the Pacific is blocked off or dried out in its passage across the mountains. The Gulf of Mexico provides a fortunate source of warm moist air that keeps these dry regions from reaching farther eastward. Similarly, the Andes Mountains of South America block the easterly trades, preventing moisture from reaching the Pacific coast of South America between the equator and latitude 30°S. Thus the coasts of Peru and northern Chile are almost completely rainless.

The tropical wet and dry climate reaches unusually far northward in Asia because of the

monsoon effect (Chapter 6). As the vast continent of Eurasia heats up in the summer, the
resulting low pressure pulls the ITC far north
of its average summer latitude, causing heavy
rain over India and Southeast Asia. Drought
follows when the ITC moves back into the tropics. But the causes of the monsoons are complex, and include the high-pressure cell over
Siberia in winter, low pressure over the Indus
Valley in summer, the presence of the Himalaya Mountains, and sudden changes in the
location of the jet stream over Asia.

Nevertheless, many climatic regions on the
continents (Figure 8.3) follow the simple pattern of climates on the hypothetical continent
(Figure 8.2). Near the equator are tropical wet
regions, with tropical wet and dry immediately
to the north and south. Subtropical deserts
sprawl across North Africa, the Arabian peninsula, Iran, and Pakistan. Smaller dry regions
occur in the southwestern United States and
western Mexico. Africa and Australia show the
expected dry regions near latitude 30°S. The
dry summer subtropical climates are found on
the western sides of the continents in five locations that would be expected to have them: California and Oregon; central Chile; Portugal,
Spain, Morocco, and the northern Mediterranean coastlands; the Cape Town region of
South Africa; and small areas of western and
southern Australia. The higher-latitude west
coasts of North America, South America, and
Europe show the expected marine climates.
The interiors of the United States and Europe
are humid continental climatic regions. Subarctic and polar climates occupy northern Asia
and North America.

Figure 8.3 correctly suggests that a resident
of Seattle would find familiar weather in such
far-off places as the British Isles and New Zealand, and an Alabaman or Georgian would find
the weather in south China or eastern Australia more like that at home than the weather
only a long day's drive to the north or west. The
occurrence of similar climates and associated
environments in widely separated locations on

the earth is one of the most important facts
to be understood in the study of physical geography.

ANOTHER VIEW OF GLOBAL CLIMATES: ENERGY AND MOISTURE INTERACTIONS AT THE SURFACE

The preceding view of climate emphasizes
the atmosphere as a delivery system for energy
and moisture but does not consider the interaction of energy and moisture with environmental systems on the surface of the earth. However, systems on the earth's surface do share
the energy and moisture that the atmosphere
delivers; plant growth, for example, depends on
the amount of soil moisture available to vegetation and not directly on delivered precipitation. So to emphasize the relationships between climate and other systems, the view of
climate can be extended to include the interactions of delivered energy and moisture at the
earth's surface.

In terms of the interaction of climate with
environmental systems, three distinct *climatic
realms* can be designated. The criteria that distinguish the three natural realms of climate
can be stated in terms of potential evapotranspiration.

The *frozen* realm embraces regions where
potential evapotranspiration equals zero. The
dry realm includes all regions where precipitation is less than potential evapotranspiration.
Characteristic of the dry realm are dry grasslands and deserts, where deficiencies of soil
moisture are common. The *moist* realm, in
which forests are located, includes all regions
where precipitation exceeds potential evapotranspiration.

Different systems of climate classification
use different methods to distinguish between
dry and moist realms. Two widely accepted
classification systems based on energy-mois-

Climates of the Great Lakes Snowbelts

Inspection of Figure 6.2 (pp. 140–141) reveals that there are six areas near the Great Lakes that on the average receive unusually large amounts of snow each winter. If we arbitrarily assume a seasonal snowfall threshold of 254 cm (100 in.), the map shows the location of six snowbelts: southeastern Lake Ontario, southeastern Lake Erie, southeastern Lake Huron, eastern Lake Michigan, eastern Lake Superior, and southern Lake Superior. In each of these areas a large proportion of the winter snows originate in spectacular and occasionally life-threatening snow squalls induced by the interactions between the "warmer" lake waters and polar outbreaks with air temperatures at −18°C (0°F) or even lower. The map also shows that snowfall averages are closely related to elevation and terrain features, but we shall exclude from discussion the heavy snowfall areas along the summits of the mountains of New England, where much smaller proportions of the snow are due to the effects of the waters of the lakes.

Within the snowbelts the largest seasonal snowfalls on the average occur just east of Lake Ontario in New York over the Tug Hill Plateau where the seasonal snowfall averages more than 482 cm (190 in.), and over the Keweenaw Peninsula of Upper Michigan. These two areas are unusually high, and orographic effects induce even greater snowfalls during squalls.

The snowbelts can be thought of as distinct climatic regions. The frequent snowfalls and deep, long-lasting snowcover have a significant impact on personal activities from November into April and on the total economy of the snowbelts. A disproportionate allocation of public funds must be dedicated to highway snow removal, and as Figure 8.5f illustrates, excessive snow loads must be removed from roofs after particularly heavy squalls. The squalls are often accompanied by strong gusty winds that reduce visibilities to near zero and build shoulder-high drifts across roads in an hour or two.

The more intense squalls are long, narrow cells, perhaps 30 km (19 miles) long and only 8 km (5 miles) wide, with the longer dimension parallel with the windflow aloft at 1,500 meters (5,000 ft) above the surface. Local motorists are very much aware of the danger of being marooned on rural roads, and some radio stations report the precise locations of the squalls at frequent intervals. In recent years radar has helped locate the squalls, and ongoing research on the dynamic interactions between the lakes, the polar air, and the terrain is helping to improve the potential to forecast them in a more timely way.

ture interactions are the Köppen system and the Thornthwaite system.

The Köppen System of Climate Classification

The most widely used system of climatic classification was developed and refined in the early decades of this century by Wladimir Köppen, a German botanist and climatologist. The *Köppen climatic classification* was influenced by the work of nineteenth-century plant geographers who had mapped the world's vegetation on the basis of extensive field studies, and by the implications of the three fundamental realms of climate.

Basic Climatic Types

Köppen's classification is related to vegetation types that he thought to be responses to climate; it recognizes five general climatic types, designated *A*, *B*, *C*, *D*, and *E*. The types *A*, *C*, and *D* represent the moist climatic realms in which precipitation exceeds evapotranspiration, so that there is an annual water surplus. Areas having these climates support forests under natural conditions. The dry realm is represented by the *B* climates where precipitation is exceeded by potential evapotranspiration, so that there is an annual water deficit. The *E* climates are polar types of the frozen realm, in which low temperatures may reduce potential evapotranspiration to near zero. The *B* climates are subdivided into the *BW*, or arid (desert) climatic type (*W* from German *Wüste,* or desert), and the *BS*, or semiarid (steppe) climatic type. The *E* climatic type includes the *ET*, or tundra climate, and the *EF*, or frost (perpetual ice) climate.

Köppen's definition of the *ET* climatic type illustrates his use of vegetation as an indicator of climate. In the *ET* climate the average temperature of the warmest month falls between 0°C and 10°C (32°F and 50°F). The 10°C (50°F) limit for the warmest month corresponds approximately to the poleward boundary of tree growth. By choosing this limit, Köppen ensured that regions with the *ET* climate would be essentially treeless. Still, the 0°C (32°F) lower limit ensures that temperatures above freezing do occur, so that some plant growth is possible, which is not true in the *EF* climate.

Subdivisions

Köppen established specific temperatures and precipitation amounts and ratios to distinguish between his principal climatic types and their subdivisions; the specific criteria are given in Appendix II. The Köppen system was an outstanding achievement for its time, and it still provides a useful introduction to world climatic patterns (Figure 8.4). The system allows additional symbols to be used to describe special features of climatic regions (see Appendix II). The symbol *f*, for example, means that precipitation adequate to offset potential evapotranspiration occurs in every month of the year. Therefore, the symbol *Af* specifies a tropical wet climate, as in the equatorial rainforests of South America. The symbol *Cf* describes the

warm, moist climate of the eastern United States. The symbols *w* and *s* signify winter and summer dry seasons within the *C* and *D* climates of the humid realm, respectively. Thus the tropical wet and dry climate is an *Aw* type, and the dry summer subtropical climate is the *Cs* type. Additional symbols are added to characterize special conditions, such as a monsoon climate or frequent fog.

Global Patterns

The Köppen climatic system represents an intermediate historical step in the evolution of climatic classification from the point of view of the availability of energy and moisture at the surface of the earth. The concept of potential evapotranspiration had not yet been developed, and Köppen vastly overestimated the areas of the continents included within the humid climatic realm. Later in this chapter we shall see that Thornthwaite's concepts of *PE* and the water budget provide more appropriate information about the climatic relationships between energy and moisture, and their interactions with plants, soils, and other environmental systems.

Nevertheless, the Köppen climatic system is comprehensive and flexible. The global patterns of climatic types in Figure 8.4 can also be used as a teaching model of climatic types associated with the atmospheric delivery of energy and moisture *to* the surface of the earth. The Köppen climatic types were developed originally in terms of interactions of energy (represented by temperature) and moisture with global vegetation patterns. Indeed, many of the climatic types are designated in terms of vegetation. However, the system is now most useful as a global scheme of dynamic atmospheric climates. These dynamic characteristics of the individual Köppen climatic types are summarized in Table 8.1, which integrates in capsule form by climatic regions many of the fundamental concepts developed in Chapters 3 through 7. The photographs in Figure 8.5 illustrate some examples of associated landscapes.

Figure 8.4
This map depicts the broad global distribution of climates according to the Köppen system of climatic classification. The definite boundary lines shown between the principal climatic regions are assigned by the Köppen system; in actuality, climatic regions shade gradually into one another. (See Appendix II for a more detailed explanation of Köppen's criteria for climate classification.) (Andy Lucas and Laurie Curran after Köppen-Geiger-Pohl map, 1953, Justes Perthes, and Köppen-Geiger in *Erdkunde*, Vol. 8; and Glenn T. Trewartha, *An Introduction to Climate*, 4th ed. New York: McGraw-Hill, 1968)

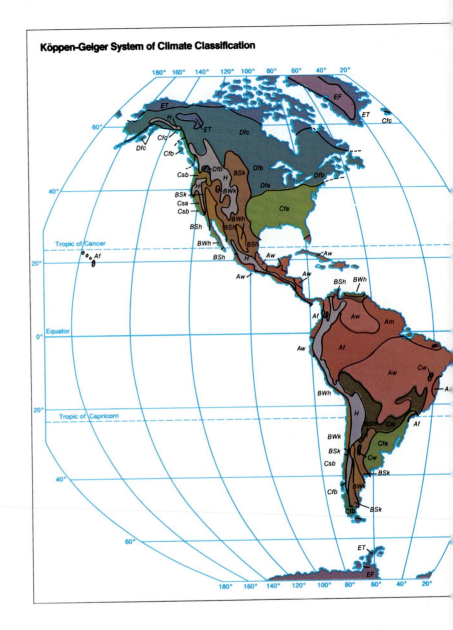

Köppen-Geiger System of Climate Classification

The Thornthwaite System of Climate Classification

In an attempt to overcome the limitations of the Köppen system, C. Warren Thornthwaite in 1948 devised a method for classifying cli-

mates according to water budget evaluations of energy and moisture.

Moisture Index

In *Thornthwaite's climatic classification*, energy is specified by potential evapotranspira-

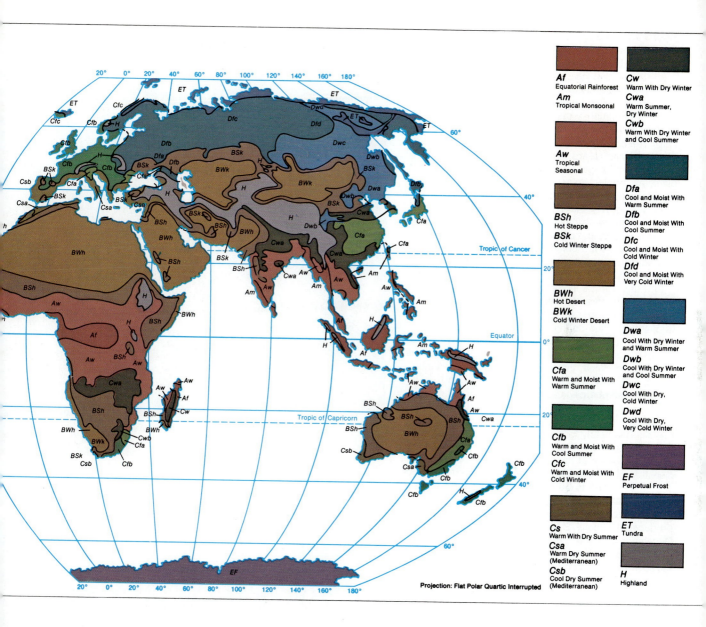

Projection: Flat Polar Quartic Interrupted

Af Equatorial Rainforest
Am Tropical Monsoonal

Aw Tropical Seasonal

BSh Hot Steppe
BSk Cold Winter Steppe

BWh Hot Desert
BWk Cold Winter Desert

Cfa Warm and Moist With Warm Summer

Cfb Warm and Moist With Cool Summer
Cfc Warm and Moist With Cold Winter

Cs Warm With Dry Summer
Csa Warm Dry Summer (Mediterranean)
Csb Cool Dry Summer (Mediterranean)

Cw Warm With Dry Winter
Cwa Warm Summer, Dry Winter
Cwb Warm With Dry Winter and Cool Summer

Dfa Cool and Moist With Warm Summer
Dfb Cool and Moist With Cool Summer
Dfc Cool and Moist With Cold Winter
Dfd Cool and Moist With Very Cold Winter

Dwa Cool With Dry Winter and Warm Summer
Dwb Cool With Dry Winter and Cool Summer
Dwc Cool With Dry, Cold Winter
Dwd Cool With Dry, Very Cold Winter

EF Perpetual Frost

ET Tundra

H Highland

tion *(PE)*, and moisture is expressed by a *moisture index.*

Thornthwaite's moisture index depends on the difference between precipitation and calculated values of potential evapotranspiration (rate of water loss to the atmosphere from land surface with enough soil moisture). The moisture index has the value of −100 when there is no precipitation, and may exceed +100 where rainfall far exceeds potential evapotranspiration. At the boundary between the dry and moist realms, the moisture index is zero.

GLOBAL SYSTEMS OF CLIMATE

Figure 8.5

Some "climatic" landscapes from west to east across the United States.

(a) Orographic lifting of persistent easterly trade winds creates a nearly continuous cloud canopy over the Hawaiian island of Kauai. Precipitation is so frequent and intense that the summit area in the background is one of the rainiest places on earth.

(b) The dry summer subtropical landscapes of coastal California are striking mosaics of dark evergreen oaks, golden brown grasses, bright green plants in irrigated areas, and—in this view from Mt. Tamalpais just north of San Francisco—a very blue lake. (Robert A. Muller)

(c) During the rainless summer seasons of the dry summer subtropical climates, the vegetation dries out and brush fires are common. The fires in this view burned for several days in August 1977, in the hills east of Oakland, California. (T. M. Oberlander)

(d) Lush tropical landscaping on this island in San Diego Bay is possible only because of the total absence of freezing temperatures and because of frequent irrigation during the dry summers. North of Mexico, similar exotic tropical landscaping is limited to southern California, lowland places in Arizona such as Phoenix and Yuma, the Rio Grande valley in southern Texas, and southern Florida. (Robert A. Muller)

(e) The steppe climates of the Great Plains are much too dry for tree growth, but dryland wheat production is very successful during years with normal rainfall. This nearly featureless plain in the Oklahoma Panhandle experiences runs of dry years, such as the Dust Bowl years of the 1930s, which can be disastrous for the regional economy. (Robert A. Muller)

(f) Roofs need to be cleared of heavy snow loads in the snowbelt regions of the Great Lakes. This scene was repeated over and over again after snow squalls produced 182 cm (72 in.) of snow in 6 days in December 1958, in a narrow band between Oswego and Syracuse, New York. Ladders to the roofs and caved-in old buildings are characteristic features of the landscapes where big snow squalls are common. (Robert A. Muller)

(g) A massive thunderstorm towers to more than 12.2 km (40,000 ft) inland from the coast near Pensacola, Florida. Some vacationers do not appreciate that these thunderstorms tend to form inland along the leading edge of the cooler sea breeze, leaving the beaches along the Atlantic Coast sunny and breezy much of the time. (Robert A. Muller)

(a)

(b)

(c)

(d)

(e)

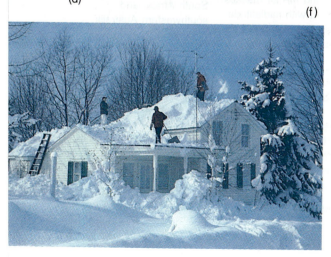
(f)

(g)

weekend in 1974 devastated crops across much of the upper Midwest. The record-breaking heat in Texas, the Plains states, and the South during the early summer of 1980 increased energy use, seared pastures, and withered crops. Fluctuations in water budgets are especially significant. Above normal deficits can greatly reduce crop yields (the "dust bowl" years of the 1930s are a dramatic example), and along the Mississippi, alternations in the surplus have resulted in both floods and low-water problems.

Only in recent decades have scientists begun to unravel the detailed outline of the earth's climatic history for the past million years, using studies of tree rings, fossil plant pollen, ancient soils, lake and deep-sea sediments, mountain glaciers, and cores from ice caps on Greenland and Antarctica. Figure 8.9 illustrates the major trends of global climate based on these analyses. Perhaps the most astounding features are the *glacial-interglacial cycles* that last about 100,000 years, with shorter interglacials separated by long periods of glacial climates, as shown in graph *a*. Glacial conditions appear to develop slowly and irregularly, with interglacial climates evolving suddenly.

Time Scales of Climatic Change

Recent studies indicate that climatic change on a time scale of about 100,000 years is associated with systematic variations of the elliptical orbit of the earth around the sun. Climatic change on a time scale of about 41,000 years is associated with systematic variation of the tilt of the earth's axis of rotation between about 22° and 24.5°. And climatic change on a time scale of about 26,000 years is associated with the wobble of the earth's axis so that the North Pole faces first toward Polaris (the North Star) and later toward Vega before returning again toward Polaris; this precession results in perihelion occurring during midwinter in the northern hemisphere, and then during midsummer in the northern hemisphere about

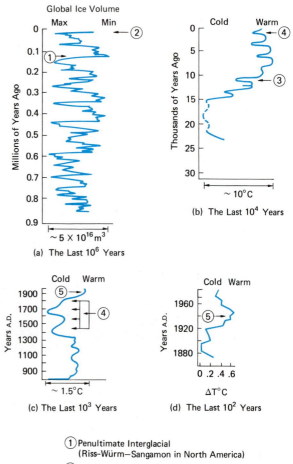

(1) Penultimate Interglacial (Riss-Würm—Sangamon in North America)
(2) Present Interglacial (Holocene)
(3) Younger Dryas Cold Interval
(4) Little Ice Age
(5) Thermal Maximum of 1940's

Figure 8.9
The main trends of global climate during the past million years. Graph *a* is based on isotope analyses of deep sea floor sediments; *b* on alpine tree lines, fluctuations of alpine glacier snouts, and fossil pollen; *c* on historical records of alpine glacier snouts and wine and grain production; and *d* on measured temperature data. (Modified from *Understanding Climatic Change: A Program for Action,* National Academy of Sciences, 1975, Washington, D.C.)

13,000 years later. These three cycles are associated with variations in the seasonal incomes of solar radiation by latitude that are closely related to the global temperature patterns during the Pleistocene illustrated in Figure 8.9.

Graph *b* in Figure 8.9 shows the rapid warming trend that began about 15,000 years ago. During the very sudden and brief Younger Dryas cold interval about 10,500 years ago, much forest in Europe was destroyed. Graph *b* also shows that the mildest post-Pleistocene temperatures in Europe occurred about 5,000 to 7,000 years ago.

Temperature variations in eastern Europe over the past 1,000 years are shown in graph *c*; the temperature reconstruction is based on historical records of the positions of the snouts of alpine glaciers and of the wine production at various monasteries. The lower temperatures associated with the *Little Ice Age* between 1400 and 1850 resulted in social and economic dislocations. The thermal maximum of the 1940s represents the highest average temperatures of the past 1,000 years. Graph *d* shows the cooling that began about 1945, but recent data suggest that the global cooling may have leveled off since the mid-1970s.

Short-term Climatic Variability

Much shorter-term climatic variability has direct impact on environmental processes and economic activities. Figure 8.10 shows monthly temperature and precipitation variability for southeastern Louisiana from 1911 through June 1982. Each month's deviation from its respective average for the 30-year period 1931–1960 is shown in graphs *a* and *d*. Graphs *b* and *c* show short-term variations in terms of twelve-month "running averages."

There are two outstanding features of these graphs for southeastern Louisiana. One is the downward trend of temperature beginning in the fall of 1957 and continuing, with the exception of several winters, to the present. The other is the tendency for clusters of months or of years to be warmer or colder, or wetter or drier, than normal. Especially obvious are the dry periods of 1915–1918, 1933–1938, and 1951–1955, all associated with droughts on the Great Plains; other intense dry periods include 1924–1925, 1962–1963, and 1967–1971. Climatic variability on these time scales significantly affects agricultural yields, flooding, and water-resource management.

Causes of Climatic Change

Explaining why climates change is extremely difficult because many interactions of the general circulation are not understood in detail. Furthermore, slight climatic changes can produce disproportionate effects in the environment. If the average global temperature were to drop a few degrees, ice and snow would cover a greater proportion of the earth's surface. Because of its high albedo, the ice would reflect more radiant energy back to space, which would cause further cooling and further spread of ice and snow.

A number of different explanations have been proposed to account for climatic variability over geologic time, but none is entirely satisfactory. Glacial climates appear to have been associated with continental drift and plate tectonic processes (see Chapter 12), which caused continents to move into polar regions, allowing ice sheets to develop. In addition to Pleistocene glaciation, there is evidence of at least three widely separated glacial periods over the last billion years. Within these glacial periods, it is now believed that systematic variations in the earth's orbit account for the 100,000-year glacial-interglacial cycle. Furthermore, mountain uplift is believed to be connected with colder temperatures and increased snowfall over uplands. Stratospheric dust from intense volcanic activity has been linked to periods of somewhat diminished solar radiation income and lower temperatures at the earth's surface. The early summer freezes and snows in northern New England in 1816 followed the massive eruption

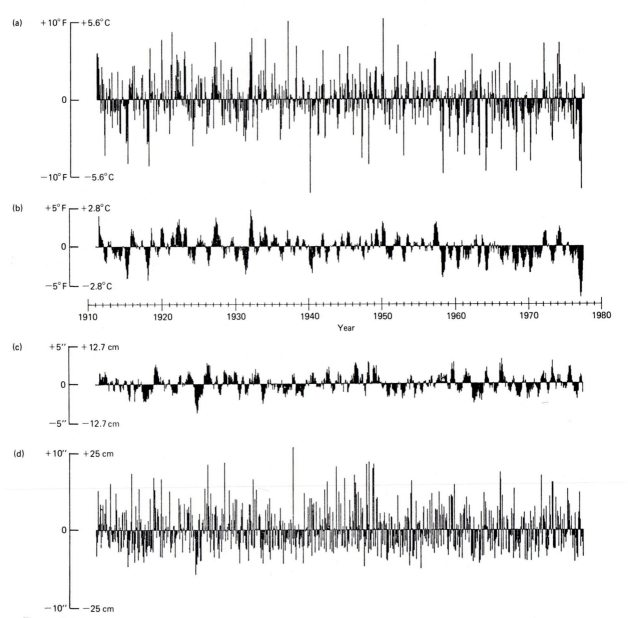

Figure 8.10

Temperature and precipitation variability in southeastern Louisiana. The temperature deviation of each month from the averages of the 30-year period between 1931 and 1960 is shown in graph *a*, and precipitation deviations are shown in graph *d*. For example, January 1940 was 7°C (12°F) below normal, and the rainfall during October 1937 was 28 cm (11 in.) above normal. Twelve-month running averages of temperature deviations are used in graph *b* to illustrate a "smoothed" interpretation of short-term temperature variation, and graph *c* is a similar smoothed interpretation of precipitation variation. Note especially the colder temperatures beginning in late 1957 and the fluctuations between warmer and colder, and wetter and drier conditions. (Adapted from R. A. Muller & J. E. Willis, 1978, "Climatic Variability in the Lower Mississippi River Valley," *Geoscience and Man*, Vol. 19, pp. 55–63)

of the volcano Tambora in the Dutch East Indies in 1815, and several years of spectacular red sunsets followed the eruption of Mt. Agung in Indonesia in the early 1960s. Some scientists believe that gases and particles ejected into the stratosphere by the Mexican volcano El Chichón in 1982 had an effect on the unusual atmospheric and oceanic circulation patterns during the winter and spring of 1982–1983.

Numerous attempts have been made to relate *sunspot activity* to climatic variation, especially drought periods in the Great Plains. Although there is some correlation between sunspots and dry periods on a time scale of several years, no cause-and-effect relationship has been demonstrated. Climatic variability on shorter time scales, such as those shown for southeastern Louisiana, are obviously related to changing patterns of the circumpolar vortex of westerlies in the upper atmosphere. Meteorologists and geophysicists continue to debate the causes of these circulation changes and the eventual return to more "normal" patterns. Precise forecasting of the circulation patterns a season or more in advance remains an elusive objective at present.

Future Climatic Change

In predicting the climates of the near future, most concern centers on the effect of human activities, particularly the use of the atmosphere as a dumping ground. The atmosphere is a sensitive controller of climate; cloudiness and dust in the air directly affect the amount of energy reaching the earth, and *carbon dioxide* in the air affects the amount of energy that is radiated away. Industry and agriculture add smoke and dust particles to the air, increasing the albedo of the atmosphere, decreasing the amount of solar radiation received at the surface, and possibly lowering surface temperatures as well.

The interconnections of the earth's systems make it difficult to predict the net effect from a particular change. The amount of carbon dioxide in the atmosphere is expected to increase by at least 20 to 30 percent in the last half of this century, primarily because of industrial activities. The direct effect of the increase in carbon dioxide will be a decrease in longwave radiation losses from the earth's surface, which will raise the average temperature. The higher temperatures will increase evaporation, perhaps leading to increased cloudiness. Because it is difficult to determine the degree to which each of these modifications will affect other environmental properties, the final temperature change caused by the increase of carbon dioxide cannot be predicted.

Any small, long-term change in the atmosphere is critical. The present state of the environment is a function of its past history; because the environment accumulates the effects of minor changes, there is no fresh start each year. Orbit wiggles, volcanic dust—these and other small and complicated interactions influence climatic trends. Now human activities can be added to the list.

KEY TERMS

teleconnections
climatic indexes
microclimate
continentality
tropical wet climates
tropical wet and dry climates

dry climates
dry summer subtropical climates
marine climates
humid subtropical climates
humid continental climates
subarctic climates

polar climates
climatic realms
   frozen
   dry
   moist
Köppen's climatic classification
Thornthwaite's climatic classification
moisture index

subhumid climates
glacial-interglacial cycles
Little Ice Age
volcanic dust
El Chichón eruption
sunspot activity
carbon dioxide

REVIEW QUESTIONS

1. What climatic indexes are generally used in physical geography?
2. What factors determine variations within a microclimate? How do they differ from those that determine global climates?
3. Why are different types of climate developed on the east and west coasts of the hypothetical continent at subtropical latitudes?
4. What are the main differences in the climatic pattern over the northern and southern hemispheres of the hypothetical continent?
5. Why does the marine climate of Western Europe extend further inland than the marine climates on the west coasts of North and South America?
6. What environmental conditions characterize the boundary between the moist and dry realms?
7. Why was dependence on vegetation as an indicator of climate a drawback to Köppen's system?
8. What are the principal ways in which Thornthwaite's system of climate classification differs from Köppen's?
9. Where would you expect farmers to depend more on irrigation—Manhattan, Kansas, or Cloverdale, California? How would a comparison of the water budgets of the two areas reflect this difference?
10. What have been the major trends of global temperature over the past 1,000 years?
11. What systematic variations of earth-sun relationships on time scales of 26,000 or more years are thought to produce major climatic variation and even cycles of glaciation?

APPLICATIONS

1. Along portions of the coasts of North and South America, Southern Africa, Portugal, and Australia there are climates that are generally regarded as "Mediterranean" in type. However, the climates of these coasts do not exactly duplicate the climates of the Mediterranean coastlands themselves. What is the difference?
2. The largest high latitude land area that was not covered by an ice sheet during the Ice Age was Siberia, which now has the coldest winters of any area outside of Antarctica. How do you explain this contradiction?
3. Suppose the Gulf of Mexico and the Caribbean Sea did not exist and the North American coastline extended continuously from Florida through the Bahamas, Puerto Rico, and the Lesser Antilles and Trinidad to Venezuela. How would this affect the climates of North America? What would happen to the geographic location of Thornthwaite's zero moisture index that

separates dry and moist climates? Draw a map showing hypothetical climatic regions on such a continent, using the Köppen system of classification.

4. The Mississippi River Valley interrupts an otherwise continuous belt of hilly terrain and low mountains extending from Oklahoma to New England. In some places these mountains were much higher in the past. If they were as lofty today as the Rockies are, how would the climates of North America be affected?

5. Climatic variability from year to year can have significant impact on environmental systems and economic prosperity. Using Appendix II and National Weather Service data on monthly temperature and precipitation for some place of interest, determine the Köppen climatic type for each of five successive years. What portions of the earth would be most likely to fall into different Köppen climates from year to year?

6. Imagine the effect on the earth's climates if the axis of the earth's rotation were perpendicular to the plane of the earth's orbit around the sun, rather than being inclined at the present angle of 23½ degrees. What differences in climatic types would result? Would the earth be more or less habitable?

FURTHER READING

Carter Douglas B., and **John R. Mather.** "Climatic Classification for Environmental Biology." *Publications in Climatology,* Vol. 19, No. 4 (1966):305–395. The evolutionary sequence of the several Thornthwaite climatic classifications and some of their relationships to vegetation distribution are presented in this technical paper. Average water-budget tables from thousands of places around the world are included in other issues of this specialized journal.

⸻, **Theodore H. Schmudde,** and **David M. Sharpe.** "The Interface as a Working Environment: A Purpose for Physical Geography." *Tech. Paper No. 7,* Comm. on College Geog., Assoc. of American Geog. (1972), 52 pp. This monograph stresses the relationship between water budget components and environment responses.

Critchfield, Howard J. *General Climatology,* 4th ed. Englewood Cliffs, N.J.: Prentice-Hall (1983), 453 pp. An introductory text that emphasizes the interrelationships of climate to global patterns of vegetation and soils, as well as to a wide range of economic activities.

Griffiths, John F., and **Dennis M. Driscoll.** *Survey of Climatology.* Columbus, Ohio: Charles E. Merrill (1982), 358 pp. This introductory text has a meteorological perspective and focuses on applications.

Hare, F. Kenneth. *The Restless Atmosphere,* rev. ed. London: Hutchinson (1956), 192 pp. This brief introductory classic contains outstanding regional and continental chapters that focus on the dynamics of the general circulation.

Lamb, H. H. *Climate, History, and the Modern World.* London: Methuen (1982), 387 pp. An up-to-date overview of the impacts of climatic variation and change by the pioneer scholar of these interesting and important topics.

Roberts, Walter O., and **Henry Lansford.** *The Climate Mandate.* San Francisco: W. H. Freeman (1979), 197 pp. This little paperback provides an outstanding current perspective on climatic variation and on some of its causes and consequences.

Thornthwaite, C. Warren. "An Approach Toward a Rational Classification of Climate." *Geographical Review,* Vol. 38, No. 1 (1948):55–94. Thornthwaite first set out the potential evapotranspiration and water budget systems in this classic paper.

Trewartha, Glenn T., and **Lyle H. Horn.** *An Introduction to Climate,* 5th ed. New York: McGraw-Hill (1980), 416 pp. Temperature and precipitation properties of climatic regions over the globe are stressed in this introductory text.

Wilcock, Arthur A. "Köppen After Fifty Years." *Annals Assoc. of American Geog.,* Vol. 58, No. 1 (1968):12–28. This article presents a concise review of Köppen's development of his climatic classification and the subsequent modifications by other climatologists.

All interactions among the parts of an ecosystem are united by the energy flow through the system. This intimate interdependence of living things implies a certain stability, a certain dynamic reciprocity, and a certain harmony of nature.

La Charmeuse des Serpents (The Snake Charmer) by Henri Rousseau. (Art Resource)

9
The Biosphere and Ecological Energetics

The miracle of our planet is its life—mysterious in origin and wondrously diverse—inhabiting every portion of the surface of the lands and the soils below, the surface waters of the continents, the oceans, and the lower atmosphere. All of these are components of the *biosphere*: the realm of living organisms on our planet.

All forms of life require energy and liquid water. The ultimate energy source for all but certain bacteria is solar radiation. On planets more distant from the sun, the low level of solar energy received causes any water present to be frozen and therefore unable to react chemically, thus precluding biological activity. The two planets nearer to the sun (Mercury and Venus) receive so much energy that their water has been vaporized and is present only in the upper atmosphere of Venus (Mercury has too little gravity to have retained an atmosphere).

The earth's nearest planetary neighbors, Venus (nearer to the sun) and Mars (farther from the sun) differ enormously. The surface temperature of Venus is high enough to melt lead. This is a result of the planet's proximity to the sun and of a "runaway" greenhouse effect produced by an atmosphere rich in CO_2. By contrast, Mars is a cold and nearly airless desert, swept by dust storms, with evidence of a permanently frozen layer below the land surface (Figure 9.1). Neither planet has any trace of liquid water. Thus it is not surprising that

cies have been named and described in some manner. Two-thirds of these are animals—most of them insects. It is estimated that as many as 8 or 9 million unnamed and undescribed additional species exist—as many as a million species of mites alone!

Plant and animal species live in associations, or *communities,* that offer mutual benefits to the individual members. Animals rely directly or indirectly on plants for food (chemical energy), but they also help to propagate and spread plants. Squirrels not only eat nuts and acorns; they also bury them for future use, thereby planting future trees. Animals that eat fruit distribute the indigestible seeds in their fecal droppings. In the wet tropics even fish eat fallen fruit, convey the seeds, and thereby help to propagate forest trees. Flowers, with their perfumes and bright colors, can have evolved only to attract animals to the plants. Insects that seek flower nectar inadvertently transport the pollen that is essential for plant reproduction. Without such insects many kinds of plants could not survive. Clearly, plants have evolved mechanisms to take advantage of animal activity, and animals have evolved processes to utilize plant resources.

Ecosystems and Biomes

The interactions of plants and animals, and their relationships with their physical environment, give the communities definite patterns of organization. However, the task of tracing and understanding all of these interactions is extraordinarily difficult. For example, a typical midlatitude forest may contain 50 or more species of plants and thousands of species of animals, primarily insects. Each seems to play a role in the functioning of the community as a whole.

Ecosystems

The study of the interactions among organisms with their particular habitat or environ-

ment is known as *ecology* (from the Greek *oikos:* house). One way to organize the character of these interactions is to focus on the basic ecological unit, which is known as an *ecosystem.* An ecosystem is a community of plants and animals generally in equilibrium with the inputs of energy and materials in a particular environment. Thus an ecosystem has both biotic (living) and abiotic (nonliving) components. An ecosystem can be as small as a tidal pool or a single tree, or as large as a lake or a forest. Its boundaries may be either sharply defined or gradational, and it may include several different biotic communities. The concept is a flexible one, defined by function rather than scale.

Biomes

An ecological system on the largest scale is known as a *biome.* Biomes are highly generalized ecosystem types recognized on a global scale, such as deserts, temperate grasslands, or tropical rainforests, with their faunas and associated physical characteristics. The concept of biomes includes broad, climatically influenced associations of plants and animals that can be portrayed on a world map. Every biome is a synthesis of many separate ecosystems, unified by some common characteristic such as the general morphology of the vegetation (e.g., forest, grass, or shrubs) and associated animal types. Biomes are designated in various ways and may be divided into subtypes. Chapter 10, which concerns global vegetation, is organized in terms of biomes.

Ecological Niches

Associated with the concept of the ecosystem is the idea of the *ecological niche,* which focuses on the specific ways an organism actually functions in its particular habitat. The ecological niche of a species is the combination of environmental factors under which the species can exploit a source of energy sufficiently to survive, reproduce, and successfully colonize other similar habitats. The scale of an ecological niche varies with the organism considered. For bacteria it may be a microscopic pit on a rock surface. A single grizzly bear's niche might be much of a sprawling mountain range and portions of surrounding lowlands.

It is interesting to note that organisms do not occupy all of the ecological niches to which they are well adapted. For example, there are no polar bears in the Antarctic region, despite the availability of appropriate climatic conditions and abundant food supplies in Antarctic waters. Conversely, penguins are not found in the northern hemisphere, though they should do well in the Arctic. Likewise, the large apes (chimpanzees and gorillas) of the equatorial forests of Africa are absent from similar environments in Central and South America, which have plentiful resources that these animals normally exploit. African giraffes would be at home in the Australian bush and the tree-studded llanos of Venezuela, but nothing like them exists in either place. These are rather glaring examples of ecological niches that seem underutilized.

Plant and Animal Dispersal

Why some ecological niches are fully exploited and others are not is a matter of interest to biogeographers. The explanations involve the geographical locations of the centers of evolution of the organisms adapted to the niche, their mode of dispersal, competition from other organisms, and geographical barriers to their diffusion. These barriers include impassable topography, large water bodies, and unsuitable climates.

Many ecological niches are more effectively occupied by plants than by animals. Animals can colonize new areas rapidly, but must do so by their own locomotion, whereas plant dispersal takes place by means of winds, running water, ocean currents, and animals. Clearly, cold-adapted polar bears and penguins could not cross the low-latitude deserts and humid equatorial region to go from one hemisphere to the other, though water barriers would present no difficulty to them. In order for gorillas to migrate outward from the African forests, they would have to swim oceans or cross deserts lacking water and suitable food resources. But had they evolved some 200 million years ago in Triassic times rather than within the past 10 million years, gorillas could have wandered overland into the Americas, which for millions of years were connected to the African continent (Chapter 12). Indeed, the dinosaurs of the Triassic period were common to both the Old and New Worlds.

Some of the mysteries of plant and animal distributions are explained by the horizontal movements of the continents discussed in Chapter 12. The continents appear to have been in motion on the earth's surface since their initial formation: colliding and separating, opening up migration routes and then breaking them—permitting multitudes of organisms to spill outward from centers of evolution at certain times, and locking them in isolation at other times.

Australia, long isolated from other landmasses, has the most primitive and unusual of all existing faunas, including the pouched *marsupial* animals epitomized by the kangaroo, and an egg-laying mammal, the duck-billed platypus (Figure 9.3). The marsupials evolved more than 100 million years ago in South America and somehow spread to Australia, perhaps by way of an ice-free Antarctica. But the highly diversified marsupials remaining in South America were unable to compete with the large numbers of more modern placental mammals entering from North America after the link at Panama was established only 2 or 3 million years ago. In isolated Australia the marsupials experienced no such competition and have lingered on to the present as a biological curiosity.

Changes of sea level also affect animal distributions. A lowered sea level can cause chains of islands to become connected and to

(a)

(b)

Figure 9.3
Two representatives of Australia's unique fauna.

(a) The semi-aquatic duck-billed platypus is an extremely primitive mammal found in the isolation of Australia and Tasmania. A furred animal hatched from soft-shelled eggs, it has webbed feet and a leathery toothless muzzle like a duck's bill. (Douglass Baglin Photography)

(b) Like the platypus, the Australian kangaroo represents an ancient evolutionary line surviving only in locations remote from the centers of plant and animal evolution. The kangaroo, a marsupial, is born the size of a honeybee and must mature in a pouch on its mother's belly. (San Diego Zoo)

link with a nearby continent, thus opening migration routes, whereas a raised sea level can isolate land areas and islands, blocking animal dispersal. The classic example of faunal discontinuities resulting from sea level fluctuations is the area of Malaysia, Indonesia, and Australia, illustrated in Figure 9.4.

More than a century ago, the English naturalist Alfred Russel Wallace chose the islands of Southeast Asia as a key area in his attempt to explain the origin and distribution of species. Wallace was the first scientist to note that

one association of plant and animal species was common to the Asian mainland and nearby islands west of the line *A* on Figure 9.4, while another association of quite different species occurred south and east of line *B*. The light blue areas were all above sea level during the Pleistocene Ice Ages when sea level was more than 100 meters (330 feet) lower than at present. This permitted free movement of plants and animals between areas that are now large and small islands in the seas of Southeast Asia. However, the much deeper water present between lines *A* and *B* has remained a permanent barrier separating the two dissimilar biotic realms representing completely separate evolutionary streams.

Biogeographical Realms

Scientific observers recognize various *biogeographical realms* that reflect the evolutionary centers and patterns of dispersal of specific plant or animal associations (Figure 9.5). These realms reveal the close proximity of the northern hemisphere landmasses to each other over a long period of time, the barrier imposed

Figure 9.4
The distributions of animal species in Southeast Asia, Australia, and the islands between reflect sea levels that were lower in the past, exposing land connections (light blue) between the Asian mainland and adjacent islands, and between Australia and New Guinea. The deep water between lines A and B prevented the interchange of land animals even during times of lowered sea level. Line A is known as Wallace's Line, in honor of the naturalist who first revealed the significance of animal distributions in this area. (After Time-Life Nature Library, *Tropical Asia,* 1969)

by the subtropical deserts, and the isolation of certain landmasses and oceanic islands.

Biological realms contain related species. They are different from the biomes resulting from the *convergent evolution* of different species toward similar forms suited to particular habitats. Thus the very similar-appearing plants and animals of the desert biome in various parts of the world are actually genetically unrelated species that have evolved virtually identical strategies to minimize heat stress and water loss.

All of the great biogeographical realms, both plant and animal, include several biome types. The *Nearctic* and *Palaearctic zoogeographic realms* are especially diverse, including the tundra, grassland, deciduous and coniferous forests, chaparral, and desert biomes. The *Neotropical* and *Afrotropical* realms are dominated by tropical forests, savannas, and deserts. In the *Australian* realm a desert core is surrounded by tropical forest and savanna, while the *Indomalayan* realm is dominated by tropical forest, as is the *Oceanian* realm. The *Antarctic* realm is one of temperate forest and grassland in New Zealand, and tundra and ice in Antarctica.

The zones of transition or interfaces between the various biogeographical realms are some of the most important influences on the distribu-

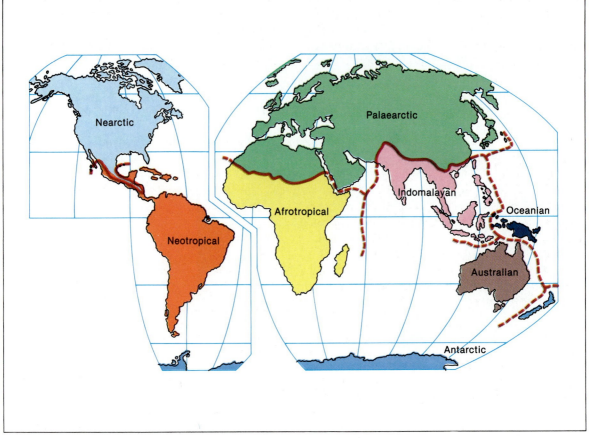

Figure 9.5
The global zoogeographical realms reflect the different evolutionary centers for animal life. The subdivisions are based on the distributions of related groups of animal species. Although certain key species remain confined to their area of origin, mixed populations appear in some of the transition zones, and some species range far beyond the realm in which they first evolved. Varying biomes composed of particular forms of vegetation and animal life are present within each evolutionary realm, as indicated in Figure 10.4, p. 256. (After Miklos Udvardy and *Coevolution Quarterly,* 1975)

tion of life on our planet. The life of the Nearctic realm is separated from that of the Neotropical realm by deserts and the absence, until recently, of a land connection between the evolutionary centers of the two biotic realms. The lifeless center of the Sahara Desert divides the Palaearctic and Afrotropical realms, which earlier were separated by an arm of the sea. Arabia is open to invasion from the north and during the important recent period of mammalian evolution has become separated from Africa by the opening of the Red Sea.

The Indomalayan realm is presently isolated from the Palaearctic realm by the lofty Himalaya Mountains. Previously there was a sea lane between the two landmasses. We have seen that the Indomalayan realm is separated from the Australian and Oceanian realms by deep ocean water. Although the Australian biota is more closely related to that of Oceania than to the Indomalayan realm, its animal life is so distinctive that it is regarded as a separate faunal realm. New Zealand's mammals have affinities with those frequenting Antarctic shores, so that it has been included in the Antarctic realm on Figure 9.5.

ENERGY FLOW THROUGH ECOSYSTEMS

Ecologists view the flow of energy in an ecosystem as a pyramid composed of several *trophic levels,* in which energy is stored in an organism that serves as food for an organism at a higher trophic level. Each successively higher level supports a smaller number and mass of organisms.

Trophic Levels

There are various types of trophic levels. At the base of the pyramid are the *primary producers* consisting principally of green plants that can convert solar energy to organic energy in the form of living tissue. This in turn is usable as an energy source by other organisms, none of which can manufacture their own food. Certain bacteria are also primary producers, having the ability to use the chemical bonds of rock and soil minerals as energy sources. All other organisms are dependent directly or indirectly upon the primary producers (Figure 9.6).

These other organisms are either *consumers,* which ingest live plant material or the prey that they or others have killed, or *decomposers,* such as bacteria, molds, and fungi, which make use of the energy stored in already dead plant and animal tissues. The animals that ingest plants are called *herbivores,* and these animals may in turn be ingested by other animals,

Figure 9.6
Flows of energy and materials in ecosystems.

(a) The flow of energy in a natural ecosystem. Following the flow from left to right, solar radiant energy input is converted to food energy by green plants. The plants use some of the energy for respiration, herbivores consume some of it, and organisms of decay feed on some of it. Part of the energy in the herbivores is transferred to successive carnivores. However, the amount of transferred energy decreases at each step because energy is used for respiration and other purposes.

(b) This diagram indicates how materials circulate between the principal units of an ecosystem and between the ecosystem and the environment. (After R. Whittaker, *Communities and Ecosystems,* Macmillan, 1970)

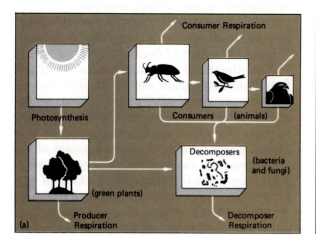

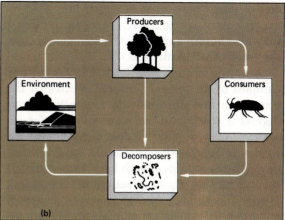

Consider a simple plant–herbivore–carnivore food chain consisting of grass plants, mice, and snakes. The mice obtain approximately 10 percent of the energy absorbed earlier by the grass plants, and the snakes obtain approximately 10 percent of the energy absorbed by the mice. Thus the carnivore receives only about 1 percent of the energy originally absorbed by the plants. The fraction of the original energy available to a succeeding carnivore stage—a hawk, for instance—is still less. The rapid decrease of available energy along a food chain limits such chains to four or five links. Large carnivores, such as lions, which are the last natural link in a food chain, obtain only a small fraction of the energy absorbed by the primary producers in their habitat. Lions must roam over large areas to obtain their food, and one region cannot support many of them.

ENERGY CAPTURE AND PLANT PRODUCTION

We can begin a closer study of the energy flow in an ecosystem by looking at the factors that determine how a primary producer captures and utilizes energy. We shall consider green plants, which support most terrestrial ecosystems. Of the solar radiation energy falling on a plant leaf, a small amount is immediately reflected and approximately 80 percent is absorbed, as Figure 9.8 illustrates. Some of the absorbed energy functions to warm the leaf and is then given off as longwave radiation, while much of the absorbed energy is used in the evaporation and transpiration of water stored in the plant. Only about 1 percent of the solar radiation is used in the process of photosynthesis to produce the chemical energy the plant requires for growth and maintenance.

The chemical processes involved in photosynthesis are complex. In terms of energy, however, photosynthesis may be thought of as a process in which simple molecules, water

(H_2O) and carbon dioxide (CO_2), are joined with the aid of solar radiant energy to form more complex carbohydrate (sugar or starch) molecules (CH_2O):

$$H_2O + CO_2 + solar\ energy \rightarrow CH_2O + O_2$$

The solar radiant energy is stored as chemical energy in carbohydrates. Further chemical reactions use the carbohydrates and nutrients from the soil to produce complex protein molecules that the plant's cells require for growth. In addition to synthesizing carbohydrates, photosynthesis produces free oxygen gas. The formation of oxygen is incidental to photosynthesis, but it is important to the ecosystem as a whole, as noted in Chapter 1. Without photosynthesis, there would be very little free oxygen in the atmosphere, and animal life as we know it could not exist (Figure 9.9).

Photosynthesis and Plant Growth

In 1926 Edgar Transeau was able to conclude that only 1.6 percent of the total solar energy incident on a field of corn during the 100-day growing season became stored chemical energy through photosynthesis. This percentage is the energy conversion efficiency of corn. No plants have energy conversion efficiencies of more than a few percent.

Photosynthesis is dependent on only the visible light portions of the total solar spectrum (see Figure 3.1, p. 44), and the rate of photosynthesis in a leaf varies with the intensity of the incoming light. If the intensity of light is increased, the rate of carbohydrate production will also increase, up to the maximum value for each plant species. Further increases of light intensity beyond this point will not result in increased photosynthesis. Leaves that are partially shaded, or that receive only indirect light, are able to carry on photosynthesis near the maximum rate. Because plants that have adapted to the tropics receive more solar energy, they generally have higher maximum

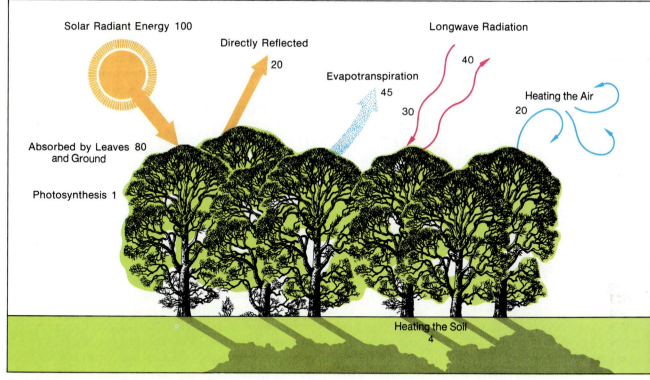

Solar Radiant Energy 100

Directly Reflected
20

Longwave Radiation

40

Evapotranspiration
45

30

Heating the Air
20

Absorbed by Leaves 80
and Ground

Photosynthesis 1

Heating the Soil
4

Figure 9.8
For every 100 units of solar radiation incident on a forest in a humid climatic region, only about 1 unit becomes stored chemical energy through photosynthesis. This diagram shows how the 130 units of energy received (100 solar radiant, 30 longwave) are utilized for various components of the local energy budget: reflection, 20 units; evapotranspiration, 45; longwave radiation, 40; heating the air, 20; heating the soil, 4; photosynthesis, 1.

rates of photosynthesis than plants native to midlatitude regions, where less solar energy is available. Graphs depicting the relationships between photosynthesis and solar radiation are shown in Figure 9.10.

The rate of photosynthesis depends also on the temperature of the leaf, which may differ from the air temperature. For plants in midlatitude regions, photosynthesis for a given light intensity reaches a maximum at a leaf temperature of about 25°C (77°F). For arctic plants,

maximum photosynthesis occurs at lower leaf temperatures. The rate of carbohydrate production decreases above and below a plant's optimum leaf temperature, and production stops if leaf temperatures rise above 40°C (104°F) or so.

If there is no wind to cool the leaves, leaf temperatures may rise so high that photosynthesis stops during the middle of the day, when solar energy input is maximum. Under such conditions, carbohydrate production is limited to a period in the morning and to a brief period in the late afternoon. It is partly for this reason that the warmest regions of the tropics tend to have lower agricultural yields than midlatitude regions.

A number of other factors also influence the rate of photosynthesis. Adequate supplies of water and carbon dioxide are necessary for efficient photosynthesis. The availability of nutri-

THE BIOSPHERE AND ECOLOGICAL ENERGETICS

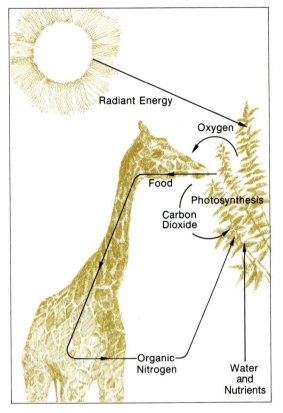

Figure 9.9
Photosynthesizing plants combine carbon dioxide from the air with water and nutrients from the soil, using radiant energy to produce stored food energy. Animals rely on plants for food, and they breathe the oxygen released as a by-product of photosynthesis. Carbon dioxide is returned to the atmosphere by the respiration and decay of plants and animals; nutrients such as organic nitrogen compounds also cycle between organisms and the environment.
Photosynthesis thus plays a significant role in the cycles of water, oxygen, carbon, nitrogen, and nutrient minerals between the soil, plants, animals, and the atmosphere.

ents from the soil, particularly nitrogen and phosphorus, also affects the rate of carbohydrate production. Nitrogen is required for the synthesis of the plant proteins necessary for cell growth. Phosphorus, a comparatively rare element in the earth's crust, is required in plants as they make chemical compounds important in the photosynthetic process. Phosphorus deficiency is often the limiting factor for plant growth in moist climatic regions, lakes, and coastal areas.

Plant Respiration

Not all of the carbohydrate a plant produces in photosynthesis is available to animals as food. In the process of respiration, carbohydrate combines with oxygen and is reduced to carbon dioxide and water, releasing stored chemical energy for use by plant cells. Hence, respiration is a necessary process that provides the energy the plant requires to maintain its vital functions.

Photosynthesis stops in the absence of light or when leaves become too warm. Respiration, however, continues both night and day, although it is greater during the day when the temperature is higher and the leaves are exposed to light. Higher temperatures increase respiration, and as respiration rises, the amount of net energy stored by photosynthesis declines. Cool nights help plants conserve the chemical energy obtained during the day. Therefore, rates of plant respiration and utilization of stored chemical energy tend to be greater in warm climates than in cool climates.

One way to find the amount of energy a plant uses in respiration is to seal the plant in a dark box and measure the amount of carbon dioxide it produces by respiratory activity. Studies show that a wide variety of plants, including phytoplankton (floating microscopic green plants in water), use approximately one-fourth of their stored energy for respiration.

Measures of Plant Production

The rate at which a plant converts light to chemical energy is called its *gross productivity*. Since the plant uses some of the chemical energy for respiration, only a portion of the total chemical energy remains stored in plant tis-

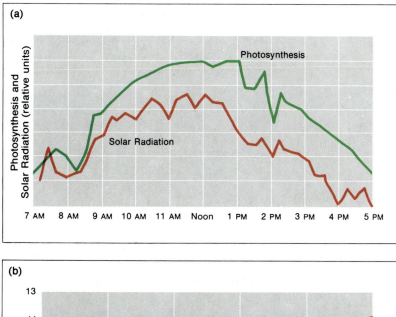

(a)

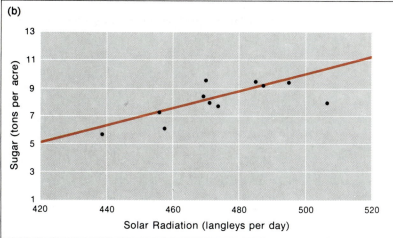

(b)

Figure 9.10
Relationship between solar radiation, photosynthesis, and yield.

(a) The course of photosynthesis through the day rises and falls with the variations of incident solar energy.

(b) This graph shows the yield of sugar from a Hawaiian sugarcane plantation during different growing seasons. Each dot shows average daily solar radiation input and associated yield of sugar during a season. The solid line represents the general trend; it shows that yield is high during seasons when fair weather causes the average radiant energy input to be high and is low during cloudy seasons. (After Jen-hu Chang, 1968)

sues. *Net productivity* equals gross productivity minus respiration over a period of time. Net productivity is the quantity important to the consumer of a plant because it represents the amount of organic material produced by the plant.

The *standing crop,* or *biomass,* represents the amount of energy stored in plants at any given time, expressed in grams of dry matter per square meter of ground surface. Biomass, which is a static measure, should not be confused with net productivity, which is the rate at which a plant produces food. A mature, unlogged redwood forest has a large biomass but a comparatively low productivity. A large amount of energy is stored in the trees, but not much additional energy is added to storage each day. Phytoplankton in the ocean have a high productivity, but their biomass is small because they are continuously being consumed by predators.

Yield represents the amount of energy stored during the growing season in the desired portion of a crop, such as the fruit. It is usually determined by weighing the harvested portion of the crop.

The productivity of a particular plant is expressed in terms of the amount of energy stored during a given time interval, such as a day or a year. When the plant is exposed to light, photosynthesis and respiration both occur. Photosynthesis causes carbon dioxide to be absorbed at the same time that respiration causes carbon dioxide to be given off. The net rate at which the plant absorbs carbon dioxide is therefore a measure of its net productivity.

Methods of Measurement

The net productivity of agricultural crops is often estimated by harvesting the complete plants, roots and all, at the end of the growing season. The plants are then dried and weighed, and the productivity is expressed in terms of the dry matter that a given area produces during a specified period of time. New productivity measured by the harvest method can be expressed as grams of dry matter per square meter of field per year, as in Table 9.1. When the harvest method is applied to natural plant communities, measurements of dry matter must be corrected for losses caused by the shedding of leaves, the depredations of insects, and the death of some plants during the growing season.

The energy content of a gram of dry matter differs among plant species, depending on the relative amounts of proteins and carbohydrates in the plant tissues. To correct for these differences, productivities expressed in terms

Table 9.1
Net Productivity of Major Ecosystems

TYPE OF ECOSYSTEM	AREA (MILLIONS OF SQ KM)*	NET PRODUCTIVITY PER UNIT AREA (DRY GRAMS PER SQUARE METER PER YEAR)† NORMAL RANGE	MEAN	WORLD NET PRODUCTIVITY (BILLIONS OF DRY TONS PER YEAR)
Tropical Forest	24.5	1,000–3,500	2,000	49.4
Temperate Forest	12.0	600–2,500	1,250	14.9
Boreal Forest	12.0	400–2,000	800	9.6
Woodland and Shrubland	8.5	250–1,200	700	6.0
Savanna	15.0	200–2,000	900	13.5
Temperate Grassland	9.0	200–1,500	600	5.4
Tundra and Alpine	8.0	10–400	140	1.1
Desert and Semidesert	42.0	0–250	40	1.7
Cultivated Land	14.0	100–3,500	650	9.1
Swamp and Marsh	2.0	800–3,500	2,000	4.0
Lake and Stream	2.0	100–1,500	250	0.5
Total Continent	149.0		773	115.0
Open Ocean	332.0	2–400	125	41.5
Continental Shelf, Upwelling	27.0	200–1,000	360	9.8
Algal Beds, Reefs, Estuaries	2.0	500–4,000	1,800	3.7
Total Marine	361.0		152	55.0
World Total	510.0		333	170.0

*One square kilometer is equal to about 0.39 square mile.
†One gram per square meter is equal to about 0.0033 ounce per square foot.

Source: Robert H. Whittaker, 1975. *Communities and Ecosystems,* 2nd ed. New York: Macmillan.

of dry matter can be converted to energy units by burning the dry matter in a device that measures the heat energy released by complete combustion.

The results obtained by the ecologist E. P. Odum demonstrate that estuaries, spring sites, coral reefs, and energy-subsidized agriculture are the richest ecosystems, having annual productivities of 25,000 to 10,000 kilocalories per sq meter. These are followed by moist forests, shallow lakes, and moist grasslands (10,000 to 3,000); mountain forests, dry grasslands, deep lakes, and continental shelf waters (3,000 to 500); deep oceans (less than 1,000); and deserts (less than 500). The two least productive ecosystems, the moisture-deficient deserts and nutrient-deficient deep oceans, cover some 80 percent of the earth's surface. However, one must keep in mind that the oceans are the source of the moisture that nourishes all terrestrial ecosystems.

Global Patterns of Plant Production

The average global net productivities of all important food crops, such as wheat, rice, corn, and potatoes, are of the order of 500 to 700 grams of dry matter per square meter per year. Surprisingly, mean agricultural productivity is no greater than that found in many "natural" ecosystems (Table 9.1). Productivities two or three times greater are attained in areas where intensive mechanized agriculture is practiced. The world average net productivity of sugarcane is high—about 1,500 grams per square meter per year. Productivities of more than 9,000 grams have been measured for sugarcane in some tropical regions, where the growing season lasts the entire year and the soil is moist and rich in nutrients.

Estimates of the mean annual net productivity of the land areas are shown in map form in Figure 9.11. The map data are estimated by means of a climatic model of the relationships among climate elements and gross and net productivity; therefore, the map patterns of net productivity are similar to the global maps of climatic regions in Chapter 8. The map patterns are also relatively similar to the net productivity data in Table 9.1.

The map and table show that productivity is greatest in equatorial and tropical regions where moisture availability meets the demands of *PE*. As both the *PE* and moisture availability decrease, so does net productivity. Because of the large areal extent and high productivity rates of tropical and midlatitude ecosystems, these regions account for more than half of the total productivity of the continents. In turn, the average annual productivity of the continents is four to five times that of the oceans, even though the oceans comprise more than two-thirds of the surface area of the globe. The most productive regions of the oceans are the shallow, sunlit waters of the continental shelves, and especially those river estuaries where nutrient-rich streamflow mixes with ocean waters across broad expanses of brackish and saline marshes. Such areas are too small to show on a world map.

Stable Climax Ecosystems

Despite the difficulties, the energy flows in several ecosystems have been studied in detail. The results of these studies are often summarized conveniently in energy flow diagrams, which show the main energy pathways through the systems.

The energy flow diagram in Figure 9.6a is that of a typical ecosystem. The diagram illustrates how the energy input to each stage in a food chain is divided among various energy flows and outputs. The principal outputs include the energy used in life processes, or respiration; the energy transferred to the next stage of the food chain; the energy transferred to the decay food chain; and any energy utilized for other purposes.

When plant communities achieve a stable, comparatively unchanging state called *climax vegetation* (Chapter 10), very little energy is used for the production of new plant material. The energy released by respiration and by the decay of dead plant and animal tissues balances the energy stored in new growth, so that the total energy stored in a climax forest remains effectively constant. Although individual trees in a climax forest use energy for growth, the forest ecosystem as a whole uses essentially all the energy input for respiration.

Researchers who studied the energy flow for an oak and pine forest near the Brookhaven National Laboratory on Long Island, New York, found that of the annual gross production of dry matter per square meter, only about one-fifth was stored as new growth, litter, and humus (Figure 9.12). Respiration accounted for more than 80 percent of the energy used, which indicates that the Brookhaven forest is in a late stage of succession toward an idealized stable climax condition.

THE PROBLEM OF DISTURBED ECOSYSTEMS

Humans have always interacted with natural ecosystems and have created artificial ecosystems to fit their needs. Throughout most of history, and in many countries today, our goal in dealing with nature has been to obtain the materials and foods that are immediately necessary for survival. Sometimes, however, these activities have made survival more difficult.

The Sahel

A vivid example is the human catastrophe centered in the semiarid transition zone between the Sahara Desert and the tropical forests of Africa, known as the Sahel. In the 1960s an unusual period of moist years there had en-

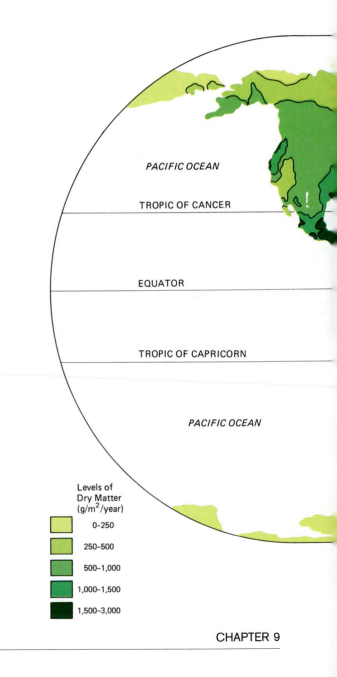

Figure 9.11
This map shows estimates of annual net primary productivity of the land areas of the earth in grams of dry matter per sq meter per year. The estimates are based on relationships of productivity to temperature and precipitation, so the patterns are similar to the climatic regions on the maps in Chapter 8. (E. Box, *Radiation and Environmental Biophysics,* Vol. 15, 1978)

Levels of
Dry Matter
(g/m^2/year)

0–250

250–500

500–1,000

1,000–1,500

1,500–3,000

couraged agriculturalists, who had already shifted from subsistence food crops to commercial crops, to expand their cropland into formerly dry areas. At the same time, wells were drilled that encouraged owners of cattle to increase their herds. When the atypical wet period ended in 1968, to be succeeded by sustained drought, local economies collapsed. Agriculture failed, and pastures that had been overstocked during the wet period could no longer support the livestock population. Across the breadth of sub-Saharan Africa some 5 million cattle died of starvation, amid a pall of dust. As many as 100,000 people perished. The

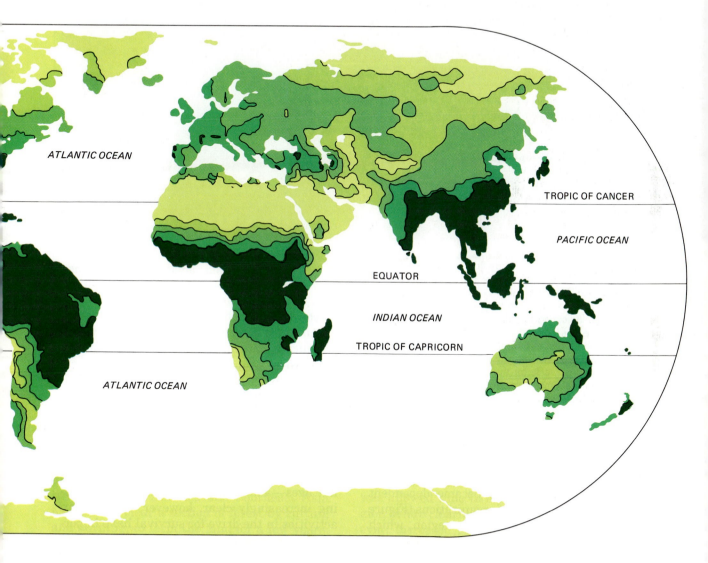

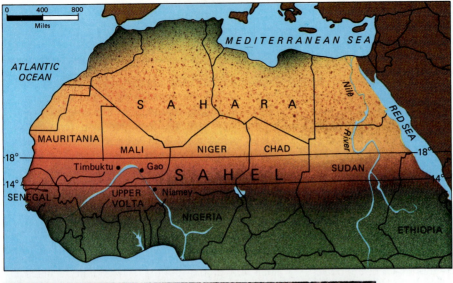

Figure 9.13
The drought in the African Sahel region, shown on the map, so reduced forage for livestock that several million cattle died of starvation. These cattle supported many populous nomadic tribes who maintained the animals for their yield of milk and blood. (Fred Sowers)

vents the reestablishment of the forest. The cattle pack the earth to a nearly impervious layer that seriously diminishes infiltration of rainfall and increases runoff and erosion. Eventually the soil becomes so depleted that even edible grasses disappear, to be replaced by coarse unpalatable grasses.

The accelerated pace of swidden cultivation and cattle pasturing may make large tracts of land in the rainforests economically useless and difficult to restore to productivity.

The Sea

The search for food has caused many countries to turn to the sea. The sea as a whole is not a highly productive region, as indicated earlier. Photosynthesis is confined to the sunlit

Tearing Up the Desert

To many Americans the desert is valueless—a wasteland useful only for the disposal of trash and the testing of nuclear weapons; a place with no owners, no restrictions, and nothing that can be damaged; a place where space-hungry urbanites can decompress harmlessly by making noise, raising dust, rutting the ground, and being the boss. The federal government, which owns the vast majority of desert lands, seems to endorse this attitude. But another group of Americans has viewed the desert quite differently. They have treasured it as a place of solitude that can be enjoyed for its changing colors, clear air, long views, and fascinating plant and animal life. The views of the two groups, both seeking their own type of escape from the pressures of everyday life, seem irreconcilable.

In fact, the first group is the smaller of the two, but its influence and visibility is out of proportion to its numbers. This group is a lucrative market; it buys expensive "dirt bikes" and "dune buggies" and the costumes and protective gear required to survive daredevil exploits in off-road vehicles or ORVs. A fully equipped ORV rider, suited-up like a space voyager, is the product of an international industry bent on fast expansion.

The best-known recurring episode in the war between those who seek escape in tearing up the desert and those desiring to preserve it in its pristine condition has been the Barstow (California) to Las Vegas (Nevada) cross-country motorcycle "race." As many as 3,000 riders have lined up for the start of this event, proceeding to ride down everything in their kilometer-wide path for the distance of 240 km (150 miles). The United States Geological Survey estimated that one such event stripped bare 9,000 acres of former desert shrubland. This is merely an exaggerated form of a year-around activity in our deserts. Small mammals, birds, and reptiles virtually disappear from areas of intensive ORV use—killed, mutilated, or driven out by noise and the crushing of the underground burrows essential for survival in the desert environment. Many bikers make a point of defying any attempt to control their activity. They see the desert as the last outpost of "freedom" and point to military and grazing use of desert land, claiming an equal right for all other (human) users. For some, this includes the right to ride down jackrabbits, to shoot tortoises, to topple cacti, to ride circles through prehistoric Indian rock intaglios, and to spray graffiti over ancient petroglyphs (rock etchings) and once-scenic rock formations.

The delicate crusts that form on sandy desert soils and the stone pavements that develop on level surfaces protect the desert against wind erosion. Vehicles disrupt these surface veneers, so that areas of ORV activity become miniature dust bowls, conspicuous even on LANDSAT images. Vehicle use also compacts desert soils, increasing water runoff and erosion. Destruction of vegetation and accelerated erosion in scattered areas increase the surface albedo, reducing surface temperature and the potential for atmospheric convection. Unfortunately, thermal convection is the main source of summer rainfall over the Southwest. Thus widespread ORV use could make the desert region still drier, weakening the vegetation and reducing animal populations even in areas not touched by ORV use. A well-known aspect of desert ecosystems is the slow rate of recovery from disturbance. Desert Indian trails unused for centuries are still clearly demarcated. The Pony Express route through Nevada, used only in 1860 and 1861, remains a conspicuous track more than 100 years later. Natural repair of a mutilated desert pavement requires hundreds of years.

The federal government promotes recreational use of its desert lands and has attracted expanding numbers of ORVs to the fragile desert environment by its designation of "open areas" for unrestricted ORV use. In the absence of any effective enforcement of the various Bureau of Land Management vehicle-use categories, and in view of the "outlaw" attitude of too many desert bikers, no place in the desert is really safe from those who constitute the greatest hazard to the scenic and psychic resources of our deserts.

upper 50 to 100 meters (160 to 330 ft) of the ocean, and life in this layer is usually limited by lack of nutrients in the water. An acre of midlatitude grassland is several times more productive than an acre of ocean. The net production of the ocean is less than half the net production of the land, and much of this production is concentrated in areas of upwelling (Chapter 4), which comprise less than 1 percent of the ocean surface.

The strategy of many nations is to maximize the investment made in modern fishing fleets and methods. Often the pressure to intensify fishing increases as the catch declines. For this reason, the oceans are being overfished, and many species, including some types of whales, have almost disappeared. In 1935 commercial fishermen in California landed 60,000 tons of Pacific sardines. The catches decreased with time, and by 1961 the sardine industry of California no longer existed. Similar overfishing is evident as regards Pacific perch, mackerel, and tuna.

The best long-term strategy for an intelligent predator is to restrict the catch to the point at which the population of prey species can remain stable. However, the desire of many nations for immediate increases in their living standards becomes more important than fear of the inevitable decline in ocean productivity.

The misuse of the Sahel and rainforests and the overfishing of the oceans are only three examples that indicate how immediate benefits can be accompanied by large hidden losses and costs. The farmers and the fishing fleets gain access to an inexpensive source of energy, but restoring the ecosystems that produced the energy will eventually be costly to society as a whole. Because the true costs of human actions are seldom considered, we may eventually suffer the dire consequences of severely depleted ecosystems. With the rapid increase in world population, humans are far from being in a stable long-term relationship with natural and agricultural ecosystems.

KEY TERMS

biosphere
biogeography
phyla
genera
species
ecology
ecosystem
biome
ecological niche
marsupial
biogeographic realms
convergent evolution
zoogeographic realms
    Nearctic
    Palaearctic
    Neotropical
    Afrotropical
    Australian
    Indomalayan
    Oceanian
    Antarctic

trophic level
primary producer
consumer
decomposer
herbivore
carnivore
food chain
grazing food chain
decay food chain
food web
biomass pyramid
energy flow pyramid
photosynthesis
gross productivity
net productivity
standing crop
biomass
yield
climax vegetation
slash-and-burn (swidden, milpa) cultivation

REVIEW QUESTIONS

1. As an environment for life, how does the earth differ from the neighboring planets, Mars and Venus?
2. Give two examples of interactions between animals and plants that benefit both.
3. Distinguish between an ecosystem, a biome, and a biogeographical realm.
4. In terms of its biogeography, what is most distinctive about the Australian continent?
5. In what island area is the distribution of animals most clearly related to past changes in sea level?
6. Explain the concept of trophic levels.
7. Why is there a loss of energy from one trophic level to the next?
8. How does photosynthesis result in the storage of solar energy?
9. Compare the diurnal cycles of photosynthesis, leaf temperature, and solar radiation.
10. What is the difference between net productivity and biomass?
11. The continents and the seas each have areas of very high and very low productivities. Where are these?
12. How is most energy utilized in an area of climax vegetation?
13. Why is the continuing problem of "drought" in the Sahel often regarded as a human-induced phenomenon rather than a natural disaster?

APPLICATIONS

1. What are the species names for your official state plant and animal? Classify each in terms of genus, family, order, class, and phylum. At each level indicate some related organisms that illustrate the degree of diversity within each category.
2. Are there important unresolved biogeographical questions pertaining to your area? Examples might concern the time or mode of arrival of particular species of plants or animals, or the effects of changing climates on species distributions.
3. Referring to Shelford's *Ecology of North America* or a similar source, determine what natural biome existed in the area of your home or campus. How important are the plants and animals of the natural biome currently? How might you label the human-dominated biome existing in the area now?
4. Describe several ecosystems on your campus. How are these present ecosystems different from "natural" ecosystems at the same sites before the time of European settlement?
5. How would you try to document ecosystem changes associated with climatic variation since 1970?
6. Some cultural geographers have specialized in the study of the domestication of wild plants and animals and their diffusion to other regions. Where are the primary regions of domestications, and how is the diffusion of plants and animals related to global patterns of climates?
7. Describe three very different food chains found in ecosystems characteristic of your state.
8. What can you learn about the productivity of land in your state? Which areas are generally believed to have the highest and lowest productivities respectively?
9. Agricultural yield data for significant crops are normally published for each county. Map the variation of the yield by counties of a significant crop in your state, and speculate about the reasons for the variation from one region to another.

FURTHER READING

Bennett, Charles F., Jr. *Man and Earth's Ecosystems: An Introduction to the Geography of Human Modification of the Earth.* New York: Wiley (1975), 331 pp. This unique text is organized by world regions in order to focus on the ecological impacts of human use of the earth. The book specifically analyzes the geographical and historical background of the environmental crisis.

Eckardt, F. E., ed. *Functioning of Terrestrial Ecosystems at the Primary Production Level.* Proceedings of the Copenhagen Symposium, July 1965. UNESCO (1968), 516 pp. This collection of technical papers by some of the world's leading scholars includes several studies that are especially useful at the introductory level.

Furley, Peter A., and **Walter B. Newey.** *Geography of the Biosphere: An Introduction to the Nature, Distribution, and Environments of the World's Life Zones.* London and Boston: Butterworths (1983), 413 pp. The first half of this large book is an analysis of the major components of the biosphere. The second half is an analysis of each of the major biomes. Quite original in its approach.

National Academy of Sciences. *Productivity of World Ecosystems.* Washington, D.C. (1975), 166 pp. This short book is a collection of papers presented at a symposium in 1972. The papers, written at beginning and intermediate levels, analyze the world's major ecosystems.

Odum, Eugene P. *Basic Ecology.* Philadelphia: W. B. Saunders Co. (1983), 613 pp. This is a comprehensive work by an outstanding American ecologist.

———. *Ecology: The Link Between the Natural and Social Sciences,* 2nd ed. New York: Holt, Rinehart & Winston (1975), 244 pp. This paperback introduces ecological principles. More emphasis is placed on human interactions with ecosystems than in the preceding work.

Odum, Howard T. *Systems Ecology: An Introduction.* New York: Wiley (1983), 644 pp. H. T. Odum is famed for his systems approach to biological modeling, using a standard set of symbols applicable in all cases. This book is his most recent effort in this direction.

Park, Chris C. *Ecology and Environmental Management: A Geographical Perspective.* Boulder, Colo.: Westview Press (1980), 272 pp. Park's book deals with the problem of environmental protection against ever more intense human pressures on natural ecosystems.

Phillipson, John. *Ecological Energetics.* New York: St. Martin's Press (1966), 57 pp. This very small book is often considered a classic; most of the basic themes of ecology are introduced succinctly.

Whittaker, Robert H. *Communities and Ecosystems,* 2nd ed. New York: Macmillan (1975), 385 pp. This is an outstanding paperback on ecology, especially pertinent to this chapter.

Where both moisture and solar energy are abundant the earth's land surfaces are cloaked with a living blanket of green vegetation. This phenomenon, unique in our solar system, varies remarkably in form and behavior in response to our planet's diverse climates and surface characteristics. The appearance of plant life on the continents made possible the emergence of all other terrestrial life forms, whose food chains are ultimately based on vegetal matter.

Studio di cielo e albieri (Study of Sky and Trees) by John Constable. (Art Resource)

10
Vegetation and Climate

When land plants evolved on earth, some 400 million years ago, they soon spread to all parts of the planet except those regions covered by glacial ice. To do so they had to evolve adaptations to every combination of energy, moisture, and soil conditions found on the earth. This was accomplished through genetic variation and natural selection. Although these adaptations have caused plant life to vary widely in form, from algae to giant forest trees, most of the higher plants (excluding algae, fungi, and lichens) have certain similarities. Most have a root system to gather chemical nutrients dissolved in soil moisture, green leaves to convert solar energy into chemical energy for plant growth, and stems or branches to support the leaves and to channel the nutrients and chemical energy throughout the plant.

PRINCIPAL ADAPTIVE STRATEGIES OF VEGETATION

The basic elements of plant structure described above must vary greatly to be effective over a wide range of environmental conditions. The depth and spread of roots reflect the supply of moisture and nutrients. Leaves may be large or small depending upon light levels and moisture stress. Leaf form varies in response to environmental constraints. Leaves are retained all year by *evergreen* plants and are dropped seasonally by *deciduous* plants to prevent frost damage or moisture loss when resupply is impossible. Under conditions of severe moisture stress plants may do without leaves altogether, with stems and branches taking over the function of photosynthesis, as in the case of the cactus family. Stems, too, are adapted to local conditions and take many forms. In later pages we shall see how these adaptations assist plants in specific environments.

In terms of their relationship to moisture supplies, plants may be classified as *hydrophytes*, which grow in water, *xerophytes*, which are structurally adapted to survive in extremely dry soils, and *mesophytes*, which grow where the water supply is neither scant nor excessive. Likewise, plants may be classified as *perennial*, persisting from year to year and enduring seasonal climatic fluctuations, and *annual*, dying off during periods of temperature or moisture stress but leaving behind a crop of seeds to germinate during the next favorable period for growth.

By their many adaptations, plants have developed the variety allowing them to grow and reproduce from the edge of wind-whipped snowfields to black-water swamps and sun-baked deserts. This chapter first summarizes some of the factors that are responsible for the development of natural vegetation types found in varying environments. Following this, the general characteristics of the principal global vegetation formations and their associated climates are described. It must be kept in mind that the natural vegetation that has evolved in equilibrium with undisturbed environments has been vastly altered by human activities. These activities have caused the virtual disappearance of some natural vegetation formations and threaten the extinction of others in the not-too-distant future.

ECOLOGY OF VEGETATION

In Chapter 9 the emphasis was on ecological systems, with attention to the energetics of photosynthesis, respiration, growth, and ecosystem productivity. In the initial portion of the present chapter we shall consider the sequential development, or succession, of plant communities with the passage of time.

Plant Communities

The associated plant species that form the natural vegetation of any one place are known as a *plant community*. In any midlatitude forest, for example, many kinds of trees, shrubs, ferns, grasses, and flowering herbs all live together in one plant community. Numerous genetically unrelated species of plants not only live in association with one another, but also with bacteria, fungi, insects, and burrowing, seed-collecting, grass- and herb-eating (grazing), and twig- and leaf-nibbling (browsing) animals. Besides providing food and shelter for animals, a plant community affects its local environment by modifying soils and moisture storage conditions, by shading the ground, and by cooling the air through the process of transpiration and latent heat transfer. In such ways, individual plants also affect one another, so that the plant community behaves as an ecological system composed of many interacting parts.

Each species in a plant community has its own particular way of utilizing energy and moisture. In a broadleaf forest, ferns are able to use the subdued light that filters through the high leafy canopy. They do not compete with trees for direct sunlight. Mosses, lichens, and fungi growing on rocks utilize a moisture supply that is different from that of ferns growing on the forest floor. Because of this principle, farmers in the tropics often successfully grow several dissimilar crops—such as bananas, a tall open plant, and cassava, a low root crop—in the same field. As long as the crops do not compete for the same moisture and sunlight, the crop yield per acre can be greater than if the fields were planted in a single crop (Figure 10.1).

Figure 10.1
Examples of two-level agriculture in the humid tropics.
(left) Coconut palms standing above rice in the
Philippines. (right) Oil palm with an understory of taro
in Malaya. (Robert Reed)

Ecotones

The boundary between dissimilar plant communities may be a gradual zone of transition and community intermingling or a relatively abrupt interface. The plant community in a distinct zone of transition between other more extensive communities is known as an *ecotone*. Ecotones can vary in scale from that between an agricultural field and a patch of forest to the general zone of contact between global vegetation regions such as the midlatitude forests and adjacent grasslands. Studies of ecotones can reveal the dynamics of vegetative change, such as expansion and contraction through time.

Plant Succession

Hikers walking through the New England woods sometimes come upon the foundation stones of farm buildings that were abandoned more than a hundred years ago. Although the land was once cleared for agriculture, it is covered by forest again. But the forest is not immediately reestablished when cleared land is abandoned. Instead, a *succession* of plant com-

munities occurs. Each succeeding community alters the local microclimate and surface condition, making possible the appearance of a community that is more demanding in terms of nutrients and specific soil and moisture conditions. The usual trend in plant succession is toward taller, more diverse, more permanent vegetation. If the natural succession is not disturbed by fire, wind damage, or climatic change, a stable community is attained, and there is no further change in its composition. However, periodic disturbance may be a normal feature of most ecosystems.

The particular order in which plant communities succeed one another depends on whether the succession begins on cleared land, filled-in marsh, burned-over forest, or some other condition. It also depends on the climate and soil conditions of the location. In a succession that eventually converts cleared land to forest, the early communities are dominated by grasses and low shrubs, as shown in Figure 10.2. In time, the shrubs increase in number, and small-statured trees appear. These, in turn, are replaced by larger trees. As more advanced communities become established, the pace of succession slows because the larger individual

(a)

(c)

(b)

Figure 10.6
These views illustrate aspects of tropical rainforest vegetation.

(a) Epiphytes with large leaves and aerial roots cling to the trunk of a massive tree in the coastal rainforest of Brazil. (Robert Voeks)

(b) A host tree is enveloped by strangler vines, created where the aerial roots of epiphytes of the fig family *(Ficus)* have reached the forest floor and rooted in the soil around the host tree. Robbed of soil nutrients and increasingly shaded by the foliage of the vigorously growing strangler, the host tree eventually dies and rots away, leaving the strangler as a hollow tree with an interlaced trunk structure. (T. M. O.)

(c) An average-size rainforest tree with a buttressed trunk. The buttressing structures help support large trees that have shallow roots. Shallow rooting is common in rainforests as a response to a range of soil conditions, including waterlogging, deficiency in nutrients below the surface layer of organic litter, or toxic concentrations of insoluble metallic oxides at depth resulting from removal of soluble nutrients by large volumes of downward-percolating water.
(T. M. O.)

CHAPTER 10

Chapter 11 we shall see that the heavy rainfall of the wet tropics dissolves soluble mineral matter in the soil and flushes (leaches) it away, leaving behind a very infertile residue in which plants must grow. The most fertile part of the soil is the surface layer, which receives nutrients from the plant litter (leaves, branches, fruit, etc.). The roots of rainforest trees are generally shallow because of this nutrient distribution and because roots will not penetrate soils that are perpetually waterlogged. Many roots do not enter the soil at all, but creep under the decaying litter on the forest floor, extracting nutrients directly from the organic debris. In all rainforests certain tall tree species have buttressed trunks that flare widely at the base to support a shallow-rooted, massive structure reaching high toward the sunlight.

It may be hard to believe that the survival of the once vast rainforests of the wet tropics is not at all certain, but such is the case. Human activities such as logging and clearing for agriculture and cattle pasturing are destroying the rainforests at an alarming pace. Reestablishment of a cleared rainforest may be next to impossible because the soil deteriorates rapidly as soon as the forest cover is removed (Chapter 11).

Regional Climate

In tropical rainforest areas, most of the precipitation falls as heavy showers and thunderstorms between late morning and early evening when solar heating at the surface makes the humid tropical air most unstable. Clusters of wet days are sometimes followed by days with only scattered showers. The wetter periods are associated with the westward passage of weak low-pressure disturbances within the trade-wind systems and intertropical convergence zone.

Air temperatures vary little through the seasons because midday solar altitudes are high throughout the year and polar air masses and

fronts are absent. For example, the average monthly temperature at Singapore, located at latitude 1°N, shows little variation from season to season, and the rainfall exceeds 15 cm (6 in.) in every month (Figure 10.7). Such a climate can be described by a *thermoisopleth diagram* (Figure 10.8), which shows average air temperatures throughout the day for each day of the year. The pattern of the temperature contour lines, or *isotherms,* on a thermoisopleth diagram reflects the climatic characteristics. When the isotherms on the graph are primarily horizontal, temperature variations during a day are greater than variations from season to season. The thermoisopleth diagram for Belém, Brazil (Figure 10.8), located in a tropical rainforest near the mouth of the Amazon River, shows that the temperature changes by 5°C (9°F) or more during each day but varies by no more than 2°C (3.6°F) at any given hour through the year.

The local water budget can be used to provide another perspective on climate. The water

Figure 10.7
Singapore is in a tropical rainforest region (Köppen *Af* climate). As the graph shows, the average monthly temperature and precipitation are high and nearly constant through the year. (Doug Armstrong after H. Nelson, *Climatic Data for Representative Stations of the World,* © 1968, by permission of University of Nebraska Press)

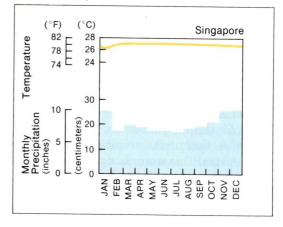

mads whose sheep, goats, and camels are at the mercy of the unreliable rains.

Regional Climate

The most common characteristic of regions of the dry realm is that precipitation is so much less than potential evapotranspiration. Most precipitation goes briefly into soil moisture storage, but then is quickly returned to the atmosphere by evapotranspiration. During heavy rainshowers, there may be some surface runoff. Most of this water drains into local basins and then evaporates, but some percolates down to the subsurface water table. Where soil conditions are favorable, irrigation agriculture can be very productive, but irrigation water must be obtained from nondesert upland areas where orographic precipitation is much greater than potential evapotranspiration.

Because there is little water available for evapotranspiration in desert areas, most of the net radiation gain goes into heating of the land surface and the lower troposphere. At Phoenix, Arizona, afternoon summer temperatures typically exceed 40°C (104°F), and soil temperatures in places like California's Death Valley can reach 90°C (194°F).

Chaparral

An almost unique assemblage of vegetation has evolved where winters are mild and rainy and summers are hot and dry. This dry-summer subtropical climatic region appears along the western coasts of continents between about 30° and 45° latitude, on both sides of the equator. Both the climate and associated vegetation are often described as Mediterranean, because the Mediterranean coasts comprise the largest area of the dry-summer subtropical climate. This climate and the vegetation associated with it are also found in smaller coastal areas of Spain and Portugal, California, Chile, South Africa, and Australia.

Chaparral Characteristics

The distinctive vegetation of the dry-summer subtropical climate is known in North America as *chaparral*. Chaparral consists of an almost impenetrable mat of brush, ranging from knee-high to twice the height of a person. The shrubs are small-leaved and deep-rooted to survive the summer drought period, and are generally evergreen (Figure 10.15). Because summer drought increases the danger of fire, most chaparral species have evolved the capability of resprouting from subsurface roots after being burned off above ground. Damp north-facing slopes may be covered by oak woodlands, and flat areas may be a virtual savanna, with large oaks rising from grass or shrub. As in the case of desert vegetation, chaparral vegetation on different continents and in different hemispheres is remarkably similar in appearance, even though the plant species are unrelated. Convergent evolution of plants is especially clear in this distinctive climatic realm.

The chaparral vegetation region in California often receives nationwide attention when late summer fires sweep over mountain slopes. Dry winds from the interior deserts can make fire fighting in chaparral extremely difficult and dangerous. If a heavy winter rain occurs before the burned chaparral has resprouted, rapid erosion sends torrents of mud and boulders rushing into valleys. This repeatedly damages communities along the California coast.

Regional Climate

The monthly temperature and precipitation regime for Palermo, Italy, shown in Figure 10.16, is representative of dry-summer subtropical climatic regions. Winter rainfall is associated with midlatitude cyclones. The mild temperatures at this location are due mostly to the protection from cold polar continental air offered by neighboring mountain barriers. Because of atmospheric subsidence associated

Figure 10.15
The shrubby vegetation covering these hillslopes in California is known as *chaparral*. The foreground area is in a stage of succession after a brush fire—a common phenomenon in summer-dry areas such as this. The person in red provides scale. (T. M. O.)

Figure 10.16
Palermo, Italy, has a Mediterranean climate characterized by moderate temperatures and a dry summer (Köppen *Cs* climate). As the graphs show, winter in Palermo is cooler and more moist than summer. The average annual precipitation is nearly 80 cm (31 in.), but for June, July, and August combined the precipitation is only 4 cm (1.5 in.). (Doug Armstrong after H. Nelson, *Climatic Data for Representative Stations of the World,* © 1968, by permission of University of Nebraska Press)

with the subtropical highs, summers are hot and dry. In other locations, such as San Francisco, upwelling ocean water along the coast keeps summer temperatures cool.

The average water budget for San Francisco (Figure 10.17) shows a large moisture deficit during summer and fall and a smaller surplus during spring, after soil moisture storage has been recharged. In most dry-summer subtropical climatic regions, local water surpluses are not adequate to sustain irrigation agriculture during the summer. Winter surpluses from mountain regions must be stored in reservoirs to be delivered to irrigated cropland in the summer.

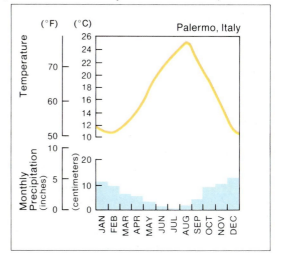

VEGETATION AND CLIMATE

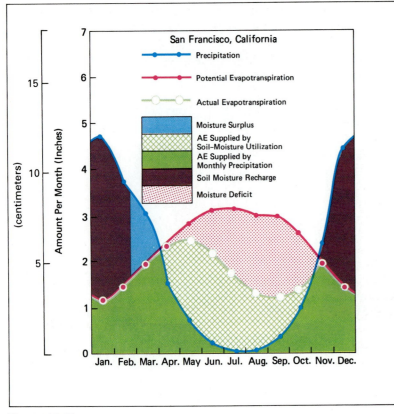

Figure 10.17
The local water budget for San Francisco, California, is characteristic of stations with a Mediterranean climate (Köppen *Cs* climate). A deficiency of moisture persists through the dry summer months. Plants that live through the summer must draw upon fog drip and stored soil moisture. From October through March, when most of the precipitation is received, soil moisture is replenished and a surplus is generated. (Doug Armstrong after D. Carter and J. Mather, *Publications in Climatology,* Vol. 19, 1966)

Midlatitude Forests

In the preagricultural period, hardwood forests were dominant across the eastern United States, western Europe, Japan, Korea, and eastern China. Similar forests occupied much smaller areas of South America, southern Africa, Australia, and New Zealand. These for-

ests are associated with humid continental climates, but they also occur in most of the humid subtropical and marine west coast climatic regions. Most of these regions have been cleared for cropland and pasture, and their upland forests have been cut over for lumber and firewood, so that very little of the original forest remains.

Broadleaf Hardwoods

The midlatitude forests on the different continents include different combinations of species. In eastern North America, oak, hickory, maple, and beech each tend to be dominant in various areas. The multistoried canopy of trees usually rises 30 meters (100 ft) or more above the ground. Some shrubs and shade-tolerant annuals occupy the ground surface, but the forest tends to be relatively open below the canopy. Most of the trees drop their leaves before the onset of winter, so the appearance of the forest changes dramatically through the seasons (see Figure 10.18).

In large portions of the northern hemisphere forests soil water in the root zone is frozen in the winter and is therefore unavailable to the trees. With no uptake of moisture from the soil, continued transpiration of moisture by plant leaves would be fatal to the plants. But even before this occurred, the moisture-laden cells composing plant leaves would be ruptured and destroyed by winter freezes, causing the death of the plant. Therefore, broadleaf plants must drop their leaves and remain dormant until the threat of frost is past. Thus the deciduous habit can result either from seasonal drought, as in the mixed tropical forests, or from seasonal cold.

In the southern hemisphere, where winters are mild because of the strong maritime influence, the midlatitude forests are dominated by broadleaf evergreens such as *Euca-*lyptus, *Podocarpus,* and *Nothofagus* (evergreen beech). In wet areas these evergreen forests include very large trees covered with mossy epiphytes that give them the appearance of tropical rainforests.

Needleleaf Evergreens

The broadleaf deciduous forests in the eastern United States and western Europe become

Figure 10.18
Seasonal change in midlatitude forests.
(left) The mixed broadleaf deciduous forests are alive with color for a few weeks in autumn; this example is from southern Wisconsin. Each species progresses through a sequence of color changes and leaf fall. Variable temperature and moisture conditions cause the date of maximum color to vary by as much as four weeks or more. (Robert A. Muller)
(right) This hardwood forest in southern Kentucky is seen in the spring before the deciduous trees produce a new set of leaves. Note the amount of sunlight reaching the forest floor. Such bright conditions are never encountered within the world's evergreen forest types. (T. M. O.)

mixed with coniferous (cone-bearing) needle-leaf evergreens on their northern margins, and with broadleaf evergreens on their southern margins. Pines tend to be dominant on the higher portions of the coastal plain of the southeastern United States, where sandy soils lack nutrients and store only limited amounts of moisture. The southern pines are an important timber resource, but also are more susceptible than deciduous trees to fires that sweep areas of the coastal plain from time to time. Some scientists believe that the southern pines are only an intermediate successional stage toward the broadleaf deciduous forest climax.

Regional Climate

Most midlatitude forest regions experience large temperature ranges from winter to summer and a relatively even distribution of precipitation throughout the year. Pittsburgh, Pennsylvania, is near the climatic boundary between Köppen's humid subtropical *(Cfa)* and humid continental *(Dfa)* climates; its monthly temperature and precipitation regimes are displayed in Figure 10.19. In general, summers are hot, but temperatures well below freezing are common during winter months. Monthly precipitation varies little. Klagenfurt, Austria, is representative of midlatitude forests with humid continental climates; the large range of its seasonal temperature is shown in the thermoisopleth diagram in Figure 10.20. This diagram should be compared with that in Figure 10.8, which illustrates the temperature regime in a tropical rainforest.

The average water budgets of most places within midlatitude forest regions show small summer deficits and relatively large winter and spring surpluses. The changing patterns of the upper-air circulation in the middle and higher latitudes, however, tend to produce clusters of wetter or drier, or warmer or colder years in each region. Figure 10.21 illustrates some of the hydrologic impacts of climatic variation during an extensive drought across the

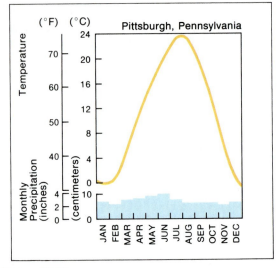

Figure 10.19
Pittsburgh, Pennsylvania, is near the boundary between the humid subtropical and humid continental climates (Köppen *Cfa* and *Dfa* climates). The graph shows the great range of the average monthly temperature between summer and winter. The monthly precipitation is nearly constant through the year, however. (Doug Armstrong after H. Nelson, *Climatic Data for Representative Stations of the World,* © 1968, by permission of University of Nebraska Press)

northeastern United States in the 1960s, as represented by water budget components.

Midlatitude Grasslands

Grasses are the dominant vegetation where precipitation does not meet the needs of trees and shrubs, or where repeated fires prevent tree regeneration from seedlings. Vast areas of continuous grasslands once extended from Texas to Alberta and Saskatchewan in central North America, and across Eurasia from the Soviet Ukraine to Manchuria. These grasslands included *tall-* and *short-grass prairies* and *steppes.* Similar extensive grasslands also were present in South America, particularly in

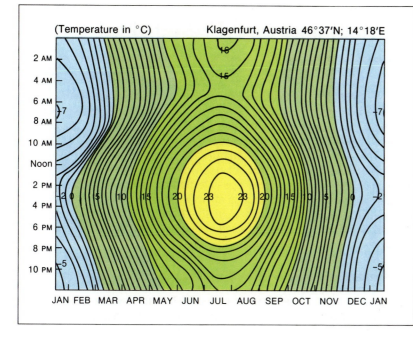

(Temperature in °C)   Klagenfurt, Austria 46°37'N; 14°18'E

Figure 10.20
Klagenfurt, Austria, has a humid continental climate (Köppen *Dfa* climate). As the thermoisopleth shows, the range of temperature through the year exceeds the range through the average day. In January the noon temperature is approximately −4°C (25°F), whereas in July the noon temperature is 22°C (72°F). Note that in November the temperature through the day varies by only a few degrees, partly because of the moderating effect of cloudy weather. (Doug Armstrong after C. Troll, *World Maps of Climatology,* © 1965, Springer-Verlag Publishing)

Argentina and Uruguay. Intense agricultural exploitation has left very few areas of natural grassland in any of these regions.

The midlatitude grasslands contained a large number of plant species that are different from the grasses of the tropical savannas. The dominant midlatitude grasses were usually perennials that lie dormant during the winter and continue their growth in the next growing season. Near the boundary between forest and grassland, where moisture is comparatively abundant, the natural grassland vegetation was usually prairie grass, one to two meters in height. Where there is less moisture, the domi-

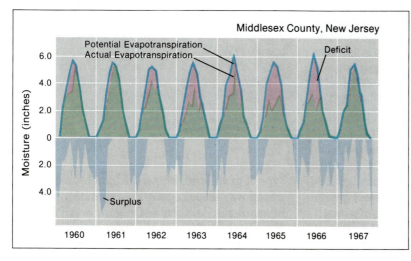

Figure 10.21
This eight-year water budget for Middlesex County, in central New Jersey, illustrates climatic variability in a midlatitude forest. During the drought years of 1962 through early 1966, deficits were large, and winter-spring surpluses were not great enough to meet the water-resource needs of the region. Water restrictions were common, and there was some loss of shrubs and trees in suburban areas. (Doug Armstrong after Robert A. Muller, 1969)

VEGETATION AND CLIMATE

transition zone centered just east of the 100th meridian from central Texas to the Dakotas, with a westward swing into Saskatchewan and Alberta. This transition zone is tempting but dangerous for agriculture. Wet years repeatedly lure wheat farmers onto the short-grass prairies only to meet disaster when the inevitable dry years follow. These transform the ploughed land into a "dust bowl," causing mass emigration from the region. The cycle has been repeated several times on the North American Great Plains.

Regional Climate

Most of the world's grasslands are located where average precipitation barely equals po-

tential evapotranspiration. In the grasslands of continental interiors in the northern hemisphere, summers tend to be hot, with periodic thunderstorms. Winters are very cold and dry.

The average water budget for Huron, South Dakota, located in the transition zone between the tall and short grasslands, is shown in Figure 10.23. Despite a summer rainfall maximum, there is still a relatively large summer moisture deficit, and precipitation is not great enough during winter to generate a significant surplus. In addition, summer rainfall from convective thunderstorms is highly variable from year to year. Some climatologists and ecologists believe that clusters of drier than normal years prevent the establishment of forests

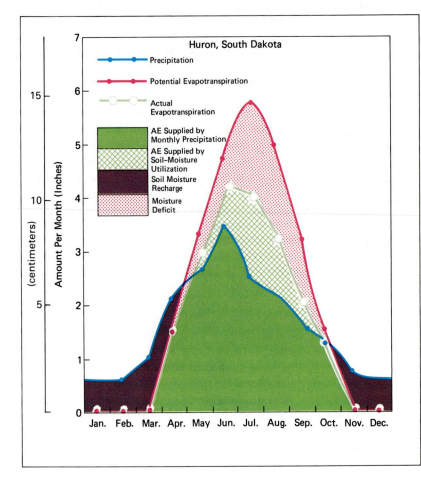

Figure 10.23
Huron is located in the prairie grasslands of South Dakota. The local water budget for Huron shows that precipitation occurs in all months and is most plentiful during early summer. However, high summer temperatures cause potential evapotranspiration to be high then as well, and moisture deficits normally occur. The soil moisture is recharged during the cooler months, when the vegetation's demand for moisture is small, but winter precipitation is not large enough, on the average, to generate surpluses to runoff. (Doug Armstrong after D. Carter and J. Mather, *Publications in Climatology*, Vol. 19, 1966)

within the wetter margins of the prairie region.

Northern Coniferous Forests

Coniferous (cone-bearing) needleleaf evergreen forests dominated by spruce, fir, and pine extend in a broad band across North America, Europe, and Asia between about 50° and 65°N latitude in the subarctic climate region. These forests, also known as *taiga*, or *boreal forests*, endure the largest annual temperature ranges encountered on the earth. Winters are bitterly cold and very dry, with only light snowfalls. These areas are dominated by the Yukon and Siberian highs and are the source regions for continental polar air masses. Summer is brief, but daylight periods are long and temperatures are mild, even warm, occasionally exceeding 25°C (77°F).

In North America the coniferous forests of the subarctic climatic region spread southeastward from the McKenzie River Valley in northwestern Canada across the Canadian border into sections of Michigan, New York, and New England.

Forest Structure

The conifers of the boreal forests are tall, slim, and tapered, as shown in Figure 10.24. Most conifers are evergreen and do not lose their leaves during winter. Their small needle-shaped leaves and thick bark resist moisture losses during the long cold winters, and their conical crowns may be adapted to intercept the oblique rays of a sun that in their latitude is never high in the sky. Since they are not required to manufacture new leaves before they can begin photosynthesis and growth in the spring, they conserve energy and are able to commence growth as soon as temperatures rise sufficiently. This is a clear advantage where the growing season is very short.

Coniferous forests contain a comparatively small number of plant species. Low sun angles and dense foliage allow little light to reach the ground, and cool temperatures also limit plant growth. Spruce is common in the coniferous forests of North America. Larch, which drops its needle-like leaves in the winter, is dominant in eastern Siberia, where winter conditions may be too severe even for needleleaf evergreens. January temperatures in the latter region plummet to −40° to −50°C (−40° to −60°F). On the southern margins of the coniferous forests, the trees are tall and densely packed. Farther poleward, the trees are smaller and the forest more open.

Pacific Coast Forests

Along the Pacific coast of North America the coniferous forest extends southward into Washington, Oregon, and northern California. Here winters are wet and summers are very dry. The virgin forests of the Pacific coast are often composed of giant trees, and the species differ from those of the taiga. Redwoods (*Sequoia*) and Douglas fir are dominant in the northern California Coast Ranges, and fir, spruce, cedar, and hemlock prevail in the forests of Oregon, Washington, and British Columbia. Moisture stress during the warm season has favored the survival of an evergreen coniferous forest in this midlatitude region of summer drought. The immense size of many trees causes the biomass in portions of the evergreen forests of the Pacific coast to be even greater than that in tropical rain forests.

Regional Climate

Figure 10.25 shows the monthly regimes of temperature and precipitation at Moose Factory in central Canada. Monthly mean temperatures range from −20°C (−35°F) in winter to 16°C (60°F) in summer. Much of the annual precipitation falls as rain during summer, when potential evapotranspiration is nearly as high as in midlatitude regions. Hence, most summer rainfall is utilized for evapotranspira-

(a)

(b)

Figure 10.24
Coniferous forest in the state of Washington.

(a) This Douglas fir forest in Washington's Cascade Range shows how a small number of tree species dominate in most coniferous forests. In this area evergreen coniferous forests are a response to occasional summer droughts that exclude broadleaf forests. Farther poleward, where a short growing season and deeply frozen subsoils are a limiting factor, spruce and pine are the dominant trees. Mt. Rainier is in the background.

(b) Pure stands of large trees that are useful for lumber have caused the coniferous forests of the middle latitudes to be intensively exploited by loggers. Here, close to Mt. Rainier, we see steep slopes recently stripped of their forest cover. Clear-cutting in such terrain greatly increases water erosion and soil slippage. The small forest remnant may have been spared to reseed the cleared area. (T. M. O.)

tion. Local water budgets show that the only moisture surplus is in the late spring at the time of snowmelt. Nearly rainless periods occasionally result in small deficits. Some summers are dry enough for the danger of forest fires to be serious.

Tundra

Trees cannot survive unless the average temperature during the growing season exceeds 10°C (50°F) for a period of two to three months. Near the Arctic Ocean, the trees of the

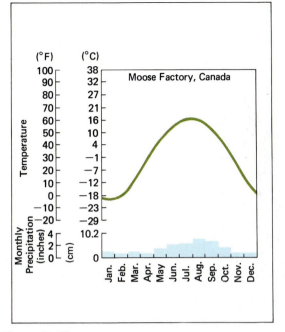

Figure 10.25
Mean monthly temperature and precipitation regimes at Moose Factory, in central Canada. The range between winter and summer temperatures in the subarctic climate is the largest of all global environments. The northern coniferous forest is well adapted to long severe winters and short but warm summers, when much of the annual precipitation falls. (Adapted from Glenn Trewartha, *Introduction to Climate,* 4th ed., 1968, McGraw-Hill Book Co.)

northern coniferous forest give way to low shrubs, grasses, and flowering herbs, with mosses and lichens on rock surfaces (Figure 10.26). This vegetation formation is known as *tundra.*

Although winter temperatures in the tundra regions do not reach the extremes experienced in the more continental taiga of eastern Siberia, they are nevertheless very low, and are frequently accompanied by gale-like winds from which there is no shelter. At −18°C (0°F), a 30-km (20-mile) per hour wind produces the equivalent of a temperature of −40°C (−40°F). This depression of effective temperature is called the *windchill factor.* At −30°C (−24°F) the same wind lowers the equivalent temperature to −55°C (−68°F). The combination of wind and low temperature causes extreme moisture stress and danger to water-bearing plant tissues. The freezing of water ruptures plant cells, and evaporation induced by the wind cannot be offset by water intake from the frozen ground. As a consequence, tundra plants are small-leaved, like desert plants, and low-growing so that they will be blanketed by snow when icy winter winds sweep over the land surface.

Figure 10.27 is a thermoisopleth diagram for Sagastyr, at latitude 73°N, in the tundra of

Figure 10.26
Low shrubs, grasses, and flowering herbs are the dominant vegetation types in tundra regions, as shown here in Norway. The tundra is wet during the period of thaw because water cannot drain through the permanently frozen ground below the surface. (Brian Hawkes, Kent, England)

VEGETATION AND CLIMATE

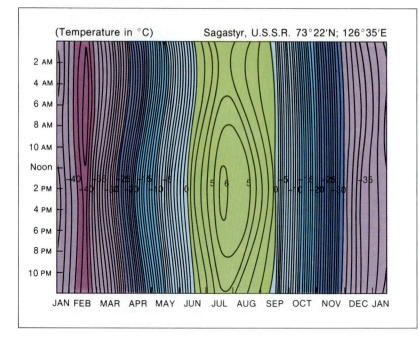

Figure 10.27
Sagastyr, U.S.S.R., is a station in the tundra region of Siberia, north of the Arctic Circle (Köppen *ET* climate). The temperature during the year varies through an extreme range because of the lack of sunlight near the time of winter solstice and the continual daylight at summer solstice. The nearly vertical temperature contours indicate that the temperature during any given day is essentially constant. (Doug Armstrong after C. Troll, *World Maps of Climatology,* © 1965, Springer-Verlag Publishing)

northern Siberia. The nearly vertical pattern of isotherms shows that the daily variation of temperature is only a few degrees because of the long daylight periods in summer and the absence of sunlight during midwinter. During a year, however, the average midday temperatures vary from 6°C (43°F) in July to −40°C (−40°F) in February.

In tundra regions precipitation tends to be low throughout the year. Winters are long, and the tundra is thinly covered by snow for 6 to 8 months. Most of the sparse precipitation falls as rain in the brief summer. Despite the relatively low summer precipitation, the tundra is usually moist and waterlogged during the warm months, when the surface layer of soil thaws but cannot drain downward because the subsoil is permanently frozen.

Highland Vegetation

Under average conditions, temperatures decrease with increasing elevation. Thus, one can progress through different climatic and vegetation zones while climbing upward in highland regions.

Figure 10.28 shows that in humid climatic regions the sequence of vegetation types encountered with increasing elevation is similar to the sequence met in traveling from the equator to the poles. In dry climates the increase in precipitation with elevation results in spectacular vegetation changes within short horizontal distances. In the San Francisco Peaks region of northern Arizona, where the vertical sequence of North American life-zones was first studied, the upward progression begins with desert shrubs at an elevation of 1,200 meters (4,000 ft) and ends in a spruce-fir forest at 3,000 meters (10,000 ft) that is not unlike the spruce-fir forests of subarctic Canada. Normally growing thousands of kilometers apart, the two completely dissimilar vegetation associations exist here within a distance of a few kilometers. In tropical uplands the diversity of contrasting vegetation is even greater. In the

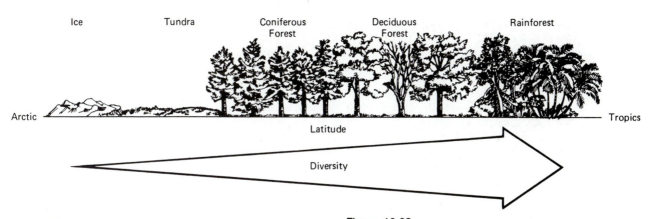

Ice   Tundra   Coniferous   Deciduous   Rainforest
Forest   Forest

Arctic ———————————————————————————————— Tropics

Latitude

Diversity

Snow

Alpine Tundra

Coniferous Forest

Diversity

Deciduous Forest

Rainforest

Elevation

Figure 10.28
Relationship of latitude and elevation to vegetation type and community diversity. Increasing latitude or elevation results in a shift to cooler vegetation types with a decrease in community diversity. Climbing 4,500 meters (15,000 ft) up a tropical mountain will reveal changes in communities analogous to those observed on a trip from the tropics to the pole. (Barbara Hoopes)

Local climates and vegetation in highland areas are very complex, however. Adjacent slopes with different orientations differ dramatically in terms of solar radiation, temperature, and precipitation. North-facing slopes of deep valleys in the middle latitudes receive little direct sunlight over the year, while nearby south-facing slopes may bake in many hours of sunshine each day. Leeward slopes may be rather dry, but precipitation on windward slopes tends to increase sharply with elevation.

For at least the first 1 or 2 km of elevation, upland areas produce much larger water surpluses than surrounding lowlands. A flourishing vegetative cover in upland regions retards water runoff, reducing soil erosion on steep slopes and decreasing flood hazards in adjacent lowlands. At the same time, forested uplands release water slowly to streams, sustaining their flow through rainless periods. Thus preservation or human management of upland veg-

Peruvian Andes, which rise to glacier-clad summits from rainforests on the east and deserts on the west, nearly all the world's latitudinal temperature and vegetation zones can be recognized in vertical succession (Figure 10.29).

VEGETATION AND CLIMATE

Ecosystem Reserves

The vegetation formations discussed in this chapter are global biomes—complete ecosystems with interacting assemblages of both plant and animal life. Only the ecosystems of extreme environments that offer limited opportunities for human activities still exist over large areas in their natural state: deserts, tundras, and the high-latitude coniferous forests. All others are under steadily increasing human pressure and are fast disappearing or being irreversibly altered.

Ecologists have proposed schemes on national and global scales to set aside remaining fragments of natural ecosystems as "biospheric reserves" or "natural areas." Such reserves would allow us to study the processes of natural ecosystems and to preserve endangered life forms included within them. These reserves would be biological islands surrounded by disturbed ecosystems, comparable to islands in oceans. We know that oceanic islands have unique ecological traits, which would likewise evolve in island-like biotic reserves.

An isolated reserve representing a global biome is only a small sample of the original: it includes only the diversity present at the sampling point itself. Species diversity is known to be considerably lower on islands than on mainland areas of similar size and type, and small islands have less diversity than large islands. Another factor in island diversity is proximity to the source of an island's biota. An island distant from the main reserve of its biota will contain fewer species than one closer to the source. Unfortunately, experience indicates that the equilibrium number of species on artificially created biotic islands is less than the initial number of species present in the same island area. The smaller the island, the more rapid the extinction of species down to the equilibrium level.

A problem in planning biotic reserves is to determine the critical size of each type of reserve that will permit it to act as a large island (or even a continent) as opposed to a small, species-depleted island. In a study of the bird life of California coastal islands, it was determined that the critical area would be about 22,500 sq km (8,100 sq miles). When rare or endangered species are involved, particularly large carnivores, the critical area increases. To avoid the genetic hazards of in-breeding, there must be room for several unrelated families. In the Rocky Mountains of Montana and Wyoming, about 300 sq km (115 sq miles) are required for each family of grizzly bears (two adults, two cubs); 600 sq km are required for two families and so on. One mountain lion needs about 95 sq km (35 sq miles); a viable wolf pack requires at least 600 sq km. But undisturbed areas of such sizes (some 600 to 800 sq km) no longer exist, except in the deserts, tundras, and coniferous forests. In normal ecosystems sizable areas temporarily lose their productivity because of natural events such as fires, tree blow-downs, infestations of parasites, and disease. Reserves must be large enough to survive and recover from such setbacks, which might become more frequent when the area is encircled by an altered peripheral environment.

It has been proposed that species survival would be enhanced by preserving natural corridors, or "stepping-stones," between the larger biotic reserves of a particular type. This would create a functional whole larger than the single reserves that could be set aside at this late date. Clearly, some such plan must be implemented soon if we are to preserve anything resembling the natural diversity of plant and animal life that preceded us on the earth.

etation, the control of soil erosion, and the stabilization of stream flow all go hand-in-hand. This is a matter of critical concern to those who live within the highland regions and far beyond them on the great river flood plains of the earth.

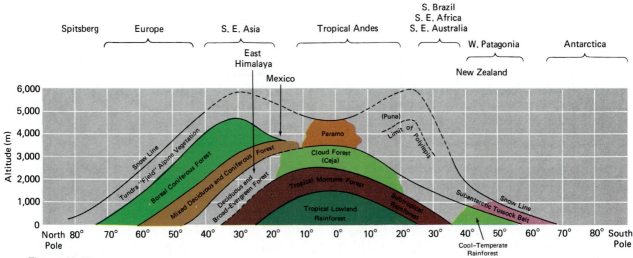

Figure 10.29

This idealized pole-to-pole transect shows the relationship between latitude and the altitudinal limits of the major nondesert vegetation zones. Note that vegetation formations in the northern hemisphere that are adapted to midlatitude continental climates are not present in the southern hemisphere, where maritime influences dominate in the middle latitudes. The páramo vegetation of the equatorial highlands (not mentioned in the text) is a grassland with unusual tree forms adapted to a cool, damp, cloud-shrouded environment. (After Carl Troll, 1968; from Larry W. Price, *Mountains and Man,* University of California Press, 1981)

KEY TERMS

vegetation
evergreen
deciduous
hydrophyte
xerophyte
mesophyte
perennial plant
annual plant
plant community
ecotone
plant succession
climax vegetation
natural vegetation

liana
epiphyte
strangler
thermoisopleth diagram
isotherm
edaphic
llanos
chaparral
tall- and short-grass prairie
steppe
taiga
boreal forest
tundra

REVIEW QUESTIONS

1. Illustrate some specific plant adaptations to varying environmental constraints.

2. In what general ways do plant communities usually change during succession to-

ward a climax vegetation?

3. How does tropical rainforest vegetation reflect the climatic characteristics of the equatorial zone?

4. What is a thermoisopleth diagram?

5. How do forests in areas of the tropics that experience a dry season differ from the rainforests of the equatorial zone?

6. What edaphic condition is widespread in savanna regions?

7. What are the two broad categories of desert vegetation? How do these two dissimilar vegetation forms survive in areas of extreme moisture deficiency?

8. What unusual problems do Mediterranean environments pose for vegetation? What is the nature of the vegetation that has evolved in such areas?

9. In what other areas of the world does one encounter forests like those originally covering the eastern United States? How do these forests reflect environmental stress?

10. What general climatic characteristic is associated with the midlatitude grasslands? Are the locations of grasslands strictly controlled by climatic parameters?

11. Why are needleleaf evergreen trees better adapted than broadleaf deciduous trees to severe winters and a short growing season?

12. How does low stature benefit tundra plants?

13. Where, over a short distance, can one view vegetation changes similar to those occurring with changes in geographic latitude?

APPLICATIONS

1. In the vicinity of your campus there is probably land, once disturbed or cleared, that fairly recently has been allowed to revert to a more natural or uncontrolled vegetation cover. Locate two such sites and compare the composition of their vegetation. Is there any good explanation for the differences you can see?

2. Analyze the map of global vegetation (Figure 10.5) in comparison with the map of global climate (Figure 8.3). Make a list of the climates of the natural vegetation types. Now make a final list of global regions where the general associations of climate and vegetation do not seem to fit, and suggest some possible reasons.

3. Are there any preserved remnants of the prehistoric natural vegetation of your region? Was the prehistoric vegetation actu-

ally regarded as a climax type, or was it part of a plant succession after some natural disturbance? In what areas of North America would you expect the prehistoric vegetation to have been other than a climax form?

4. How does the size of plants appear to be related to the availability of energy and moisture?

5. There are similarities in the appearance of the vegetation in desert and tundra regions. What are the resemblances? Is this a result of the same or different climatic stresses?

6. One peculiar aspect of the climatic regime of tundra regions is also seen in the tropical rainforests. This climatic characteristic is especially evident in a comparison of thermoisopleth diagrams from different vegetation regions. What is the similarity? Are its causes the same in the two regions?

FURTHER READING

Billings, W. D. *Plants, Man, and the Ecosystem,* 2nd ed. Belmont, Calif.: Wadsworth (1970), 160 pp.

This brief paperback, part of the Fundamentals of Botany series, includes many succinct sections that supplement this chapter.

Eyre, S. R. *Vegetation and Soils: A World Picture,* 2nd ed. Chicago: Aldine (1968), 328 pp. This book, written from the perspective of the British Isles, focuses on relationships between vegetation and soils. It is organized by global vegetation types and is especially useful for Chapters 9 and 10 of this text.

Johnson, Hugh. *Hugh Johnson's Encyclopedia of Trees.* New York: Gallery Books (1984), 336 pp. The best illustrated compendium on trees seen in their natural settings and in gardens. Nearly 1,000 color photographs and drawings.

Milne, Lorus, and **Margery Milne.** *Living Plants of the World.* New York: Random House (1975), 336 pp. Nontechnical description of 150 of the most significant plant families, well-illustrated in color. The authors are botanists and expert natural history writers.

Shelford, Victor E. *The Ecology of North America.* Urbana: University of Illinois Press (1963), 610 pp. Shelford's book is a detailed account of North American biomes. It is packed with information on plant and animal interactions, and is a valuable reference book, available in a paperback edition.

Tivy, Joy. *Biogeography,* 2nd ed. New York: Longman (1982), 459 pp. Plant distribution and vegetation formations are explained in terms of environmental factors.

Vankat, John L. *The Natural Vegetation of North America: An Introduction.* New York: Wiley (1979), 261 pp. This very informative paperback presents the ecological basis of the major vegetation formations of North America from tundra to tropical rainforest.

Walter, Heinrich. *Vegetation of the Earth in Relation to Climate and the Eco-Physiological Conditions.* Translation of 2nd German ed. New York: Springer-Verlag (1973), 237 pp. This work stresses relationships between climate and vegetation and also includes short descriptions of the vegetation formations of each natural vegetation type.

The thin layer of soil that forms at the interface of the atmopshere and lithosphere supplies the nutrients that sustain all natural and cultivated plant life. The earth's soils are the product of thousands of years of interactions between energy, moisture, minerals, and living organisms.

The Gleaners by Jean François Millet, 1857. (Scala/Art Resource)

11
The Soil System

A vital factor in the natural productivity and human utilization of any locality is the nature of its soils. Soils supply the water and nutrients for land plants, and plants support the food chains that sustain the web of life on earth. Thus it is important for geographers to understand the processes of the soil system. In fact, soil is a much more complex phenomenon than is generally realized. It is not just loose "dirt" that one can dig into with a shovel or push around with a bulldozer. A true soil is the product of a living environment. Where there is no life, as on the moon, there is no true soil. Since the earth teems with life, soils of some type are present almost everywhere.

In this chapter we shall examine the most important characteristics of soils, the reasons for geographical variations in soil types, the current system of soil description and classification, and problems of soil management.

SOIL-FORMING PROCESSES

For soil to develop, two things must happen. First, water percolating down through loose rock material must cause physical and chemical modifications in the original material. Second, the activities of living organisms must bring about further changes. The less water and organic activity present, the weaker the development of the soil. The more strongly developed the soil, the more it is apparent that percolating water and active organisms have caused the original material to develop visible layers with varying physical and chemical characteristics. The presence of these layers, produced by soil-forming processes, identifies a true soil.

Soils develop on rock material that has already been reduced to fine fragments. This material may be either a mass of decomposed rock, or sediment that has been transported and deposited by an agent of erosion, such as running water, wind, or glacial ice. Thus, the story of soil development begins with the initial fragmentation of the solid rock of the earth's crust, a process known as *weathering*.

Weathering

Solid rock can be fragmented by both mechanical and chemical processes. All rock masses are brittle; they break to form organized systems of deeply penetrating cracks, called *joints,* as a result of slow stretching or twisting motions of the earth's crust (Figure 11.1a). When molten lava cools and solidifies, it breaks into prismatic columns with hexagonal or pentagonal cross sections (see Figure 12.12d, p. 337). Rock masses also expand somewhat when erosion of the overlying land reduces the downward pressure that confines them. The release of pressure allows massive, sparsely jointed volumes of rock to dilate by the separation of sheets that are parallel to the land surface in a process called *exfoliation* (Figure 11.1b). Joint production and exfoliation seem to be the only rock fragmentation processes that do not in some way involve water.

Mechanical Weathering

Water enters rock masses by way of joint systems, as well as through tiny pores between the mineral grains composing most rocks (see Chapter 12, p. 331). When water freezes in joints and pore spaces, it splits rocks into smaller fragments, just as freezing water bursts pipes when a heating system fails in winter. This is because the phase change from liquid water to ice produces an increase in volume of nearly 10 percent. Wherever winters are severe, this so-called *frost weathering* is an important agent of rock disintegration (Figure 11.1c). The prying action of plant roots and the burrowing activities of animals also help to widen the joints in rocks.

In dry climates the evaporation of water carrying dissolved salts results in the growth of crystals of evaporite minerals on rock surfaces and in pore spaces between rock particles. As the crystals grow they wedge apart the mineral grains composing the rock. Subsequently, expansion and contraction of the evaporite crystals due to daily changes in temperature and humidity help to loosen adjacent particles.

The products of all these forms of mechanical weathering are chemically unchanged rock debris of all sizes, from giant blocks and slabs to microscopic silt particles.

Chemical Weathering

Except in very cold climates, chemical weathering is more important in the fragmentation of rock than is mechanical weathering. The positively charged hydrogen ions and negatively charged hydroxyl (OH) ions in water are quick to react with the chemical compounds (minerals) of which rocks are composed (see Chapter 12, p. 331). Chemical weathering is assisted by plants, whose roots and litter release carbon dioxide and organic substances that modify downward-percolating water into a weak acid. This acidic soil moisture is more reactive with most rocks than is natural rainwater. Some of the products of the chemical reactions between soil moisture and subsurface rock material also are seen at the surface, such as the orange rust that forms on iron and the green coating that appears on copper roofs and leaky copper plumbing. Both of these are products of the chemical reaction known as *oxidation*. Most soils and decayed rocks are shades of brown, due to the oxidation of iron-bearing minerals. In general, chemical reactions cause rock disintegration by swelling and softening

(a)

(b)

(c)

Figure 11.1
Three ways in which massive granite rock is fragmented.

(a) This road cut near Prescott, Arizona, reveals a typical weathering profile in granitic rock. Chemical weathering gradually converts masses of jointed granite to isolated *corestones* surrounded by decomposed rock. If these corestones do not disintegrate first, they may in time be exposed at the surface as piles of boulders. (T. M. O.)

(b) The reduction of confining pressure due to erosion into masses of granitic rock causes the rock to expand by breaking into parallel sheets. These are often curved and resemble an onion structure. This process, called *exfoliation,* affects only sparsely jointed rock, as here in California's Yosemite region. (T. M. O.)

(c) This photo illustrates the effect of frost weathering in cold climates in high latitudes and at high altitudes. The angularity of the forms produced here in the Teton Range of Wyoming is characteristic of frost weathering. (T. M. O.)

some minerals and altering others to less stable types.

Chemical weathering is most effective where moisture is plentiful and high temperatures accelerate chemical reactions. The result is a *weathered mantle* composed of fine particles of chemically altered material overlying the solid bedrock (Figure 11.2). This mantle, which may be several meters deep, often has a high clay content. Although only certain minerals decompose to clay, these minerals are abundant in many rock types. The proportion and type of clay formed indicate the kind of environment in which the weathering occurred.

Translocation

The mantle of fine particles produced by chemical and mechanical weathering is not yet a true soil. Additional changes are required. Some of these result from the process of *translocation,* in which both solid and dissolved materials are moved downward by water sinking into the ground. Translocation involves both the loss of material from the upper part of the soil and the arrival of translocated substances in the lower part of the soil. The process of *eluviation* is the downward flushing of both solid and dissolved (or *leached*) matter. Every soil has a somewhat porous *eluvial layer* (or zone) from which translocated material has been lost.

The deposition of translocated material at a lower level in the soil is known as *illuviation.* The fine particles arriving in the *illuvial layer* (or zone) of the soil make this layer denser than the eluvial zone above it. The deposition in the illuvial zone of iron and aluminum oxides leached from the eluvial layer by acidic waters sometimes imparts a yellow or orange color to the illuvial zone. In dry regions calcium carbonate is translocated into the illuvial zone, producing whitish flecks or veins of lime.

It is the translocation of material that causes well-developed soils to become differentiated into layers of varying density, color, texture, and chemical composition. These layers are known as *soil horizons.*

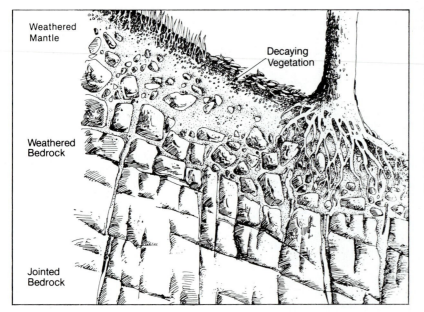

Weathered Mantle

Decaying Vegetation

Weathered Bedrock

Jointed Bedrock

Figure 11.2
Chemical and physical weathering processes break down massive rock into the small particles that form the *weathered mantle,* which is the inorganic component of soil. In warm, moist climates, weathering proceeds most actively along rock joints. (John Dawson after Arthur N. Strahler, *Physical Geography,* 3rd ed., © 1960, John Wiley & Sons, by permission)

Organic Activity

Soil horizons also reveal the influence of plant and animal activity. Organic activity is crucial to soil development, and organic matter is a vital component of soils. Several types of organisms play a role in soil formation: plants contribute vegetative matter such as leaves, twigs, flowers, seed pods, and dead roots; fungi and bacteria reduce this vegetative matter to *humus;* and ants, termites, earthworms, and other soil inhabitants (which also alter plant detritus) mix the humus downward into the mineral matter of the soil.

Humus

Humus is the dark brown to black organic substance that makes the upper portions of the best soils so much darker than the deeper subsoils. Organically generated humus becomes so mixed with the mineral matter of the soil that it is almost impossible to separate. It has many beneficial properties. It increases the soil's ability to retain both moisture and soluble plant nutrients. It is an important source of the phosphorus and nitrogen required by plants. And it maintains a soil structure that is neither too compact nor too porous for plant growth.

Nitrogen Fixation

Aside from releasing nutrients and producing humus from vegetative debris, soil bacteria are vital in *nitrogen fixation*—the conversion of gaseous nitrogen to forms that can be utilized by plants. Nitrogen, which is essential to plant growth, is present in the air that occupies pore spaces in the soil, but it cannot be taken up directly by plants in gaseous form. Usable nitrogen is made available to plants by the action of bacteria that inhabit the soil or are parasitic on the roots of the large family of plants known as *legumes,* which include peas, soybeans, clover, and alfalfa, as well as many uncultivated plants. Progressive farmers rotate their other crops with legumes to add extractable nitrogen to the soil. It is estimated that in the eastern United States organic activity fixes more than 135 kg of nitrogen per hectare (or 120 lbs per acre) each year.

The abundance of organisms in the soil is far greater than one would imagine. There may be a million earthworms and 25 million insects in 1 hectare (2.47 acres) of pasture land. As many as a million bacteria can inhabit 1 cu cm of soil. All are functioning components of the soil system.

PROPERTIES OF SOILS

It is possible to compare soils of various types by focusing on several properties, including texture, structure, chemical characteristics, color, and profile development. These properties reflect the environment of a soil and the parent material from which it was derived. They are the products of a number of soil-forming factors to be discussed subsequently.

Soil Texture

Soil texture refers to the size distribution of the mineral particles composing the soil. Soil particle sizes are classified into four general categories: gravel, sand, silt, and clay—with further subdivisions within each category (Figures 11.3 and 11.4). Table 11.1 indicates the size ranges included in the categories and their subdivisions.

The United States Soil Conservation Service has established a classification of soil textures that describes the various mixtures of different particle sizes. In dealing with gravelly soils, it is the finer matrix that is classified, with the term "gravelly" or "gravel" appended. Figure 11.4 shows the standard classification scheme. The term *loam* indicates a mixture of sand, silt, and clay, which has a more favorable structure for root growth than does a soil composed of a

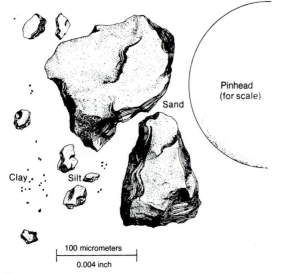

Figure 11.3
The particles that constitute the inorganic component of soil are classified according to size. Particles smaller than 2 micrometers in diameter are called clay, particles from 2 to 50 micrometers in diameter are called silt, and larger particles are considered sand or gravel. (John Dawson)

Table 11.1
Particle Size Classification

| | DIAMETER | |
PARTICLE CLASS	MILLIMETERS	MICROMETERS
Gravel	greater than 2.0	greater than 2,000
Very Coarse Sand	2.0–1.0	2,000–1,000
Coarse Sand	1.0–0.5	1,000–500
Medium Sand	0.5–0.25	500–250
Fine Sand	0.25–0.10	250–100
Very Fine Sand	0.10–0.05	100–50
Silt	0.05–0.002	50–2
Clay	less than 0.002	less than 2

Source: U.S. Department of Agriculture, *Soil Taxonomy,* 1975.

more uniform particle size, such as sand or clay. Not shown in Figure 11.4 are the gravelly loams. Coarse-grained sandy soils are permeable and absorb water easily, but dry out rapidly. They are said to be "light." Their permeability also permits rapid leaching of soluble nutrients. Because individual clay particles (called micelles) are so small as to be visible

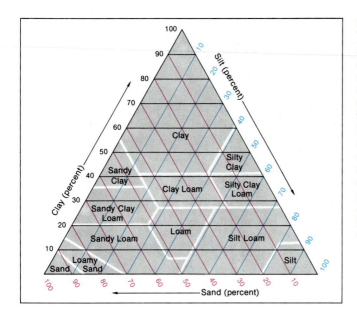

Figure 11.4
The texture of a soil is determined by measuring the proportions of clay, silt, and sand in the inorganic part of the soil. This is done by sifting the soil sample through a series of screens graded from coarse to fine. The soil texture triangle shown in the figure can be used to classify the texture of a soil sample once the percentages of the components are known. Texture classes are separated by white lines. If a soil sample contains 30 percent clay and 40 percent sand, for example, it would be classified as a clay loam. (After E. M. Bridges, *World Soils,* © 1970, Cambridge University Press)

only under an electron microscope, clay soils are very dense, or "heavy," making them hard to work. They accept moisture very slowly and then hold it tenaciously.

Soil Structure

The particles composing most soils clump together in characteristic small masses called *peds*. The sizes and forms of peds determine the *structure* of the soil. As shown in Figure 11.5, soil peds vary considerably in form. This is largely a consequence of variations in the type and content of clay. Most plants grow best in loamy soils with a granular or crumb structure, having peds measuring 1 to 5 mm (0.04 to 0.2 in.) in diameter. Soil structure determines the rate of water absorption and the ease of root penetration.

Another aspect of soil structure is soil *bulk density*—mass per unit of volume, including pore space. Bulk density increases with clay content and is a measure of soil compactness. The greater the bulk density, the less the pore space between soil particles, and the more compact the soil. High bulk density is undesirable because diminished pore space reduces the rate of water acceptance by the soil. Unfortunately, pore space is often reduced unintentionally by the use of heavy agricultural machinery on clay-rich soils. This results in soil compaction and can lead to increases in surface runoff, soil erosion, and gully development.

(a)

(d)

(b)

(c)

Figure 11.5
Four common soil ped structures: (a) platy, (b) prismatic, (c) blocky, and (d) granular. Scales are in inches. (Roy W. Simonson, Courtesy U.S. Department of Agriculture)

THE SOIL SYSTEM

Soil Chemistry

The chemical behavior, or "fertility," of soil is closely related to the *clay–humus complex* in the soil. The clay–humus complex consists of microscopic particles of humus and clay bound together so that they behave like large molecules (Figure 11.6). The extremely small (less than 2 micrometers) clay and humus particles carry negative electrical charges and remain in suspension in soil moisture. As a result they attract positively charged ions, or *cations,* such as those of the chemical bases calcium, magnesium, potassium, and sodium. These essential plant nutrients are easily dissolved and would be leached from the soil were it not for their retention by the clay–humus complex.

Cation Exchange

In a fertile soil the clay–humus complex maintains a delicate balance. It must hold nutrients strongly enough to keep them from being leached away but not so strongly that plants cannot extract them from the soil. Many different types of cations are attracted and held in the clay–humus complex, with various degrees of strength. Weakly bound cations may be replaced by others that are more strongly attracted. Basic cations, such as the sodium ion, are easily replaced by metallic cations, such as iron or aluminum ions, or by hydrogen ions from water. The ability of a soil to absorb and retain exchangeable cations is known as the *cation exchange capacity,* or *CEC*. A high CEC indicates a fertile soil. A soil's CEC is related to the types and amounts of clay and organic matter present.

Not all soils have an active clay–humus complex. Desert soils are rich in soluble nutrients because so little water moves through them, but they contain almost no humus and very little clay. When such soils are watered artificially, the lack of a clay–humus complex permits rapid leaching of their soluble nutrients. This is a major problem of desert irrigation agriculture.

Figure 11.6
The finest inorganic particles and humus in a soil bind together to form the clay–humus complex. On a submicroscopic scale, the particles of the clay–humus complex act like giant molecules with the power to attract cations electrically. Acidic soil moisture contains an excess of hydrogen ions that can replace basic cations on the surface of the complex, as the figure shows (H^+ replacing Na^+). Hence acidic soil moisture removes basic inorganic nutrients from the soil. (John Dawson)

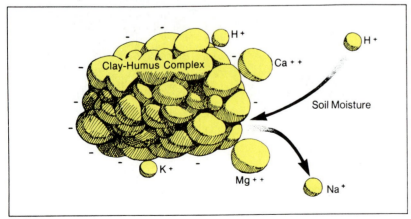

Soil Acidity

The hydrogen ions in a soil may present a problem, because they can displace plant nutrients. Hydrogen ion concentration, known as *acidity,* is measured in terms of the *pH scale,* ranging from about 3 to 10 in soils. Pure water has a pH of 7, which is regarded as neutral. This means that pure water contains 10^{-7} grams of hydrogen ions per 1,000 cm³. The lower the pH, the more acidic the soil.

The best agricultural yields are obtained from soils with pH values between 5 and 7 (Figure 11.7). Acidic soils with low pH values occur in wet areas having abundant partially decayed vegetation on the soil surface. It is possible to raise the pH of acidic soils by adding lime ($CaCO_3$) to them. However, soils with pH values greater than 7, called *alkaline* soils, can also present problems. They may contain amounts of sodium that exceed plant tolerances and contribute to high bulk density and thus very poor soil structure. The alkalinity of such soils, which are common in semiarid regions, can be diminished by adding a source of hydrogen, such as ammonium sulfate ($[NH_4]_2SO_4$).

Soil Color

An obvious characteristic of any portion of a soil is its color. Soil color often changes as one digs downward. It also varies from place to place. Soils of the humid tropics commonly are orange or red. In the temperate grasslands soils are dark brown to black. Soils under coniferous forests tend to be gray near the surface, with an orange or yellow zone below.

Soil color results almost entirely from the amounts of organic matter and iron present, and from the chemical state of the iron. Organic matter colors the soil dark brown to black. Iron that has been oxidized (combined with oxygen) contributes reds, yellows, and browns. Where oxygen has been excluded, iron compounds have greenish and gray-blue hues. This occurs most often where waterlogging has kept air from moving through the soil. Other coloring matter is sometimes present, espe-

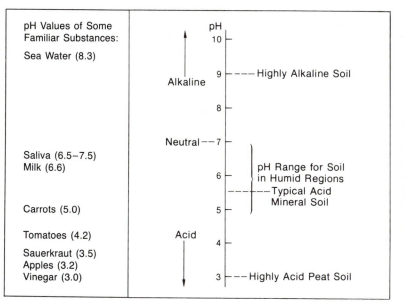

Figure 11.7
The pH value of a soil, an indication of its hydrogen ion concentration, is one of the measures that can be used to estimate a soil's suitability for agriculture. A low pH indicates that a soil is acidic and may have lost many of its nutrients by exchange with hydrogen ions. A high pH indicates that a soil contains strong alkalis, which may be damaging to plant root tissues. (After Lyon and Buckman, *The Nature and Property of Soils,* 4th ed., © Macmillan Co.)

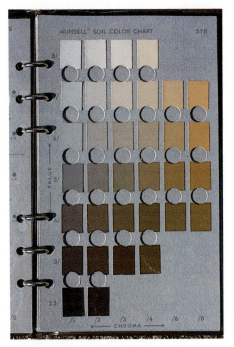

Figure 11.8
Soil color is determined by the use of a Munsell Soil Color Chart bound into a standard field reference booklet. The page with the color chips is placed over the soil to be described, so that the soil can be seen through the holes below each of the color chips. The color that best matches the soil is characterized in terms of *hue* (position in the color spectrum), *value* (degree of lightness or darkness), and *chroma* (color saturation, or departure from a gray of the same value). (T. M. O.)

cially the white of calcium carbonate (lime). Soil color identifications are made by comparing soil samples with color charts designed specifically for the purpose of soil description (Figure 11.8).

Soil Profiles

Every soil has its own distinctive sequence of layers, or horizons, resulting from the processes of translocation and organic activity. These horizons constitute the *soil profile*. Six

separate layers, termed the *O, A, E, B, C,* and *R* horizons, can often be distinguished. These are further subdivided, as shown in Figure 11.9.

The *O* horizon consists of undecomposed plant litter or raw humus at the soil surface. Below it is the *A* horizon, in which humus is mixed with mineral particles. The presence of organic matter makes the *A* horizon dark in color at the top; farther down the loss of fine mineral and soluble substances by eluviation (Figure 11.10) causes the *E* horizon to be lighter in color and relatively porous and light in texture (sandy or silty) compared to the underlying *B* horizon.

The *B* horizon receives material translocated from the *A* and *E* horizons, giving it a higher clay content and a greater bulk density than either the *A* or *E* horizons. It is in the *B* horizon that we encounter blocky and prismatic soil structures and sometimes dense clay "hardpans" as much as a meter thick. The *B* horizon may also be vividly colored by iron oxide leached from the *E* horizon. The *O, A, E,* and *B* horizons collectively are known as the *solum*—the portion of the soil displaying the effects of pedogenic processes.

The *C* horizon is composed of weathered material or loose deposits that have not yet been affected by soil-forming processes. It may be somewhat colored by oxidation of iron, and in dry regions may contain translocated calcium carbonate (lime). The *R* horizon, where present, consists of unweathered bedrock.

From one soil profile to another the characteristic horizons vary greatly in thickness, depth below the surface, and strength of development. Laboratory procedures are necessary to determine such important aspects of soil profiles as bulk density and the concentrations of organic matter, carbonates, clay, oxides, and other constituents at various levels. Soil profile characteristics are routinely used to identify and map specific soil types for purposes of crop selection, irrigation scheduling, fertilization, and other aspects of land-use planning.

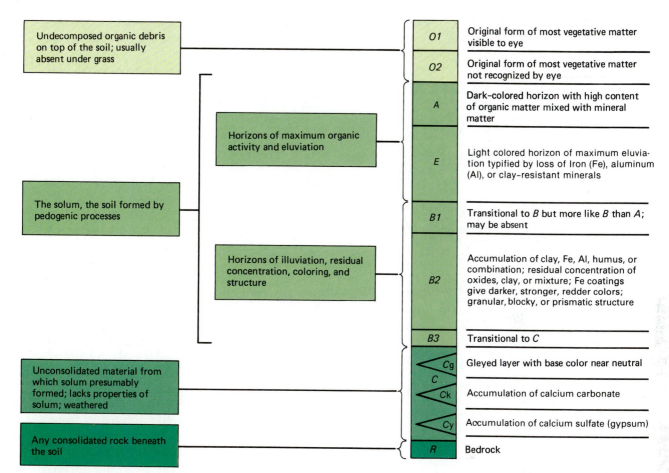

Figure 11.9
Standard horizons in soil profiles. No single profile contains all of the horizons shown. Additional subhorizons similar to those indicated for the C horizon include: *B2t*—illuvial clay; *B2s*—illuvial sesquioxides (iron, aluminum); *B2h*—illuvial humus; and many others. (Modified from Robert Ruhe, *Geomorphology*, 1975, Houghton Mifflin, and Soil Survey Staff, 1951, 1962)

FACTORS AFFECTING SOIL DEVELOPMENT

Our present understanding of soils has developed largely out of work begun by Russian soil scientists more than a hundred years ago. Over the vast area of Russia, soil characteristics seem closely associated with large-scale patterns of climate and vegetation. Research by the early Russian soil scientists and their more recent successors around the world has indicated that soil profiles show the influences of five separate factors: (1) parent material, (2) climate, (3) site, (4) organisms, and (5) time. Soil scientists today call these the *factors of soil formation.*

Parent Material

The inorganic material on which a soil develops is called the soil's *parent material*. This can

THE SOIL SYSTEM

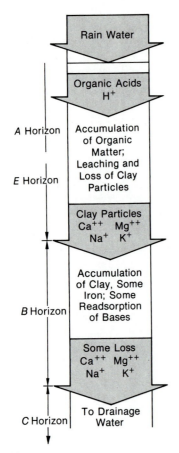

Figure 11.10
In the eluviation process, rainwater, which has been made slightly acidic by dissolved carbon dioxide or organic acids in humus, infiltrates the soil. The hydrogen ions in the acidic water displace basic cations from the clay–humus complex, causing a downward movement of soluble nutrients. (After E. M. Bridges, *World Soils,* © 1970, Cambridge University Press)

be rock that has decomposed in place, or loose material that has been deposited by streams, glaciers, rockfalls, landslides, or the wind. The parent material determines what chemical elements are initially present in the soil. Some parent materials have an abundant supply of the nutrients most needed by plants; others lack them in varying degrees; and some contain substances that are toxic to certain types of vegetation.

If the parent material is rich in soluble bases—calcium, magnesium, potassium, and sodium—which are easily dissolved by water, it can continually supply these nutrients despite constant leaching from the soil. Limestone and basaltic lava have a high content of soluble bases and in humid areas produce the most fertile soils. Outstanding examples are seen in southeastern Pennsylvania near Lancaster and Harrisburg and in Alabama's Black Prairie, which are areas of limestone bedrock; in regions of basaltic rock in Virginia, Maryland, Delaware, New Jersey, and Pennsylvania; and on the calcareous (lime-rich) glacial deposits of the so-called Corn Belt, extending from Iowa to Ohio.

If soluble nutrients are not abundant in the parent material, water moving through the soil removes bases and replaces them with hydrogen ions (Figure 11.6). The soil thus becomes increasingly acidic and decreasingly suitable for agriculture. Soils developed on sandstone are often infertile, being poor in nutrients and coarse in texture, which facilitates leaching. Thus the natural vegetation of the sandy Coastal Plain from Delaware to New Jersey consists of pine forest, which has very low nutrient demands. The soils of this region are too poor for any type of agriculture. Similar "pine barrens" occur on sandy soils in the Carolinas, Georgia, and Florida.

In addition to soluble bases and nitrogen, crop plants require iron, phosphorus, and sulfur, and trace amounts of such elements as boron and copper. If these are not available in the soil's parent material, farmers must supply them artificially.

Abrupt changes in natural vegetation commonly reflect a change in soil due to variations in parent material. Even so, soils from dissimilar parent materials may become quite similar with the passage of time if other soil-forming factors are the same.

Climate

On a global scale, major soil types show a close relationship to climatic zones. The energy and moisture delivered by the atmosphere influence many aspects of soil formation. These include translocation, the rates of chemical reactions, and organic activity in the soil. Abundant rainfall aids translocation, and warm and wet conditions favor chemical reactions and organic activity.

Both vegetative production and the activity of soil bacteria and of larger organisms are reduced in desert areas and in tundra regions at high latitudes or lofty elevations. In the dry desert environment, both plant litter and soil organisms are minimal, and in the tundra, organic litter decays slowly, often forming acidic peat rather than a clay–humus complex. Plant production is at a maximum in the warm and wet tropics, but here the destruction of litter by organisms is so rapid and thorough that the soil is actually poor in organic matter.

Climate affects the chemistry of soil moisture, which, in turn, affects the solubility of various substances in the soil. For example, iron can be removed only by acidic water. Soil water tends to be acidic in cool, wet areas, which are normally covered by coniferous forest. Therefore iron is leached from the eluvial horizon in such areas (see p. 300). In dry regions, lime leached from the upper portion of the soil is redeposited at a lower level where the moisture evaporates rather than moving through to the water table.

Many soils contain features formed thousands of years ago under different environmental conditions. Such "relict" features are important indicators of past climates and vegetation. Soil features have given evidence of shifts in the forest–tundra boundary in high latitudes and in the forest–grassland boundary in the midcontinent region. Soils also reflect expansion and contraction of the world's deserts, as well as less severe climatic fluctuations in nearly all parts of the world. The interactions between climate and soil formation will be made clearer in the discussion of major pedogenic regimes later in this chapter.

Site

The specific location of a soil helps determine the soil type. Since water drains downward, soil at the foot of a slope will evolve in a wetter environment than soil on the hillcrest or on the slope itself. The material at the slope foot is finer than that upslope because fine material is washed downslope on the surface, while the greater dampness at the slope foot causes clay formation by chemical weathering within the soil itself.

Generally, soils on slopes are thinner, stonier, lower in organic matter, and less well developed than those on level or low-lying land. On level surfaces the effects of translocation are much stronger, producing an eluviated E horizon resting on a more massive clay-rich B horizon.

Organisms

The type and intensity of organic activity vary geographically. Since much of this variation is due to climate or microclimate, the climatic and organic factors in soil development are sometimes hard to separate. We have already noted the general influences of organisms in the discussion of soil-forming processes, but there remains one aspect of special importance—the *nutrient cycle*.

Nutrient Cycling

Wherever plants and animals exist, there is a constant cycling of material and energy between life forms and their environment. Organisms need the nutrients in soils to carry out their life-sustaining processes. They return these same nutrients to the environment as waste products, or as litter, or in the form of their own bodies when they die. This establishes a nutrient cycle. Without constant uptake and return of nutrients, soluble compounds would soon be leached out of the soils of humid regions. This would cause a steady decrease in the soils' capacity to support life.

Nutrients not used by organisms are indeed gradually flushed from the soil. Which nutrients are lost depends upon the climate and life forms present (Figure 11.11). We have noted previously that pines do not have high nutrient requirements. Therefore the soils under pine forests gradually lose their soluble nutrients and become acidic. On the other hand, tropical forests have high nutrient demands. Despite the heavy rainfalls of the humid tropics, the forest soils are stabilized by rapid nutrient cycling between the vegetation and the soil.

Human removal of such forests, in which the bulk of the nutrients are stored at any time, is followed by rapid soil deterioration. The soil can be restored for agricultural use only by constant application of costly fertilizers.

Figure 11.11
The nature of the nutrient cycle helps determine soil fertility.

(**left**) Plant species with high nutrient demands prevent soluble compounds from being leached from the soil. The plant extracts nutrients and then returns them to the soil in the form of plant litter. The constant two-way exchange between the plant and the soil maintains a high concentration of nutrients in the soil.

(**right**) Plant species with low nutrient demands permit unused soil chemicals to be leached away, and themselves return few nutrients to the soil. Thus the soil's fertility declines.

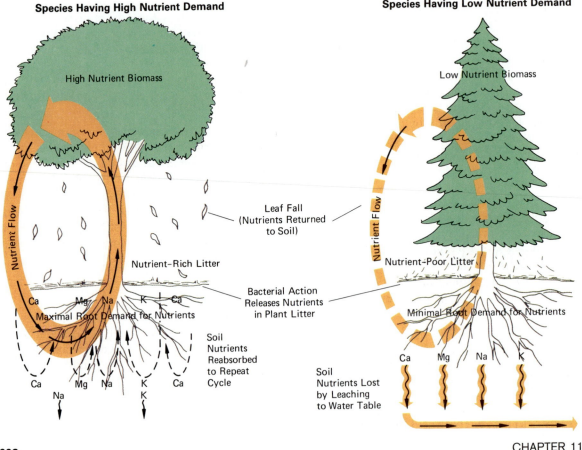

Time

Time is required for translocation and organic activity to produce strong horizon development in soils. But do soil horizons continue to strengthen with the passage of time, or does the soil eventually reach an equilibrium with its environment and cease to be altered further? Because the environment and soil-forming factors frequently change, the answer to this question is uncertain.

It has been possible to study the rate of soil formation on parent material that was deposited in historic time. When the volcano Krakatau, between Java and Sumatra, exploded in 1883, large amounts of volcanic ash fell on the surrounding land. Under the moist tropical climate prevailing there, soil development on the ash was rapid—the soil thickened at a rate of about 1 cm (0.4 in.) per year. In drier areas in Central America, it has taken a thousand years to produce a soil about 30 cm (1 ft) thick on volcanic ash.

Many of the world's most valuable soils are fertile because they are young. They have formed too recently to be strongly affected by chemical and mechanical eluviation. The soils of river flood plains are usually very productive. These soils have developed upon *alluvium* composed of sand, silt, and clay deposited by streams during floods. Soils on recent volcanic ash and glacial deposits are also normally fertile because they have not yet been strongly modified by translocation.

By contrast, flat land surfaces of great age tend to have infertile soils. In such environments thin impoverished *A* and *E* horizons overlie impermeable *B* horizons composed of dense clays or cemented into a rock-like mass by translocated lime, iron oxide, or silica. Here and there on the plains of Australia, Africa, and South America, and also in the southwestern United States, the land surface has been eroded down to these "hardpans," which produce resistant crusts over the weathered parent material. These so-called *duricrusts* are best developed in areas that have well-defined dry seasons. Figure 11.12 shows a duricrust layer cemented by calcium carbonate.

MAJOR PEDOGENIC REGIMES

Most of the earth's soils have been created by one of a small number of distinctive *pedogenic*, or soil-forming, *regimes*. These are related to climate both directly and indirectly through the influence of vegetation. Each of the major pedogenic regimes produces a soil of a distinctive general type that reflects major geographic variations in energy and moisture budgets.

Laterization

The unique feature of most tropical soils is their high content of iron and aluminum oxides in relation to silica, giving them a brick-red color. Such soils result from intensive chemical weathering and leaching under hot, wet conditions. Silica is normally the most abundant mineral in decomposed rock because it is a common constituent of many rock types and is highly resistant to solution. To remove silica from soils requires a very aggressive regime of weathering and soil leaching. This distinctive soil-forming regime is known as *laterization* (from the Latin word *later,* meaning brick or tile). The process is diagramed in Figure 11.13.

In the humid tropics, decomposition and insect consumption of plant litter are usually too rapid to permit the formation of abundant humus. The paucity of humus decreases the soil's ability to retain the soluble nutrients that are not immediately taken up by plants. When forests are cleared and the nutrient cycle is broken, leaching removes nearly all soil cations, and in extreme cases even the silica from decomposing clays. The residue of the process

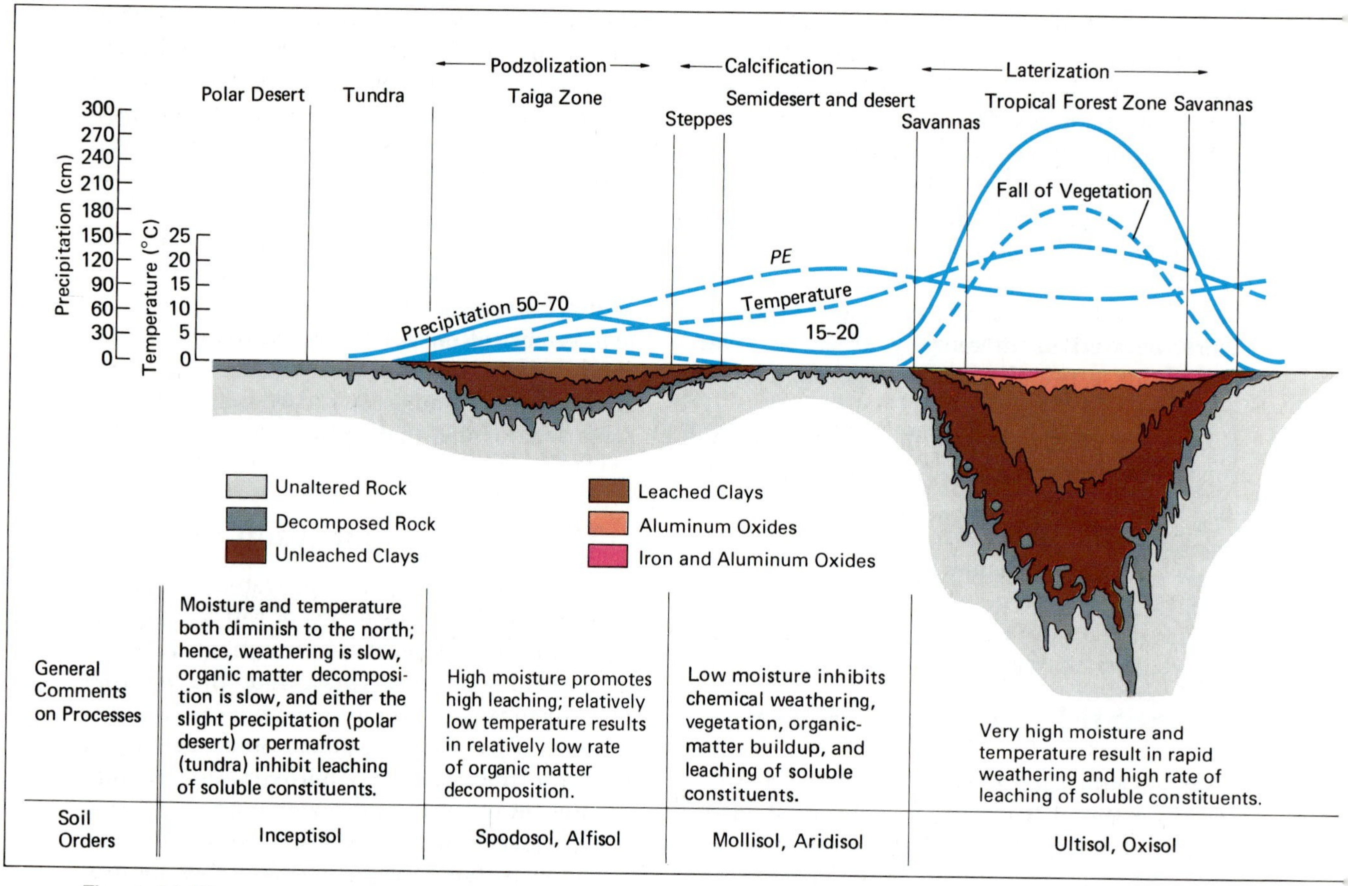
Podzolization
Calcification
Laterization
Polar Desert
Tundra
Taiga Zone
Steppes
Semidesert and desert
Savannas
Tropical Forest Zone
Savannas
Precipitation (cm)
300
270
240
210
180
150
120
90
60
30
0
Temperature (°C)
25
20
15
10
5
0
Fall of Vegetation
PE
Temperature
Precipitation 50-70
15-20
Unaltered Rock
Decomposed Rock
Unleached Clays
Leached Clays
Aluminum Oxides
Iron and Aluminum Oxides
General Comments on Processes
Moisture and temperature both diminish to the north; hence, weathering is slow, organic matter decomposition is slow, and either the slight precipitation (polar desert) or permafrost (tundra) inhibit leaching of soluble constituents.
High moisture promotes high leaching; relatively low temperature results in relatively low rate of organic matter decomposition.
Low moisture inhibits chemical weathering, vegetation, organic-matter buildup, and leaching of soluble constituents.
Very high moisture and temperature result in rapid weathering and high rate of leaching of soluble constituents.
Soil Orders
Inceptisol
Spodosol, Alfisol
Mollisol, Aridisol
Ultisol, Oxisol

Figure 11.18
Latitudinal distribution of influences on soils. This diagram summarizes how precipitation, energy, potential evapotranspiration (*PE*), and vegetation affect rock weathering and soil formation. Irregular boundaries indicate uneven depths of the zones indicated. (Modified from N. M. Strakhov, *Principles of Lithogenesis*, Vol. 1, 1967, Plenum Publishing Corp., and P. W. Birkeland, *Pedology, Weathering, and Geomorphological Research*, 1974, Oxford University Press)

orders, described here in order of increasing level of development.

Entisols (from the word "recent") are soils with poor horizonation (Figure 11.21), because of any of several factors: youth of the soil, rapid erosion during soil formation, or human interference, such as plowing. Obviously, these conditions differ considerably, and thus the range of Entisol types is wide. Suborders make the necessary distinctions.

Histosols (from the Greek *histos,* tissue) are composed primarily of plant material (see Figure 11.21). They develop in waterlogged environments, where organic matter decomposes slowly due to the lack of oxygen required by bacteria (Figure 11.17b). Suborders specify drainage conditions and the degree of decomposition of the plant material. Histosols are usually acidic, and their horizons are based on the degree of compaction and state of decomposition. They can develop almost anywhere, from the high Arctic to equatorial forests, but they are most extensive in tundra regions and glaci-

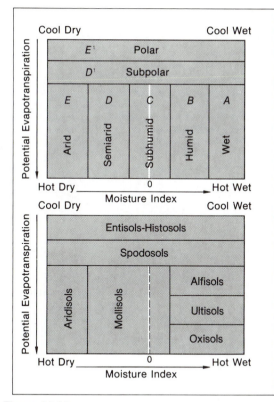

Figure 11.19
Climate is one of the important factors in soil formation. The top diagram shows the principal climate regions according to the Thornthwaite system. Potential evapotranspiration, a measure of energy, increases downward from cool to hot on the vertical scale. The moisture index, a measure of the moisture available, increases toward the right along the horizontal scale. The bottom diagram classifies the principal soil types according to the same criteria.

ated landscapes of higher latitudes. Histosols can be very productive when drained; however, the subsequent oxidation and drying of the organic material cause compaction and subsidence of the land surface, and increase the hazards of fire, flooding, and wind erosion.

Vertisols (from the Latin *verto,* to turn) are clay-rich soils in which horizon development is impeded by the churning effects of expansion and contraction due to seasonal wetting and

drying. Such soils develop in semiarid regions and are dominated by clay minerals that have exceptional water-absorbing capacity. When dried, Vertisols shrink, harden, and develop systems of cracks as much as 2.5 cm (1 in.) wide and 50 cm (20 in.) deep. These cracks collect organic debris that falls into them. Wetting causes the soil to swell, closing the cracks and churning sticky soil masses against one another.

In much of California, Vertisols are the normal soil type and are known as "adobe" from their use in the making of sun-dried brick in the Spanish period of California history. Vertisols are extensive in Texas, Australia, and India. These soils are rich in plant nutrients but are difficult to use agriculturally because of their poor structure and problems induced by the chemical effects of excessive sodium.

Inceptisols (from the Latin *inceptum,* beginning) are young soils that have developed in humid regions on recent alluvium, glacial or aeolian (wind-produced) deposits, or volcanic ash. Although some soluble compounds have been removed from the A horizon, these soils show no clear illuvial horizon and can be very productive if not too permeable or stony.

Aridisols (from the Latin *aridus,* dry) are the soils of deserts and semideserts around the world. Aridisols have the thinnest profiles of any regional soil type, due to shallow and infrequent penetration of water. They contain a minimum of organic matter, have maximum stoniness, and often include concentrations of calcium carbonate (lime), calcium sulfate (gypsum), and sodium chloride (salt), either in discrete masses or in surface or subsurface layers (see Figures 11.12 and 11.17). Aridisols are usable with irrigation, but are easily leached and require regular fertilization to remain productive.

Mollisols (from the Latin *mollis,* soft) have dark, humus-rich A horizons (mollic epipedons) with high base saturation (see Figure 11.16). Mollisols rarely, if ever, form on bedrock. They are derived from alluvium, glacial

THE SOIL SYSTEM

deposits, and loess (fine material deposited by the wind). They develop under midlatitude grasslands in the transition zone between arid and humid climates. The vegetation cover generally maintains a rich clay–humus complex due to the extremely dense mass of roots, which saturates the upper soil with organic decay products. Moisture is adequate to encourage a flourishing soil biota, but not sufficiently abundant to cause vigorous soil leaching. Mollisols show varying degrees and depths of calcification.

Unlike Vertisols, which may also be dark and rich in organic matter, Mollisols do not experience major volume changes, nor do they harden when dry. Thus, provided there is sufficient moisture, the best Mollisols are the "cream" of agricultural soils. Their natural cover of wild grasses has been replaced everywhere by crop grasses: wheat, barley, rye, corn, and sorghum. Unfortunately, they extend into areas that experience periodic droughts, leading to crop failures and recurrent "dust bowl" conditions. Mollisols vary significantly in character westward from Illinois to the much drier high plains of eastern Colorado. This is reflected in the soil suborders.

Spodosols (from the Greek *spodos,* wood ash) are soils in which a leached and eluviated light-colored *E* horizon overlies an illuvial *B* horizon that is colored by translocated iron or aluminum compounds or related organic carbon (Figure 11.15). An iron-cemented clay hardpan may be present. Spodosols are most common in forested cold-winter areas and wherever the parent material is nutrient-poor sand. The major soil-forming process is podzolization.

Figure 11.20
World-wide distribution of soil orders according to the U.S. Comprehensive Soil Classification System (7th Approximation). (Adapted from U.S. Department of Agriculture, Soil Conservation Service, 1972)

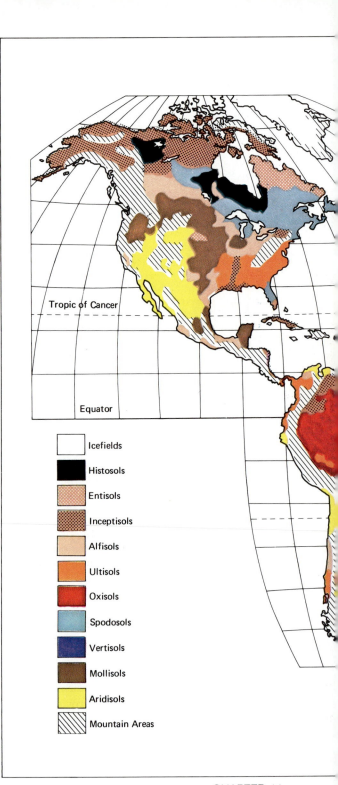

Icefields

Histosols

Entisols

Inceptisols

Alfisols

Ultisols

Oxisols

Spodosols

Vertisols

Mollisols

Aridisols

Mountain Areas

Tropic of Cancer

Equator

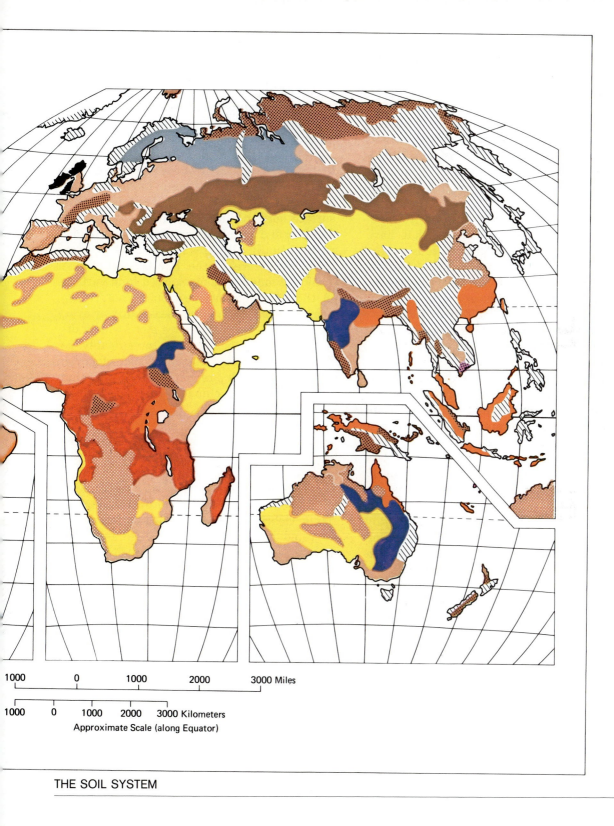

1000   0   1000   2000   3000 Miles

1000   0   1000   2000   3000 Kilometers

Approximate Scale (along Equator)

THE SOIL SYSTEM

311

Spodosols are notoriously infertile, most of their soluble bases having been replaced by hydrogen ions. In addition to their acidity, their sandy texture causes problems of moisture retention. However, some root crops, such as potatoes, thrive in these soils. To be productive for grains, spodosols must be neutralized by treatment with calcined lime (quicklime), which may increase grain yields by as much as 30 percent and pasture production by 300 percent.

Alfisols (from aluminum, *Al,* and iron, *Fe*) have yellowish-brown *E* horizons that have been partially leached of bases, causing the upper soil to be colored by iron and aluminum compounds. A clay hardpan is usually present in the illuvial zone. Alfisols are transitional between the Mollisols of the lime-accumulating *pedocal* zone and the Spodosols and Ultisols of the leached, iron- and aluminum-enriched, *pedalfer* regions.

In North America, Alfisols are found mainly in the southern Great Lakes area; however, they are widely distributed, occurring from middle to tropical latitudes. The retention of a significant proportion of bases allows the better Alfisols to be very productive agriculturally, and they support part of the intensively farmed American Corn Belt.

Ultisols (from the Latin *ultimos,* ultimate) are similar to Alfisols but more thoroughly leached of bases. Typically they are found on land surfaces older than those occupied by Alfisols. Thus Ultisols may be an advanced state of Alfisol development. Ultisols occur in climates that are warmer and wetter than those of Alfisols. They are redder due to a greater proportion of iron and aluminum oxides in the *A* horizon, and they are significantly more leached, poorer in humus, and less productive than Alfisols. The bases present in the upper soil have been brought up by deeply penetrating tree roots and are fed into the *A* horizon by way of plant litter. Forest cutting interrupts this nutrient cycle and results in rapid leaching of the bases remaining in the upper soil.

In North America, Ultisols are found throughout the southern Atlantic states and lower Mississippi Valley. These soils were severely damaged by 150 years of intensive cotton farming, which resulted in erosional losses of nearly the entire solum in some localities.

Oxisols (from the word "oxide") are even more thoroughly leached than Ultisols and are the soils characteristic of the wet tropics (Figure 11.13). They do not occur in North America. The diagnostic feature of Oxisols is a subsurface horizon consisting of a residue of clay and iron and aluminum oxides and hydroxides, with virtually all bases removed. Oxisols are the consequence of the laterization process. They are found on long-exposed land surfaces in those portions of the tropics that receive heavy precipitation either seasonally or every month. However, oxisols are more widely distributed than the climates presumably required for their development. Oxisols found in such relatively dry regions as the Australian interior indicate climatic change in these areas.

Oxisols are primarily encountered in the equatorial forest, savanna, and scrub forest regions of Central and South America, Africa, and Southeast Asia. As in the case of Ultisols, the bases present in Oxisols are a consequence of nutrient cycling by the natural vegetation. Thus, clearing the natural vegetation results in rapid leaching of soluble plant nutrients, leaving a severely impoverished soil.

SOIL MANAGEMENT

The soils in which plants grow are one of our planet's most priceless resources, equalled in importance only by air and water. In their natural state many soils seem to reach an equilibrium condition. Over the space of a year, the nutrients taken from the soil by its vegetative cover are returned to the soil as litter in the process of nutrient cycling.

Clearing the land for agriculture interrupts the cycling of nutrients. Since we remove most of the useful products of our croplands, only a portion of the nutrients that plants remove from the soil are recycled into it in the form of organic litter. Thus agriculture without artificial fertilization inevitably results in some loss in soil fertility. At the same time, soil erosion is increased by exposing bare ground to the impacts of rain, wind, and flowing water, and by disturbing the soil through plowing and cultivating.

Soil Erosion

In the United States, annual losses to erosion in agricultural areas average from 9 to 12 tons per acre, whereas new soil forms at a yearly rate of only about 1.5 tons per acre. The U.S. Department of Agriculture estimates that eroded soil is currently flushing out of the mouth of the Mississippi River at an average rate of 15 tons per second. Erosion by water and wind removes the vitally important *A* horizon that contains the humus needed for moisture retention and soil fertility. Loss of the *A* horizon reduces the soil's water-accepting capacity; this increases the proportion of water that runs off on the surface, which results in still more erosion.

In eroded soils, plant roots must seek nutrients in dense subsoils instead of in the loose, humus-rich *A* horizons to which field crops are adapted. As soils become thinner and less fertile, plants become increasingly sensitive to periodic moisture variations. Thus many "droughts" are actually normal events that have a disastrous impact because the soil system has been weakened by human activities.

Sedimentation

Loss of soil by erosion creates other problems. The erosion of deforested hillslopes has choked streams with sediments, making them shallower. This increases the hazard of flooding on the plains beyond the uplands. Deforestation and soil erosion in upland areas have become an acute problem in the foothills and high valleys of the Andes and Himalaya mountains and in the East African Highlands, due to rapid population growth that is pushing agriculturalists onto ever steeper land. This is increasing flooding in the nearby heavily populated lowlands. Soil erosion also shortens the life of reservoirs, many of which are filling with silt much more rapidly than predicted.

Remedial Measures

Of course, measures are being taken to preserve our soil resource. The cost of fertilizers of various types is a major expense on any modern farm. Crop rotation can stave off soil exhaustion. To reduce wind and water erosion, crop stubble and litter are left on the fields, or cover crops are planted after harvesting. In rolling country, plowing along the contour rather than up and down the slope will hold back runoff and reduce soil erosion by as much as two-thirds. Contour plowing can be combined with strip cropping, in which different crops are alternated in bands along the contour, as shown in Figure 11.21. A new erosion-reducing method is "minimum tillage," in which the land is disturbed as little as possible in agricultural operations, with weed control by herbicides, or natural sod being left between the crop rows. This can reduce erosion to half its usual rate.

In deforested uplands, massive plantings of trees can reduce erosion and sediment input into streams. In irrigated areas, deep drains and more carefully controlled water applications will prevent the artificial rises in water tables that cause salinization of dryland soils.

All these procedures are being carried out in many parts of the world, but only after centuries of neglect. Unfortunately, in some regions it is already too late, causing an exodus of people from rural areas. Ironically, the deterioration of soils around the world has become

To Plow or Not to Plow

The very word *farmland* brings to mind the straight furrows of a freshly plowed field, free of crop residues or intruding green weeds. For thousands of years seed agriculture consisted of preparing the field with a simple wooden plow that merely broke the soil surface and never turned a furrow. Then in the early 1800s plows were introduced that not only broke the ground but partially turned it over, inverting the surface. This buried the residue of the preceding crop, turned weeds under, and mixed fertilizer down into the soil—all to the good it seemed.

But such "moldboard" plowing also leaves a loose, unprotected surface that takes the full impact of raindrops during storms and the full force of the wind during dry periods. While it is obvious that deep plowing results in the loss of soil moisture, it is not so apparent that rainsplash erosion in a plowed field seals the soil surface and increases water runoff and sheet erosion. Although plowing along the contour reduces erosion on sloping land, loss of the most productive part of the soil by erosion has become so widespread that new erosion-control approaches seem required.

One solution is a type of agriculture known as "minimum tillage." The strategy is to minimize soil disturbance and to keep old crop residues on the fields where they can partially absorb the impact of rain and wind. Deep moldboard plowing is avoided. There is no harrowing to make the entire field a receptive seed bed. Instead, crops are planted in linear slots cut in the soil to receive seeds. This requires new types of agricultural machinery. In the midwestern Corn Belt, crop residues left on fields can reduce water erosion by 50 to 75 percent, and wind erosion in the wheat regions of the Great Plains can be reduced from as much as 30 tons per hectare to as little as 2 tons. Slot-planting also allows more efficient application of fertilizers and improves soil structure by decreasing compaction by heavy machinery.

But not all of the results are positive. A common scene on farms around the world is a flock of birds following the plow, feasting on insects and larvae exposed by cultivation and reducing the need for insecticides. In the absence of plowing, control of weeds, insect pests, and plant pathogens (fungi and harmful bacteria) must be achieved by increased use of chemical herbicides, pesticides, and fungicides. Surface and ground water quality reflect the increase in toxins applied to fields. These toxins build up through food chains and could reach levels dangerous to higher life forms. However, slot-planting should diminish contamination of water by chemical fertilizers. Compared to plowed fields, fields under minimum tillage retain more moisture, have a higher albedo, and are better insulated by the remaining surface layer. But such fields are slower to warm in the spring. The temperature differential with plowed fields may be 3° to 4°C (5° to 7°F). This will delay crop germination in cool regions, possibly extending the maturation period long enough to increase the risk of insect or pathogen damage.

Experiments with minimum tillage are now underway in a variety of agricultural regions, and will undoubtedly be more successful in some than in others. Each region will have to work out its own answers to the twin hazards of soil erosion and chemical pollution as a result of agricultural land use.

evident at the very time that the world's population and food demands are increasing explosively. While crop yields have increased markedly in many areas, the cost of producing such yields has risen even faster, and crop failures have become more frequent. Quite possibly the future of humanity will depend less on world political events than on how we handle the fragile soil resource that sustains life on our planet.

(a)

(b)

Figure 11.21
Land management problems and solutions.

(a) Gullying of this cornfield in Missouri resulted from the improper practice of plowing and harrowing up and down the slope. The damage was done in only a few days by several heavy rains. At the time of the photo, this land, which originally had a natural cover of grass, had been used agriculturally for only three years.

(b) This scene in southern Wisconsin illustrates contour strip cropping, which retards runoff and greatly reduces soil erosion. According to the Soil Conservation Service, agricultural yields have doubled where contour strip cropping has been introduced in this area. (U.S. Department of Agriculture, Soil Conservation Service)

KEY TERMS

rock joints
exfoliation
mechanical weathering
frost weathering
chemical weathering
oxidation
weathered mantle
translocation
eluviation
leaching

eluvial layer
illuviation
illuvial layer
soil horizons (O, A, E, B, C, R)
humus
nitrogen fixation
legumes
soil texture
loam
ped

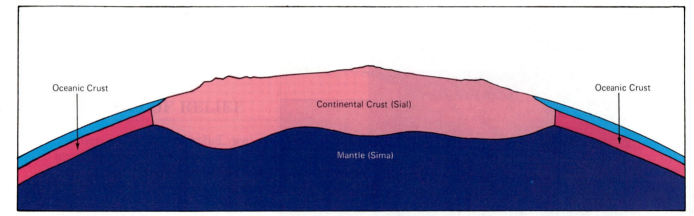

Oceanic Crust

Continental Crust (Sial)

Mantle (Sima)

Oceanic Crust

Figure 12.3
Because the sial of the continents is lighter than the sima of the mantle, the continents float like rafts in the denser mantle material. To be in bouyant equilibrium, the continental crust must project downward into the mantle under large mountain ranges, as indicated here. Each segment of the continental and oceanic crusts may be imagined displacing an equal weight of sima in the upper mantle. The vertical scale of the diagram is greatly exaggerated. The average thicknesses of the oceanic and continental crusts are 10 km (6 miles) and 40 km (25 miles), respectively, with mountain "roots" reaching as deep as 70 km (40 miles).

gested that these continents had broken apart and had somehow moved to their present positions. The hypothesis of "continental drift" was first set forward in a formal way in 1912 by the German meteorologist Alfred Wegener (Figure 12.4). Wegener pointed out that coal beds in temperate regions are composed of tropical vegetation types, suggesting that today's mid-latitude areas had once been located near the equator. Likewise, ancient glacial deposits are seen in regions that are now tropical; thus low-latitude landmasses once had polar or subarctic locations. Furthermore, geological structures and rock types along the west coast of Africa and the east coast of South America seemed to match. Finally, the fossilized remains of ancient plants clearly indicated former connections among all the widely separated southern hemisphere continents.

For almost half a century most geologists scoffed at Wegener's ideas, for there seemed to be no conceivable force that could move an entire continent. The fact that the sial of the continents projects downward into the denser sima beneath made the possibility of horizontal movement of the continents seem all the more unlikely.

Sea-Floor Spreading

In 1944 the English geologist Arthur Holmes proposed a solution for the puzzle. He suggested that it is the sea floors that are moving—and dragging the continents along with them. Holmes hypothesized that the driving force is a process of convection ("boiling") in the upper part of the earth's mantle. This would cause slow horizontal circulations of molten material below the earth's crust, which would drag the crust this way and that. The only feasible location for the subterranean "flows" is in the asthenosphere (Chapter 1)—the mantle layer having temperatures high enough to keep rock material melted and pressures low enough to allow the molten material to circulate at a rate of centimeters per year.

During World War II, it was discovered that all the oceans are divided by continuous undersea ridge systems. The best example is the Mid-Atlantic Ridge. In addition, trenches deep

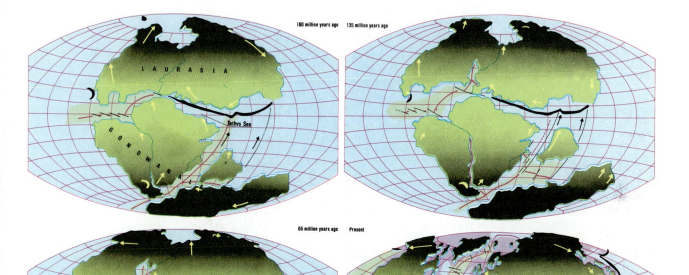

Figure 12.4
Alfred Wegener suggested in 1912 that about 200 million years ago all of the landmasses of the earth were united in one supercontinent, Pangaea. Pangaea subsequently separated into two continents, Laurasia in the northern hemisphere and Gondwanaland in the southern hemisphere. The rifting continued, with crustal plates and associated continents moving in the directions indicated by the arrows. The drift of the continents is now thought of in terms of plate tectonics, sea-floor spreading, and subduction. In this figure, subduction zones are shown in black and spreading centers in red. Thin black lines indicate transform faults; purple areas are present continental shelves. (*The Atlas of the Earth,* p. 36, © 1971, Mitchell-Beazley, Ltd.).

enough to hold the world's highest mountains almost completely ring the Pacific Ocean. Holmes suggested that volcanic eruptions along the oceanic ridge systems create new areas of ocean floor, while old areas of ocean floor eventually disappear by descending into the oceanic trenches, a process later termed *subduction* ("underflow").

In the next two decades a mass of new evidence emerged that supported Holmes's hypothesis. The continents do indeed move, riding passively like rafts on lithospheric plates that seem to move like conveyor belts from the oceanic ridges toward the oceanic trenches in the process now known as *sea-floor spreading.* The continents cannot themselves be sub-

THE LITHOSPHERE

ducted downward because of their low density relative to the material beneath them in the crust and mantle. The global pattern of crustal plates, spreading centers, and subduction zones was established by oceanographers in the 1960s and is shown in Figure 12.5. The movement of these crustal plates and the inter-actions between them are known as *plate tectonics*.

The sea-floor spreading hypothesis has been supported by many findings since the 1960s (Figure 12.6). Scientists have drilled into the floors of the oceans hundreds of times, principally from the research vessel *Glomar Challenger,* recovering cores of sediment and rock for analysis. These have shown that the sea floors are extremely young compared with the continents. Whereas the earth itself is almost 5 billion years old, and 3.8 billion-year-old rocks have been found on the continents, the most ancient portions of the sea floors have an age of only about 170 million years. Furthermore, the thickness and age of oceanic sediments increase with their distance from the oceanic ridges. This supports the idea that the sea floors are being formed at the ridges and are moving away from them (Figure 12.7).

Some of the most convincing evidence for sea-floor spreading comes from the record of

Figure 12.5
The lithospheric plates compose the earth's crust and the relative motion of the plates. The red lines represent spreading centers where new crust is being formed by volcanic activity and the blue lines mark regions where plates are descending into the mantle in the subduction process. Blue barbs are on the upper (overriding) plate and point in the direction of subduction of the descending plate. Red arrows indicate the relative directions of plate motion. The seven major plates are identified in bold type and several of the smaller plates in lighter type. Note that the Pacific basin is largely rimmed by subduction zones. Along the Atlantic coasts, however, there are no plate boundaries except those of the Caribbean Sea. (Calvin Woo from John F. Dewey, "Plate Tectonics," *Scientific American,* copyright © 1972 by Scientific American, Inc. All rights reserved.)

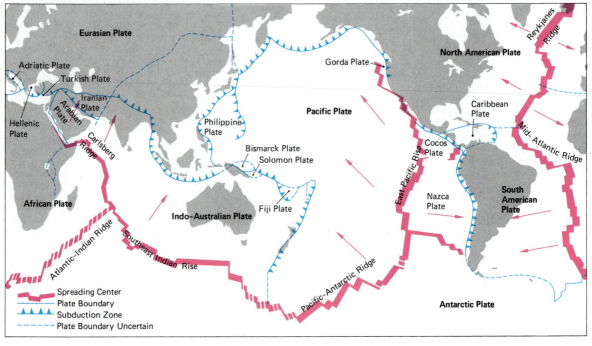

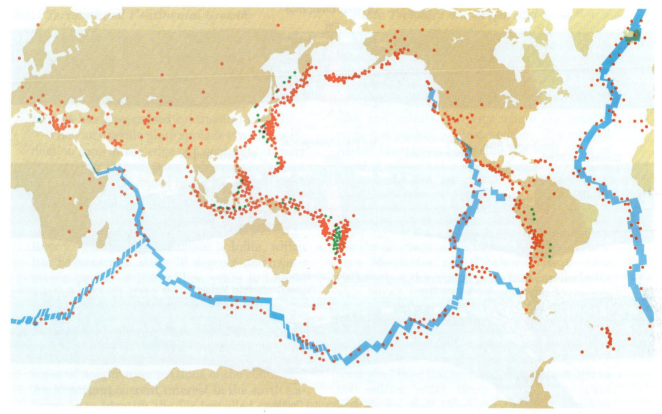

Figure 12.6
The distribution of earthquakes (red dots) coincides with the major discontinuities of the earth's surface, including oceanic ridges and trenches, and continental boundaries. These bands of seismicity mark the edges of lithospheric plates that are either diverging or converging. Green dots indicate earthquakes originating at deeper levels than those shown by red dots.

ancient magnetism in the volcanic rocks of the ocean floors. When volcanic rock solidifies from a molten condition, tiny grains of the iron-bearing mineral magnetite align with the earth's magnetic field. Studies of volcanic rocks reveal that the global magnetic field has changed in strength and reversed in polarity (so that a compass needle would point south instead of north) at intervals of hundreds of thousands of years. The last major *magnetic reversal* occurred about 700,000 years ago. A map of the lavas on the ocean floors shows alternating stripes of normal and reversed polarity. These stripes parallel the oceanic ridges and their general pattern is strikingly symmetrical on the opposite sides of the ridges. Changes in the direction of the stripes indicate changes in the direction of sea-floor spreading.

In general, the sea floors appear to be moving laterally 1 to 10 cm (0.5 to 4 in.) each year, carrying the continents with them. The Atlantic Ocean is widening by this process, whereas continents are being forced toward the Pacific Ocean from two sides. The floor of the Pacific Ocean is detached from the adjacent continents and is forced to descend under them in the oceanic trench system that almost completely encircles the Pacific (Figure 12.5).

color, hardness, and specific gravity (mass relative to that of an equal volume of water). Many minerals are extremely beautiful, and perfect specimens are sought by collectors. Certain minerals are quite familiar, such as *halite* (common salt, NaCl), *ice* (solid H_2O), and *geothite* (iron rust, FeO[OH]).

The high-density metals, such as iron, aluminum, copper, manganese, lead, zinc, nickel, tin, and silver, are rarely found free as separate minerals. They are usually combined with other elements to form minerals that must be artificially decomposed, or "refined," to extract the metal. Such minerals are known as the *ores*

of the metals (Figure 12.9). Gold and platinum are unusual in that they normally occur in pure form. Occasionally, "native" (uncombined) copper is present in a natural deposit, such as that on the Keweenaw Peninsula of Michigan, which was worked by Indians in the pre-European period. Iron meteorites have also been a source of metal for preindustrial peoples.

The elements combined in minerals are held together by electrical bonds. Thus all multielement minerals are composed of combinations of positive and negative ions. The most abundant ions in minerals are those of silicon (Si), which

Figure 12.9
These minerals are important ores. Pyrite, goethite, and hematite are all iron ores. Chalcopyrite and malachite are copper ores. Galena is the major ore of lead, sphalerite of zinc, cinnabar of mercury, and cassiterite of tin.

Pyrite and Chalcopyrite

Goethite

Malachite

Hematite

Sphalerite

Galena

Cinnabar

Cassiterite

Uranium Minerals

CHAPTER 12

Table 12.1

Major Elements of the Earth's Crust

ELEMENT	WEIGHT (PERCENT)	VOLUME* (PERCENT)
Oxygen (O)	46.60	93.77
Silicon (Si)	27.72	0.86
Aluminum (Al)	8.13	0.47
Iron (Fe)	5.00	0.43
Calcium (Ca)	3.63	1.03
Sodium (Na)	2.83	1.32
Potassium (K)	2.59	1.83
Magnesium (Mg)	2.09	0.29
Totals	98.59	100.00

*Computed as 100 percent, hence approximate. After Brian Mason, *Principles of Geochemistry*, John Wiley & Sons, Inc., 3rd ed., 1966.

has a positive charge, and oxygen (O), which carries a negative charge (Table 12.1). Most rock-forming minerals are silicates, in which positively charged ions of the hard and soft metals are combined with silica (SiO_2). About 92 percent of the earth's crust is composed of silicate minerals (Table 12.2). Oxides, which form another large family of minerals, result from the chemical decay of silicate minerals, which releases metallic ions that combine with the oxygen ions in water. Since oxygen is much more abundant than silicon in the silicate minerals, and is also an important constituent of the carbonate minerals (which combine metallic ions with negatively charged CO_3), almost 94 percent of the volume of the earth's crust consists of oxygen ions (Table 12.1). Of the vast number of different minerals known to occur on the earth, only those few silicates and carbonates shown in Table 12.2 are common constituents of unaltered rocks.

Igneous Rocks

Much of the rock of both the continents and the ocean basins has solidified from a molten

Table 12.2

Major Rock-forming Minerals

MINERAL GROUP	MINERAL	GENERALIZED CHEMICAL COMPOSITION	
		POSITIVE IONS	NEGATIVE GROUP
Silicates	Olivine	Mg, Fe	(SiO_4)
	Garnets	Mg, Al, Ca, Fe	
	Pyroxenes	Na, Mg, Al, Ca, Fe	(SiO_3)
	Amphiboles	Na, Mg, Al, Ca, Fe	(SiO_{11}), (OH)
	Micas	Mg, Al, K, Fe	(Si_2O_5), (OH)
	Clay Minerals	Al, K	
	Plagioclase Feldspar	Na, Al, Ca	(SiO_2)
	Orthoclase Feldspar	Al, K	
	Quartz	Si	O
Carbonates	Calcite	Ca	(CO_3)
	Dolomite	Ca, Mg	

After A. Lee McAlester, *The Earth,* Englewood Cliffs, N.J.: Prentice-Hall, 1973.

condition. Molten rock-forming material, or *magma,* is present everywhere on earth below a depth of about 70 km (40 miles). Occasionally this fluid material, with a temperature of 900° to 1,200°C (1,600° to 2,200°F), forces its way through the crust and spills out at the surface as red-hot *lava* (Figure 12.10), which cools rapidly to form volcanic rock. Volcanic eruptions also hurl rock particles and bits of lava into the air. This *pyroclastic* ("fire-broken") material rains down to form loose deposits of *volcanic* *ash,* also known as *tephra.* Extremely violent eruptions vent great volumes of fine glowing particles that are welded together by heat when they settle to the ground. This produces the rock known as *tuff.* Much larger volumes of magma solidify slowly below the surface, deep within the crust. In all cases the product is *igneous rock* (from the Latin *ignis,* fire). Igneous rocks that have solidified below the land surface are termed *intrusive,* while those that solidify at the surface are *extrusive.*

Intrusive Igneous Rock

As fluid magma cools, various silicate minerals "precipitate" out in crystal form at successively lower temperatures. The mineral compo-

Figure 12.10
Red-hot lava erupted at the earth's surface at temperatures of 900° to 1,200°C (1,600° to 2,200°F) rapidly congeals into extrusive igneous rock. (Charles A. Wood)

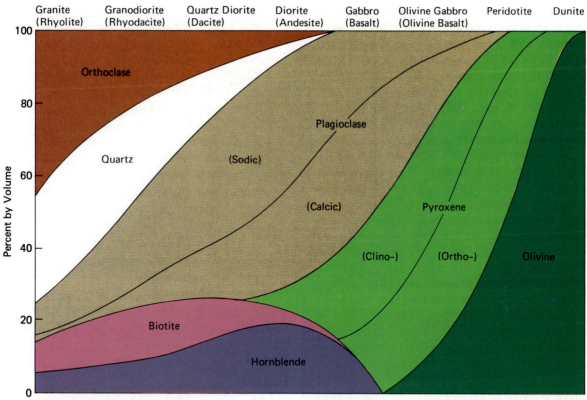

Granite (Rhyolite)   Granodiorite (Rhyodacite)   Quartz Diorite (Dacite)   Diorite (Andesite)   Gabbro (Basalt)   Olivine Gabbro (Olivine Basalt)   Peridotite   Dunite

Figure 12.11

Igneous rocks are classified by the proportions of their constituent minerals. Here the approximate mineralogical compositions of the more common igneous rocks are shown in terms of the percentage of the total volume occupied by each mineral. Thus a granite should contain approximately 40 percent orthoclase, 30 percent quartz, 15 percent plagioclase, 10 percent biotite, and 5 percent hornblende. Intrusive rock types are labeled across the top, with their extrusive equivalents in parentheses. Peridotite and dunite are rare, probably being formed only in the earth's upper mantle. (Doug Armstrong)

sitions of the more common igneous rocks are shown in Figure 12.11. Magma that crystallizes in the high-temperature environment beneath the earth's surface cools much more slowly than magma that pushes closer to the surface. The greater the time required to solidify, the larger the mineral crystals composing the resulting rock. Igneous rock formed by cooling over periods of thousands of years at a depth of many kilometers is coarse-grained *plutonic rock* (after Pluto, the Roman god of the underworld). The rock mass itself is called a *pluton.* Where many individual plutons have formed a large volume of igneous rock surrounded by other rock types, the result is a *batholith,* such as the Idaho batholith or the Sierra Nevada batholith. Deep-seated plutonic rocks, such as granite, become exposed at the surface by crustal upheaval and the erosion of covering rock masses.

Extrusive Igneous Rock

When magma forces its way to the land surface, the result is a volcanic eruption. These eruptions vary in nature (Chapter 15) but produce both lava and tuff. Lava is fluid or semifluid magma that flows out at the surface, where it normally solidifies into rock in hours

Figure 12.13

Granite is the most common intrusive igneous rock on the continents.

(a) The thin section (enlarged about ten times) shows that granite is a coarse-grained rock consisting of different minerals that have an interlocking structure. (M. E. Bickford, University of Kansas)

(b) Stone Mountain, Georgia, is a core of unjointed granitic rock that rises boldly above weathered granite in the Piedmont region of the Appalachians. (Warren Hamilton/U.S. Geological Survey)

(c) Chemical weathering in jointed granite in arid eastern California creates a jumble of exposed rock. In the distance is the steep wall of the Sierra Nevada, its crest splintered by frost weathering; Mt. Whitney is in the center. (T. M. O.)

of tiny marine animals. Under the deeper water of the continental rise areas we once more encounter sands and silts, emplaced by great submarine slides and flows of sediment down the continental slopes. When sediments of any type become deeply buried under subsequent deposits, compaction and the deposition of cementing substances cause them to become *lithified,* or converted into *sedimentary rock* of either clastic or chemical origin (Table 12.4).

Sedimentary rocks form in layers (Figure 12.14). These may be millimeters to hundreds of meters thick. Each distinct layer, called a *bed* or *stratum,* indicates a period of sediment deposition. The separations between strata are *bedding planes,* which indicate a period of no deposition at that location. Frequently the strata above and below a bedding plane are dissimilar, indicating a change in the conditions of sediment delivery. Individual beds may also vary laterally due to differences in energy con-

Table 12.4

Common Sedimentary Rocks

	UNCONSOLIDATED SEDIMENT	GRAIN SIZE	LITHIFIED ROCK
CLASTIC ORIGIN	Angular boulders, cobbles, pebbles	> 2 mm	Breccia
	Rounded boulders, cobbles, pebbles	> 2 mm	Conglomerate
	Sand	0.02–2.0 mm	Sandstone
	Silt	0.002–0.02 mm	Siltstone (Mudstone)
	Clay	< 0.002 mm	Shale

	UNCONSOLIDATED SEDIMENT	MINERAL COMPOSITION	LITHIFIED ROCK
CHEMICAL ORIGIN	Calcareous parts of marine organisms and direct calcium carbonate precipitates	Calcite ($CaCO_3$)	Limestone
	Magnesium replacement of calcium and direct magnesium carbonate precipitates	Dolomite ($CaMg[CO_3]_2$)	Dolomite
	Amorphous silica	Chalcedony, Quartz (SiO_2)	Chert (Flint)
	Compacted plant remains	Carbon (C)	Bituminous Coal
	Salt left by evaporation of sea or saline lake water	Halite (NaCl)	Rock Salt
	Gypsum left by evaporation of sulfate-laden water	Gypsum ($CaSO_4 \cdot 2H_2O$)	Gypsum

ditions and distance from the original source of the sediments. A single layer deposited in a certain time interval can change in lateral succession from *conglomerate* (cemented gravel) to *sandstone* (cemented sand), *siltstone* (cemented silt), *mudstone* (cemented silty clay), *shale* (cemented clay), and finally *limestone,* which is formed either as a chemical deposit or as an accumulation of the skeletal remains of tiny marine animals.

Nonclastic Sedimentary Rock

Limestone, which is composed of calcium carbonate, has considerable economic value. From it is made the cement used to construct highways, buildings, sidewalks, patios, and swimming pools. *Dolomite* is a calcium-magnesium carbonate rock that forms as a chemical precipitate or by chemical alteration of limestone. Limestone is peculiar among common rock types in that it dissolves completely where there is abundant moisture and vegetation. This creates very distinctive landscapes, as we shall see in Chapter 15. Dolomite too is soluble, but much less so than limestone.

Marine organic deposits that have an exceptionally high carbon content form *hydrocarbons*—the principal constituents of petroleum

(a)

(b)

Figure 12.14
Sedimentary rocks consist of fragments of rock debris or organic material cemented together by various substances, most commonly silica, calcite, and iron oxide.

(a) *Sandstone,* one of the most common sedimentary rocks, is usually composed of grains of silicate minerals cemented together by other minerals, as shown in this enlarged thin section. Sandstone is formed by the consolidation of beds of sand deposited by wind or water both on the land and in the sea. (M. E. Bickford, University of Kansas)

(b) This photograph illustrates the laminated nature of sedimentary rocks, which form *strata* that vary in thickness and physical and chemical characteristics. (Warren B. Hamilton/U.S. Geological Survey)

(c) This thick series of sedimentary rocks is exposed above the Colorado River near Moab, Utah. The ledge-forming strata are sandstone, with weaker shale producing the slopes between successive ledges. (T. M. O.)

(c)

and natural gas. Due to their low density, petroleum and natural gas migrate upward to fill openings in the more porous rocks, especially sandstone. These are the "reservoir rocks" in oil and gas fields.

Economically, the most valuable of all sedimentary rocks is *coal,* which originates as luxuriant vegetation growing in freshwater lagoons and swamps. To be preserved, this organic material must accumulate in a stagnant-water environment and be acted on by bacteria that can thrive without oxygen. To be transformed into coal, the resulting organic complex must be compressed by deep burial. The first stages of coal formation are occurring today in the swamps of the southeastern United States.

Two other sedimentary rock types of economic value are *rock salt* and *gypsum.* These are chemical deposits formed on the beds of evaporating lakes and inland seas in dry regions. The uses of salt are too numerous to mention. Gypsum likewise has many uses, among the most important being the manufacture of plaster of Paris and gypsum board, which is the standard material used to sheath the interior walls of houses.

Submarine volcanic activity can saturate sea water with silica ions, resulting in the chemical precipitation of silica as thin beds of *chert.* Podlike masses of chert, called "flint," are frequently found in limestone and, due to their hardness and ability to retain sharp edges, were used in the manufacture of stone tools by primitive peoples.

Metamorphic Rocks

The subduction process causes rock masses to be forced deep down into the lowest portions of the earth's crust. Here pressures and temperatures are hundreds of times those at the earth's surface. This causes the rock material to deform and flow in a plastic manner or to melt and recrystallize in different minerals. This process of *metamorphism* of rock material transforms limestone to *marble,* shale to *slate,* sandstone to *quartzite,* granite to *gneiss,* and lava to *schist* (Table 12.5).

In many metamorphic rocks the minerals are "smeared out," or oriented along visible planes of flow (Figure 12.15). Where the original rock contained a mixture of minerals, as in granite, metamorphism may segregate them in wavy bands of contrasting color. The result is gneiss. Complete melting of the original rock generates new fluid magma, which may force its way upward in the crust to become a pluton of igneous rock. The process of metamorphism, then, is transitional to complete melting.

Crystalline Shields

Near the centers of all the continents are rigid areas of very ancient *crystalline rocks*—rocks clearly displaying mineral crystals—principally granite, gneiss, and schist. These ancient rocks may be exposed at the surface, as around Hudson Bay in Canada and in New York's Adirondack Mountains, or they may be "basement rocks" covered by younger sedimentary rocks. These oldest portions of the continents are known as *crystalline shields.* Most of the rocks of the crystalline shields were subjected to extreme metamorphism in pre-Cambrian time (Figure 1.8), and many of the plutonic rocks appear to have been created by the melting of even more ancient sedimentary rocks. The crystalline shields represent the deep roots of ancient mountain systems that were erased long ago by the processes of erosion. They contain the earth's oldest known rocks (about 3.8 billion years in age), all of which are metamorphic types exposed where erosion has removed thicknesses of tens of kilometers of overlying younger rocks.

The Rock Cycle

During the course of time, rock materials pass from one form to another in the *rock cycle* (Figure 12.16). Our understanding of the rock cycle has been greatly advanced by the discovery of sea-floor spreading and the crustal subduction process. We have long known that most of the rocks exposed at the earth's surface are sedimentary types composed of the debris of older rocks of all types. We know too that despite the nearly 5-billion-year history of the earth, all the ocean floors have ages of less than 200 million years. Thus, in the rock cycle the older oceanic crust is swallowed by subduction, descending deep into the lithosphere where it becomes metamorphosed. At depths of 70 to 90 km (45 to 55 miles), it begins to melt into new magma. This molten material rises into the continental crust in the form of igne-

Table 12.5

Structure and Composition of Metamorphic Rocks

	METAMORPHIC ROCK	TEXTURE	MINERAL COMPOSITION	DERIVED FROM
FOLIATED	Slate	Fine-grained; smooth, slaty cleavage; separate grains not visible	Clay minerals, chlorite, and minor micas	Shale
	Schist	Medium-grained; separate grains visible	Various platy minerals, such as micas, graphite, and talc, plus quartz and sodium plagioclase feldspar	Shale, basalt
	Gneiss	Medium- to coarse-grained; alternating bands of light and dark minerals	Quartz, feldspars, garnet, micas, amphiboles, occasionally pyroxenes	Granite
NONFOLIATED	Quartzite	Medium-grained	Recrystallized quartz, feldspars, and occasionally minor muscovite	Sandstone
	Marble	Medium- to coarse-grained	Recrystallized quartz or dolomite plus minor calcium silicate minerals	Limestone or dolomite

ous intrusions. Some breaks through to the surface and erupts as volcanic ash and lava. Erosion of the surface ash, lava, sedimentary rocks, and the deeper metamorphic and plutonic basement rocks generates new sediments. These sediments are gradually transformed into new sedimentary rock, which in time may be subducted, metamorphosed, melted, and recycled as new igneous rock. In this rock cycle the same atoms that have been part of the earth since its formation are used over and over in successive generations of rock material.

Differences in the rock types resulting from the rock cycle play a very important role in the appearance of landscapes. We shall explore this topic in Chapter 15. But first it is necessary to look at the geological structures created by the deformation of rocks as crustal plates move and interact.

CRUSTAL DEFORMATION

In addition to rock type, the "architecture" of rock masses plays an essential role in the appearance of landscapes. The varying structural configurations of rock bodies are the result of past tectonic activity, including volcanism,

(a)

(b)

Figure 12.15

Metamorphic rocks form by the transformation of preexisting rocks under conditions of heat and high pressure.

(a) *Schist* is a crystalline rock dominated by a layered arrangement of platy minerals, as seen in this thin section enlarged ten times. Because weathering tends to be most effective between the sheets, schists tend to break into thin flakes. (Warren Hamilton/U.S. Geological Survey)

(b) This outcrop of *gneiss* exhibits the swirled patterns often seen in metamorphic rock. (Warren Hamilton/U.S. Geological Survey)

broad vertical warping, and crustal compression and extension.

Volcanism

Volcanism refers to the intrusion of magma into the earth's crust and the extrusion of volcanic gases and molten material at the earth's surface. Although volcanism's greatest effect is in the oceans, the floors of which are composed of basaltic lava, its more visible displays on the land are among the earth's most awe-inspiring phenomena.

Large-scale volcanism is related to lithospheric plate boundaries (Figure 12.17). Three types of plate boundaries exist: those where plates are pulling apart, those where plates are sliding past one another horizontally, and those where plates are pushing together with one being subducted under the other. The first two are zones of nonexplosive outpourings of lava; the third is the location of violent tephra-producing volcanic eruptions.

Magma Sources and Eruptive Styles

Eruptions of the earth's most active volcano, Hawaii's Kilauea, can be viewed from close at hand in complete safety. Other eruptions devastate large areas with amazing suddenness and spare only those eye-witnesses who are far away. This fact was reaffirmed by the May 1980 eruption of Mt. St. Helens in the state of Washington, which took more than 60 human lives in a matter of minutes. Volcanoes take many different forms, and so do volcanic eruptions. This is a consequence of variations in the source of volcanic energy—the magma feeding the volcano.

THE LITHOSPHERE

Surface Mining and the Western Environment

Traditionally, useful minerals have been extracted from underground mines that burrow into the earth from surface veins or outcrops. However, mining technology has been changed by the appearance of powerful excavating equipment and new refining processes that make massive low-grade mineral deposits economically useful. There is an increasing trend away from costly underground mines and toward large-scale surface mines. The metallic ores, such as those of iron, copper, and even gold, are quarried out in vast open pits, and the great energy resource, coal, is mined in shallower open cuts known as "strip mines." While the economics favor such mining, it creates troublesome environmental problems.

The most visible impact of surface mining is the wholesale destruction of the land surface and the soil resources in environments ranging from tropical rainforests to deserts and arctic tundras. In particular, strip mining of coal chews away vast expanses of land, gnawing at an open face of rock on one side and leaving acre after acre of "tailings" behind. In the United States, it has recently become mandatory to renovate the land literally turned upside down by strip mining. This involves saving topsoil and rebuilding and revegetating the land surface. Nevertheless, such massive disturbance of the surface generates enormous volumes of sediments that are washed into streams. The stream waters become opaque, degrading or destroying the aquatic ecosystem and causing increased downstream flooding as stream channels become choked with sediment deposits. The reaction of water with the substances in metallic ores and coal beds creates acidic water that few organisms can tolerate. In strip mine areas this acid water feeds directly into surface streams, degrading water quality far beyond the mine areas themselves. Such highly toxic substances as mercury and cadmium are often released by mining metals or coals. Not only do these toxins pollute streams, but they can also become con-

Basaltic Volcanism

There are two general sources of magma. One is the portion of the mantle that lies below the rigid crust of the earth. The permanent layer of fluid rock material here is chemically basic sima, having a silica content of 50 percent or less. When simatic magma reaches the surface it is extremely hot—about 1,200°C (2,200°F)—and the magmatic gases are still dissolved in the melt, making it very fluid. It pours out freely as basaltic lava. There is little explosive activity, despite the presence of impressive steam clouds.

Basaltic lava, which forms the floors of the oceans, sometimes floods out in great volumes on the land, as in the states of Washington, Oregon, and Idaho. Such a phenomenon requires special circumstances, to be discussed in Chapter 15.

Silicic Volcanism

The second source of magma is the melting of rock masses that have been subducted downward into the earth's furnace-like interior. Magma produced by the melting of older rocks contains more than 65 percent silica and is "cooler" than basic magma—about 900°C (1,600°F). It has a lower density than basic magma, and its gases have separated out below the surface, where they are under enormous pressure. Therefore, when it is able to break through to the surface, silicic magma always erupts violently. The first eruption of gases and steam reduces the pressure confining the

centrated in the sediment deposits created by streams draining the mine areas, waste dumps, and ore-processing plants. These sediments become a further source of stream contamination.

The ecological problems associated with surface mining loom large in western North America. This is the location of the world's major untapped energy resources, which consist of tens of thousands of square km of bituminous coal, oil-bearing shales, and tar sands: the energy equivalent of several trillions of barrels of oil. The principal fossil fuel reserves are in Alberta, Montana, Wyoming, Colorado, and Utah. The energy reserves contained in these near-surface deposits vastly exceed those of all the earth's petroleum reserves and remain unexploited only because they are much more costly than oil to extract and refine.

Should the price of oil rise severely in response to political developments or the inevitable depletion of reserves, massive exploitation of western coal, oil shale, and tar sands is sure to begin. Tar sands are already being mined in Alberta, Canada, in a tundra setting. All of the western deposits will be attacked by open pit or strip-mining procedures. In the arid West, where the land is poorly protected by vegetation and is very slow to recover from disturbance, the large-scale onslaught of surface mining could create an ecological catastrophe. In this setting, revegetation and erosion control on reconstructed land will be especially difficult. Streams could change character completely, affecting urban and agricultural users of water. The landscape that will be lost is one of exceptional scenic beauty.

It seems inevitable that the day will arrive when the rising price of oil will cause the great power shovels to begin tearing into the cliffs of the arid West. Temporarily, at least, this will mean an economic boom in the energy-rich western states. Unfortunately, it is feared, the cost will be the permanent destruction of large expanses of the Western environment. Nevertheless, this will be one of the rare instances in which the environmental problems will have been well studied, and options fully evaluated, before the onslaught begins.

magma, which allows more magma to flash into gas, causing more explosions, more pressure release, more gas, and so on, in a chain reaction that may continue for hours or even days.

Silicic volcanism generates much smaller amounts of lava than does basaltic volcanism, but ejects great volumes of coarse and fine pyroclastic tephra in explosive eruptions. The tephra (volcanic ash) and lava are principally andesitic to rhyolitic in composition. Explosive volcanism occurs at the continental edges, areas of crustal plate convergence, and in volcanic *island arcs,* such as the West Indies and the Japanese, Philippine, and Indonesian archipelagos (island chains)—all of which lie above presently (or recently) active subduction zones.

Broad Warping and Isostacy

Over much of the earth's surface the only crustal deformation occurring at present is broad uplift or subsidence. Although these motions do not create striking geological structures, they are important because they help control erosion and deposition. Rising areas are attacked by stream erosion, forming regions of hills or plateaus gouged by canyons. Sinking areas are usually regions of low relief in which sediments are deposited.

Broad uplift may be a result of localized crustal heating due to mantle convection. Smaller areas can rise because of a large subsurface intrusion of magma. Such an event may be occurring at present near Mono Lake in California, where recent earthquake activity

THE LITHOSPHERE

forms. The clearest examples are the landforms produced in high mountains throughout the world by glacial erosion and deposition that ended more than 10,000 years ago (Chapter 16). To erase the evidence of glacial modification of landscapes will require perhaps a million years of gradation by other processes.

SLOPES: THE BASIC ELEMENT OF LANDFORMS

The chief problem in landform analysis is to explain the conditions that produce the various types of slopes and flat surfaces that make up the earth's landscapes.

Most of the slopes we see around us were initiated by downward erosion by channeled flows of running water. The same process that cuts gullies in a sloping cornfield like that in Figure 11.21 (p. 315) has excavated the Grand Canyon of the Colorado River. Stream incision provides new vertical surfaces that are quickly transformed into slopes by other erosional processes. These processes loosen material and move it down the slopes and into streams, ever widening the excavation made by vertical stream erosion. Rock weathering, mass wasting, and water-assisted erosion all play a part in the development of slopes.

Mass Wasting

Gravity provides the energy for all slope-forming processes. Gravity itself can cause the material composing hillslopes to move downward without the assistance of any moving fluid. This general process is known as *mass wasting*. Mass wasting movements range from sudden catastrophic rockslides to the slow downslope creep of soil and rock fragments over hundreds of years.

Solid rock is held together by the attractive forces between neighboring atoms, but slopes composed of soil and rock fragments remain in place because of the friction between solid particles. Without these frictional and cohesive forces, hillsides would collapse under the pull of gravity. Weathering of the rock forming a slope, or water saturation of the weathered rock, may occasionally reduce friction and cohesion to the point where rocks break loose or a portion of the hillside avalanches downward.

Angle of Repose

For any loose material on a slope, whether soil or a layer of rock debris, there is a maximum angle of inclination, known as the *angle of repose,* that the material can maintain without slipping downward. At this threshold angle, gravitational stress is just balanced by the cohesion or friction of the material on the slope. If new material, or a large volume of water, is added to a slope that is already at the angle of repose, a portion of the slope may fail and slide downward. Soil that has become water-saturated is most likely to slide, because the absorbed moisture lessens the friction between soil particles at the same time that it increases the bulk weight of the soil.

Talus and Scree

At the bases of rock cliffs there are normally cones of debris consisting of loose rock fragments that have fallen from the face of the cliff (Figure 13.7). This material, called *talus* (large chunks) or *scree* (small particles that shift underfoot), accumulates at an angle of repose of 34° to 39°. Talus slopes resulting from rock falls are present wherever there are rock cliffs but are most common in alpine areas and dry regions where bare rock exposures are most plentiful.

Soil Creep

In most areas, where there are no rock cliffs, mass wasting acts slowly and invisibly. The soil of every sloping pasture or forested hillside moves downhill a fraction of a centimeter per

Figure 13.7
Frost weathering of well-jointed granite rock at elevations above 3,500 meters (11,000 ft) in California's Sierra Nevada loosens rock masses and produces rock falls, particularly during the spring, when ice is melting. This creates cones of rock rubble, known as *talus,* at the base of frost-shattered cliffs. (T. M. O.)

year (Figure 13.8). This form of mass wasting is aptly known as *soil creep.* Soil creep causes fence posts and tombstones on sloping ground to tilt conspicuously.

The creep process is related to expansion and contraction of the soil by wetting and drying and by freeze and thaw. Figure 13.8 shows how changes in soil volume always result in slight downslope displacement of the soil mass. The rate of displacement is related to the slope angle but is generally less than 1 cm a year. The formation of *needle ice* (Figure 13.8) on cold nights can cause the surface layer of the soil to travel much faster than the bulk of the soil. Since creep velocity is a function of the slope angle, the transported material (known as *colluvium*) accumulates at the base of the slope, thus gradually elevating the slope foot

and causing the slope angle to become less steep with the passage of time. As creep-transported colluvium moves into depressions, the slope also becomes progressively smoother (Figure 13.9).

Gelifluction

A distinctive type of mass wasting is seen in subarctic and highland tundra regions where soils that are silty and moisture-retentive cover ground that remains frozen throughout the year. We have seen that in many tundra regions only the upper meter or so of the soil thaws during the spring and summer. The water released by thawing cannot drain downward through the still-frozen subsoil. For a few days the water-saturated soil loses its cohe-

Figure 13.13

(a) Viscous flows resulting from high-intensity rains onto surfaces lacking vegetative protection.

(a) Mudflows, such as this example seen in the California desert, often transport large boulders many kilometers. In the center of the view the mudflow deposit has been eroded by later flows of water.

(b) Debris flows transport a higher proportion of coarse material than do mudflows; consequently, they are more viscous and less free-flowing. This debris cone in the California desert has been built by the gradual accumulation of angular rock fragments carried out of a canyon by thousands of years of debris flow activity. (T. M. O.)

(b)

in terms of their evolutionary development through time. Thus Davis classified landforms according to their stage in a theoretical cycle of development, using the terms "youth," "maturity," and "old age." He then inferred the processes by which the landforms of each stage would slowly evolve into those of the next stage.

Youthful Stage

Davis focused first on the evolution of landscapes dominated by the effects of stream erosion. To explain his *cycle of erosion* in the simplest way, he visualized a sudden uplift of the land surface from a lower to a higher elevation, after which erosion begins to affect the uplifted mass. Rainfall and runoff on the raised area initiate erosion as streams flow down the newly created slope toward the sea. At first these streams rapidly erode their beds downward, cutting narrow valleys that are V-shaped in cross section. As long as areas of the original uplifted land surface remain visible between the new valleys, the landscape would be in the *youthful stage* of landform develop-

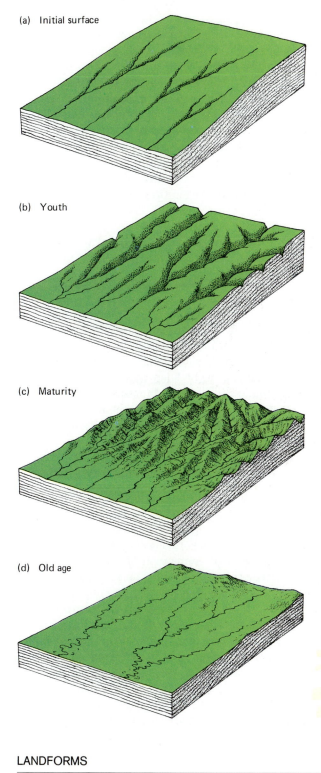

(a) Initial surface

(b) Youth

(c) Maturity

(d) Old age

Figure 13.14

This sequence of diagrams illustrates stages of the Davisian cycle of landscape evolution. This cycle applies to moist regions where erosion is accomplished primarily by flowing water. The diagrams assume that the underlying rock is uniform and exerts no controls over landform development.

(a) The initial surface after uplift of the region is a landscape of low relief. Streams have energy to begin cutting downward.

(b) In the stage of youth, the streams have cut narrow steep-sided valleys, and much of the initial surface is preserved between the valleys.

(c) In the stage of maturity, the uplifted area has been eroded into a mass of hills. The streams have stopped incising and have widened their valleys. Little or no trace of the initial plain remains.

(d) In old age, mass wasting and flowing water have eroded the region to a plain of low relief *(peneplain)* with isolated hills, or *monadnocks.* (T. M. O.)

ment (Figure 13.14b). In this stage, settlement and human activity take place on the plateau-like surfaces between the narrow, steep-sided valleys.

Mature Stage

As streams deepen their valleys, they decrease their altitude above sea level. This decreases their potential energy and, therefore, the kinetic energy available to do erosional work. Finally, in "late youth" the streams become "graded," with no excess energy to convert to the work of downward erosion, and valley deepening ceases. This permits slope gradational processes and lateral stream erosion to widen the valleys and create flat floodplains. In the *mature stage* the original uplifted land surface is converted entirely into hillslopes leading down to flat-floored valleys (Figure 13.14c). In the valleys, streams meander back and forth over continuous floodplains veneered with stream-deposited alluvium. Since the only level land is in the valleys, human activity is concentrated there.

Old Age

Further evolution from the mature stage to *old age* is assumed to proceed very slowly, requiring tens of millions of years. Hillslopes are imagined as becoming less steep with the passage of time. The landscape "flattens down." Finally, the surface is worn down almost to sea level, producing a lowland with faint relief, called a *peneplain* ("almost a plain"). Isolated areas of higher ground that are remote from the larger valleys are called *monadnocks* (Figure 13.14d), after solitary Mount Monadnock in New Hampshire. Davis stressed that his simple model could include variations, such as further uplift and return to the youthful stage during any of the intermediate stages in the cycle of erosion.

Davis's Contribution

Davis and his students devised other cycles of erosion for landscapes in which stream erosion was not the dominant process. There were cycles of evolution for deserts, coasts, glaciated areas, and regions of soluble limestone bedrock. In general, Davis viewed landforms as products of *geologic structure, geomorphic process,* and *stage of evolution.* Stage did not imply any fixed amount of time, but referred only to the development of forms in a sequence. Form evolution is slow where rocks are resistant or processes weak, and rapid where materials are weak and processes are vigorous.

Davis's work is important because it provided an easily understood basis for the organization and classification of landforms and caused geographers to begin to focus on the relationships between different landforms. However, the cycle of erosion concept originated at a time when the processes by which landscapes are transformed had only begun to be investigated. Today the Davisian cycle of erosion is regarded as being of value mainly as a way of introducing students to the concept of landscape change. The terms "youth," "maturity," and "old age," introduced into geomorphology

by Davis, have been retained, but only as descriptive terms for stream-dissected landscapes. Even this involves problems, as the youthful valley form is common in maturely dissected ("ridge and ravine") landscapes, and mature valleys are not uncommon where the dissection is still in a youthful stage. Although the cycle of erosion concept is a useful introduction to landforms, and is no doubt valid over very long periods of time, it does little to increase our understanding of geomorphic processes and the forms they create.

The Equilibrium Theory of Landform Development

Since about 1950 geomorphologists have increasingly emphasized the actual mechanics of landform development. This approach focuses directly on the detailed relationships between form and process in the landscape. It emphasizes the general tendency for the form of the land surface to be such that the energy of the erosional process is balanced to the resistance of the material it affects. In the following paragraphs we can see how this operates in the case of a hillslope.

Downslope Gravitational Stress

Two forces cause material to be put in motion on a slope. One is *downslope gravitational stress,* which is proportional to the slope angle. If you tilt a table, an object on it slides off when the downslope component of the pull of gravity exceeds the friction between the object and the table top. More specifically, the object starts to slide at the threshold when the component of gravitational force parallel to the slope exceeds the component of gravity normal (perpendicular) to the slope, plus a frictional component related to the roughness of the slope and the object.

In Figure 13.15, gravitational force, *A,* exerted on similar particles on slopes of different angles can be resolved into a downslope force,

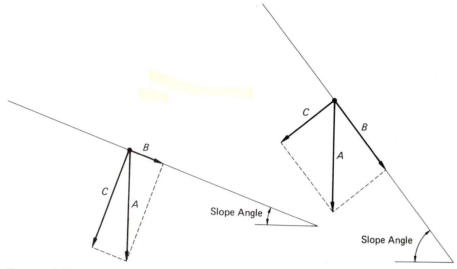

Figure 13.15
Effect of slope angle on downslope gravitational stress. Particles of the same weight *A* are shown resting on slopes of differing inclines. The downslope component of particle weight is *B*; the component of particle weight directed into the slope is *C* (particle resistance).

B, which tends to cause downhill motion, and a force directed into the surface, *C*, which tends to hold the material against the surface. The steeper the slope, the greater the downhill force *(B)* compared to the force resisting movement *(C)*. The loose particle at the left will remain in place because *C* is greater than *B*. If disturbed, the particle at the right will move downhill because *B* is greater than *C*.

Shear Stress

A second force that causes material to move is *shear stress.* This is the oblique downward and forward force exerted by one material rubbing against or flowing over another material. Shear stress is proportional to the density and velocity of the moving material. Water, ice, or wind moving over any surface exerts shear stress on that surface and may cause loose particles on it to move. The shear stress exerted by water or ice is also related to slope angle, because water and ice are usually moving due to

downslope gravitational stress. The magnitudes of both of these stresses, as well as the resistance of the material affected, can be expressed numerically and used in equations that explain or predict geomorphic processes and forms.

Controls of Slope Angles

If erosion reduces the angle of a slope, it has the effect of reducing the downslope gravitational stress on the material forming the slope. It also reduces the shear stress exerted by moving erosional agents. Then how far can erosion reduce a slope angle? Certainly not below the minimum that gives the erosional agent enough energy to continue to remove material from the slope.

On the other hand, any increase in the slope angle increases both the downslope gravitational force and the shear stress imposed by an erosional agent. This results in more vigorous erosion, which tends to reduce the slope angle.

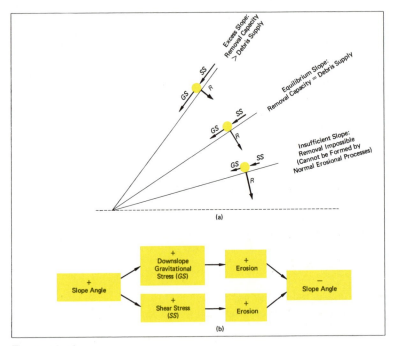

Figure 13.16
Principles of equilibrium slopes.

(a) Hypothetical slope angles indicate downslope gravitational stress *(GS)*, shear stress *(SS)* exerted by the erosional agent (such as slopewash), and resistance of slope particles *(R)*. Resistance increases with decreasing slope angles (see Figure 13.15), while *GS* and *SS* diminish with decreased slope. The steepest slope is unstable, as the sum of *GS* and *SS* far exceeds *R*, causing rapid erosional removal. The intermediate slope is stable, as the sum of *GS* and *SS* just balances *R*, so that removal can occur without changing the slope angle. The lowest slope cannot be produced by the stresses shown, as particle resistance *(R)* exceeds erosional stresses, so that erosional removal is impossible.

(b) Diagrammatic representation of a negative feedback relationship that causes slopes to be self-adjusting toward angles that equate stress to resistance. Any increase in slope increases the erosional stresses, which in turn decrease the slope angle until equilibrium is reestablished.

Here is an example of *negative feedback,* in which disturbance of a system that is in equilibrium triggers changes that tend to restore the original system (Figure 13.16). Thus, when a natural hillslope is made steeper due to highway construction or some other artificial modification, a destructive slump may follow. This is the way the hillslope returns to its original equilibrium angle.

Negative feedback is the principal means of maintaining equilibrium in physical systems and is a normal feature of many process/form systems. Negative feedback causes many landform systems to be self-regulating, so that they tend to maintain a steady state with the passage of time. In such a system, changes in form occur only when there is a change in the material or in the nature or intensity of the gradational processes.

Slope Angle Variation

Wherever the relief is high and the materials composing a slope vary in resistance to fragmentation and removal, there are corresponding variations in the slope angle (Figure 13.17). Where the rock is massive and resistant, slopes created by erosion will be steep,

T
G

At
rapic
tonic
the
excee
land
veys
gion:
ers (
regic
tatio
tion
year:
the r
the
Rive
(2 in.
the r
been
Th
duce
occu
Mou
centu
rare
ing
time
term
1 cm
more
subm
the h
So
ceed
the s
that
porte
flood
tiona
years
store
into
scape

Figure 13.17
Slope angle changes related to variations in rock type. Changes in rock resistance are especially clear in arid regions. Here in Monument Valley, Utah and Arizona, massive sandstone forms vertical walls. The lower layers of thinly bedded sandstones and shales produce much gentler slopes, although even here the more resistant layers make small cliffs. As these slopes are eroded, they wear back at a constant angle—the equilibrium angle for the material involved. (T. M. O.)

which maximizes downslope gravitational stress and shear stress. Where material is easily removed by the active agents of erosion, slopes are gentle, thus moderating the erosional stresses. Since variations in slope angles tend to equalize the erosional stresses on different rocks, it is possible for the whole landscape, tough rocks and fragile ones, to wear away at the same rate over very long spans of time. If this were not so, the local relief in all landscapes composed of varying rock types would become progressively greater with the passage of time, which is not the case.

Human Alteration of Landforms

The delicate adjustment of surface forms to local materials and gradational processes is, indeed, the most important thing to understand about landforms. It is impossible to modify either landforms or geomorphic processes artificially without impinging on process or form thresholds that trigger reactions in the natural system. The reaction will be one that tends to restore an equilibrium between process and form. Too often, however, this response is a costly or destructive event in human terms. We cannot make a significant change in a natural system and expect the system to remain passive; sooner or later it will respond with a change of its own to absorb the effect of the artificial change.

For example, artificially straightening a river to aid navigation or to prevent flooding also shortens the river and increases its slope

3. How does tectonic activity influence landforms?

4. In what two ways is water the chief agent of gradation?

5. What are the other agents of gradation? What determines which one will be dominant in an area?

6. There are several ways in which time is of importance in landform studies. Explain.

7. How is the "angle of repose" relevant to the concept of geomorphic thresholds?

8. What is the difference between "creep" and gelifluction?

9. How are large rockslides usually triggered?

10. Outline the sequence of events leading to slope erosion by surface water.

11. What conditions favor the occurrence of mudflows or debris flows?

12. Characterize the landscape (valleys and interfluves) in each stage of the erosional cycle proposed by William Morris Davis.

13. Why is the Davisian cycle of erosion concept in disfavor among contemporary geomorphologists?

14. In terms of the equilibrium theory of slope development, why do resistant rocks normally form slopes that are steeper than those of easily eroded material?

15. What problem commonly arises when landforms or geomorphic processes are artificially modified by human activity?

16. How do magnitudes of local and regional rates of denudation compare with local rates of uplift in tectonically active regions?

17. What unique landforms are associated with areas of tundra vegetation?

18. How does the creep process affect the appearance of hillslopes in humid midlatitude areas?

19. How does the seasonality of the Mediterranean type of climate create distinctive landforms or geomorphic processes?

20. Why are the effects of running water more apparent in deserts than in areas of higher annual rainfall?

21. What is the most aggressive geomorphic process in the wet tropics and tropical savannas?

APPLICATIONS

1. What are the landforms of your area on the macro, meso, and micro scales? Are there any unusual landforms or "textbook examples" of particular landforms in your vicinity?

2. Look at the slope forms and slope angles in your area. What do you think explains the differences from place to place?

3. What energy transformations occur (a) in the case of a rockslide that blocks a valley, damming the stream in the valley? (b) in the case of a wave that removes enough material from the base of a cliff to leave the higher part of the cliff without support?

4. It has been noted that for a beach to persist, particle input must be equivalent to erosional loss. What landforms in your region can be thought of in terms of a similar material budget?

5. W. M. Davis's cycle-of-erosion concept has been criticized on grounds that a full cycle from youth to old age could rarely be completed. What different factors could disturb the course of an erosion cycle?

6. Evaluate the relative roles of geologic structure, tectonic activity, gradational process, and time in the creation of the scenery of your region. Name other regions in which each of these factors is dominant.

FURTHER READING

Bloom, Arthur L. *The Surface of the Earth.* Englewood Cliffs, N.J.: Prentice-Hall (1969), 152 pp. This well-written paperback outlines the processes of landform development—brief but unusually good.

Bradshaw, Michael J., A. J. Abbott, and **A. P. Gelsthorpe.** *The Earth's Changing Surface.* New York: John Wiley (1978), 336 pp. An abundantly illustrated account of landform development; unusual and interesting.

Brunsden, Denys, and **John Doornkamp, eds.** *The Unquiet Landscape.* Bloomington: Indiana University Press (1974), 171 pp. This is a magnificently illustrated collection of articles on various aspects of landform development, from a series appearing in the British periodical *The Geographical Magazine.*

Butzer, Karl W. *Geomorphology from the Earth.* New York: Harper & Row (1976), 463 pp. Different climatic regions are discussed in this textbook on landform development. It is not highly technical and uses a geographical approach.

Chorley, R. J., S. A. Schumm, and **D. E. Sugden.** *Geomorphology,* London and New York: Methuen (1984), 605 pp. A state-of-the-art treatment of all aspects of landforms by three leaders in the field of geomorphology. Very well illustrated with diagrams and graphs.

Hunt, C. B. *Natural Regions of the United States and Canada.* San Francisco: W. H. Freeman (1974), 725 pp. This is a well-illustrated introduction to the regional landforms of North America. Fairly complete, but nontechnical.

Lobeck, A. K. *Geomorphology: An Introduction to the Study of Landscapes.* New York: McGraw-Hill (1939), 731 pp. Although dated, this text is included here because of its excellent photographs, maps, and diagrammatic illustrations of landforms of all types.

Ritter, Dale F. *Process Geomorphology.* Dubuque, Iowa: W. C. Brown (1978), 603 pp. As the title suggests, this is a modern process-oriented text. Illustrations consist largely of graphs and diagrams, and some portions are at a relatively advanced level.

Shelton, John S. *Geology Illustrated.* San Francisco: W. H. Freeman (1966), 434 pp. This book is unsurpassed for crisp photographic illustrations of landforms. The text and organization are more imaginative than most—highly recommended.

Twidale, Raoul C. *Analysis of Landforms.* New York: John Wiley (1976), 572 pp. The best illustrated treatment of landforms of all types; by a leading Australian geomorphologist.

Utgard, R. O., G. D. McKenzie, and **D. Foley.** *Geology in the Urban Environment.* Minneapolis: Burgess (1978), 355 pp. This paperback is a collection of articles concerning the significance of landforms and geomorphic processes in urban settings, demonstrating that landforms are not merely a "rural" topic.

View of Niagara Falls in Moonlight by Herman Herzog, 1872. (Museum of Fine Arts, Springfield, Massachusetts, James Philip Gray Collection with additional funds from the bequests of Richards Haskell Emerson, Ethel G. Hammersley, and Henry Alexander Phillips)

Flowing water is the most visible agent of landscape change, even in arid regions. The energy in moving water entrains, transports, and redeposits rock waste, constantly moving material from high to low places on the earth's uneven surface.

14
Fluvial and Aeolian Landforms

From 100 miles above the earth's surface, the organized patterns created by human activity are almost invisible. However, only from such an altitude can one begin to appreciate the organization of the earth's natural features. Two kinds of phenomena in particular exhibit remarkable regularity. One is erosional, the other depositional. First, and far the more widespread, are the systems of valleys carved by water runoff from the land surfaces. Second, and much more localized, are the patterns of sand dunes built by wind in deserts that seldom experience rainfall or runoff.

Valley systems and dune systems are related. Both result from the friction of a fluid (something that "flows") passing over the material of the earth's surface. The flow of water, however, is confined to channels, whereas the flow of air that concentrates sand and creates dunes has the whole landscape for its bed. The waves of sand raised by the wind as it passes over the land sometimes resemble water waves raised by the wind as it blows across the sea. Despite the difference in the natures of the "channels" for running water and for wind, the two fluids erode, transport, and deposit material in similar ways. For this reason we shall consider them together in this chapter. Since the domain of running water greatly exceeds that of wind erosion and deposition and is much more the realm of human activity, our attention will center there.

Processes related to channeled flows of water are known as *fluvial processes,* from the Latin *fluvius,* meaning river. Flowing water makes gradation possible by providing an effective transportation system connecting high and low places on the earth's surface. From a high altitude the effects of fluvial processes are conspicuous over nearly all ice-free land areas. The only exceptions are the areas covered by sand dunes. From space one can see intricate valley systems carved by running water, as in Figure 14.1; large streams collecting the water and sediment delivered by smaller streams; and the mouths of the earth's great rivers, issuing plumes of sediment that discolor the sea. Clearly, rivers are the essential disposal system in the process of gradation.

In this chapter we shall look first at the way channeled flows of surface water are organized in drainage networks. Then we shall examine the mechanics of streamflow. The vertical and horizontal patterns that fluvial channels make and the landforms that result from fluvial processes will be considered next. Finally, we shall look briefly at the quite different effects of wind on the land surface.

STREAM ORGANIZATION

Only the few streams that cross very dry regions, such as the Nile River in Egypt, flow far without being joined by other streams. All natural streams in a region are part of a *drainage network* that removes surface runoff from a *drainage basin,* which is the area drained by a system of connected stream channels. Every stream has its own drainage basin, or contributory area (also called its *catchment*), which can range in size from a fraction of a square kilometer to a sizable portion of a continent (Figure 14.2).

The specific geometrical arrangement of streams within a drainage network is the *drainage pattern,* which can take many forms. These are usually related to the local geological framework. The most common drainage

Figure 14.1
This side-looking radar image of a 32-km (20-mile) wide portion of eastern Kentucky portrays the nature of stream dissection of the land surface where there is little variation in rock type or geological structure. Note the large dendritic (branching) stream systems, the smaller tributaries, and the uniform density of channels. (Raytheon Company and U.S. Army Engineering Topographic Laboratories)

patterns are illustrated in Figure 14.3. Figure 14.1 also demonstrates the remarkable uniformity of the drainage pattern in an area undergoing dissection by fluvial processes.

Stream Order

In all drainage networks, small streams feed into successively larger streams. The characteristics of drainage networks can best be analyzed by using the idea of *stream order.* The

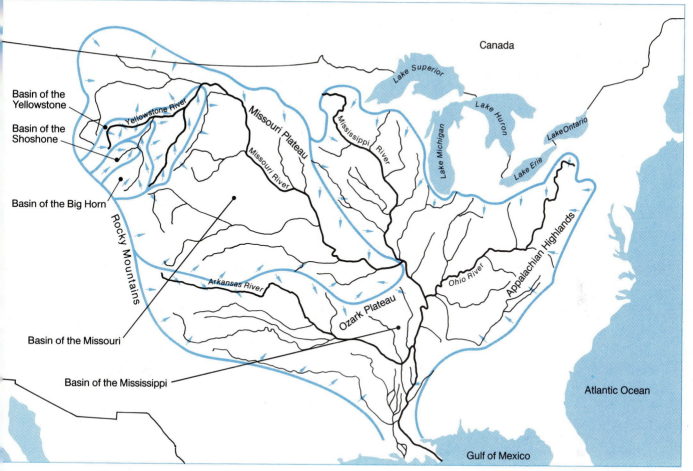

Figure 14.2

The drainage basin of a large river such as the
Mississippi contains a hierarchy of smaller nested
drainage basins, outlined in blue. In this figure, the
basin of the Shoshone River is part of the basin of the
Big Horn, which is part of the basin of the
Yellowstone, which feeds into the Missouri, which is
the major tributary of the Mississippi. The arrows
indicate the downhill direction of water flow into the
various basins. (Doug Armstrong)

smallest streams, which have no tributaries
feeding into them, are *first order* streams.
Where two first order streams join, the resulting larger single channel is a *second order*
stream. Similarly, *third order* streams begin at
the junction of two second order streams, and

so on, as shown in Figure 14.4. A stream of
higher order is formed only when two streams
of the next lower order join, but not every time
a tributary of any size enters. Counting the
streams of different orders in any drainage network, one finds that the number of streams in a
particular order is from 3 to 5 times the number in the next higher order. Thus a drainage
network is a pyramid-like phenomenon, supported by many low order streams whose waters eventually funnel into the single highest
order stream at the basin outlet.

Within any drainage network the streams
of each order usually have a characteristic

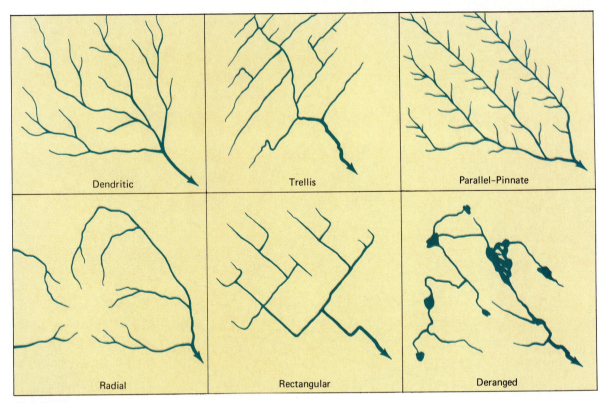

Figure 14.3
These are six of the most frequently encountered drainage patterns.

Dendritic patterns are found in areas that lack strong contrasts in bedrock resistance, such as flat-lying sedimentary rock or massive crystalline rock that is deeply weathered.

Trellis patterns develop where inclined layers of sedimentary rock of varying resistance to erosion are exposed at the surface. The parallel segments develop along the outcrops of the erodible layer of rock.

Parallel-pinnate drainage reflects a topography of long parallel ridges. The short segments drain the flanks of the ridges, and the long segments drain the troughs between.

Radial drainage indicates the presence of an isolated high mountain area and is common on individual volcanoes or dome-shaped mountainous uplifts.

Rectangular patterns reflect strong jointing of resistant bedrock, with streams incising along the joint planes.

Deranged drainage shows no geometrical pattern or constant direction and usually includes numbers of lakes; this pattern indicates destruction of prior drainage by the erosive effects of continental ice sheets. Such areas lack true valleys developed by fluvial erosion.

length, slope, and drainage basin area. The average values for these parameters, like the number of streams in each successive order, change in a very regular manner as the stream order increases, as indicated in Figure 14.4 b, c, and d. Studies of drainage networks suggest that the nature of their development is remarkably consistent from place to place. They seem to have evolved in a way that tends to maximize system efficiency with a minimum expenditure of energy both in the formation and operation of the system.

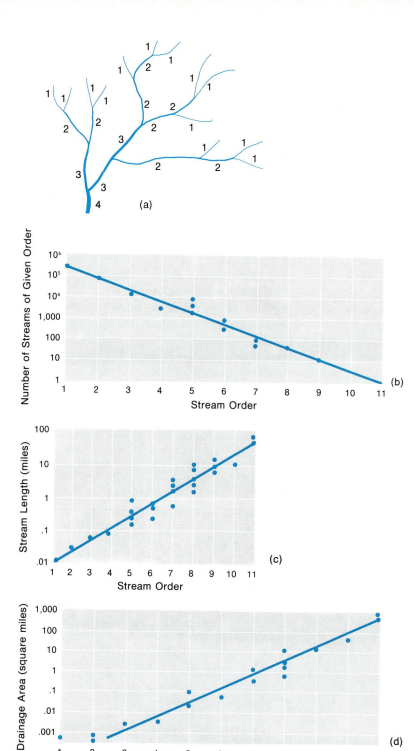

Figure 14.4
The concept of stream order and its relationship to drainage basin parameters.

(a) The system of stream ordering devised by the engineer Robert Horton and simplified by the geomorphologist Arthur Strahler. A second-order stream arises at the junction of two first-order streams; a third-order stream is produced by the junction of two second-order streams, and so on.

(b) (c) (d) These three graphs represent a Horton analysis of a drainage basin near Santa Fe, New Mexico. The graphs show the regularities that are typical features of drainage development, as revealed by a Horton analysis.
(b) The numbers of streams of a given order decrease regularly with increased stream order. (c) The average length of streams of a given order increases regularly with increased order. (d) The average area drained by streams of a given order increases regularly as stream order increases. (Doug Armstrong after *Fluvial Processes in Geomorphology* by Luna B. Leopold, M. Gordon Wolman, and John P. Miller. W. H. Freeman and Company. Copyright © 1964)

FLUVIAL AND AEOLIAN LANDFORMS

THE MECHANICS OF CHANNELED FLOW

To understand fluvial processes and the landforms they create, it is necessary to understand how water flows in channels. Take the example of water in a trough. The water does not flow unless one end of the trough is lifted higher than the other end. This permits the potential energy of the water at the high end of the trough to be converted to kinetic energy as the force of gravity draws the water to the low end of the trough. But not all potential energy is converted to energy of motion. Some energy is lost in overcoming friction between the water and the trough walls as well as friction within the flow itself. Similarly, water flowing in a natural channel encounters friction with the stream bed and banks. This friction has a significant retarding effect and causes the velocity of flow near the channel margins to be less rapid than in the center of the flow (Figure 14.5). Boaters heading upstream know that near the banks the downstream current is slower and easier to overcome.

Stream Energy

The way water flows in its channel determines the amount of energy it has for erosional and depositional work—meaning the deepening of valleys, the transport of solid and dissolved matter, and the deposition of sediments to create flat valley floors or floodplains. Stream discharge and turbulence, as we shall see, play important roles in providing this energy.

Stream Discharge

The volume of water a stream carries past a given point during a specific time interval is called the stream *discharge*. Stream discharge is measured in cubic meters (or cubic feet) per second, and is equal to the cross-sectional area of the flow times the flow velocity (distance traveled per unit of time). The average discharges of some well-known streams are given in Table 14.1. The United States Geological Survey maintains more than 6,000 stream gauging stations to measure stream discharges in the United States. The data are used to fore-

Figure 14.5
This diagram shows the average measured flow velocities, in feet per second, at various points in a cross section of Baldwin Creek, Wyoming. Note that the velocities tend to be lowest near the sides of the channel and highest near the center of the channel. The stream is therefore able to carry material suspended in its waters. Moderate velocities near the stream bed enable some material to be transported over the bed. (After *Fluvial Processes in Geomorphology* by Luna B. Leopold, M. Gordon Wolman, and John P. Miller. W. H. Freeman and Company. Copyright © 1964)

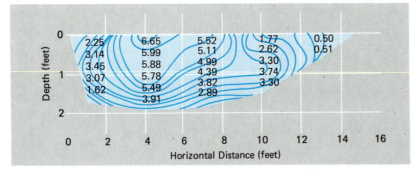

Table 14.1
Characteristics of Selected Rivers

RIVER AND LOCATION	AVERAGE DISCHARGE AT MOUTH		LENGTH, HEAD TO MOUTH		AREA OF DRAINAGE BASIN		AVERAGE ANNUAL SUSPENDED LOAD (MILLIONS OF METRIC TONS)	AVERAGE ANNUAL SUSPENDED LOAD PER SQ KM OF BASIN (METRIC TONS)
	M³/SEC	FT³/SEC	KM	MILES	KM²	MILES²		
Amazon (Brazil)	180,000	6,400,000	6,300	3,900	5,800	2,200	360	63
Congo (Zaire)	39,000	1,400,000	4,700	2,900	3,700	1,400		
Yangtze (China)	22,000	800,000	5,800	3,600	1,900	700	500	260
Mississippi (U.S.)	18,000	650,000	6,000	3,700	3,300	1,300	296	91
Yenisei (U.S.S.R.)	17,000	600,000	4,500	2,800	2,100	800		
Irrawaddy (Burma)	14,000	500,000	2,300	1,400	430	170	300	700
Brahmaputra (Bangladesh)	12,000	415,000	2,900	1,800	670	260	730	1,100
Ganges (India)	12,000	415,000	2,500	1,600	960	370	1,450	1,520
Mekong (Vietnam)	11,000	390,000	4,200	2,600	800	300	170	210
Nile (Egypt)	2,800	100,000	6,700	4,200	3,000	1,200	110	37
Missouri (U.S.)	2,000	70,000	4,100	2,500	1,370	530	220	160
Colorado (U.S.)	200	6,000	2,300	1,400	640	250	140	210
Ching (China)	600	2,000	320	200	57	22	410	7,200

Curtis, W. F., Culbertson, J. K., and Chase, E. B. 1973. "Fluvial-Sediment Discharge to the Oceans from the Coterminous United States," *U.S. Geological Survey Circular 670.* Washington, D.C.

Sources: Holeman, John N. 1968. "The Sediment Yield of Major Rivers of the World," *Water Resources Research,* 4 (August): 737–747. Fairbridge, Rhodes W. (ed.) 1968. *The Encyclopedia of Geomorphology.* Vol. III, Encyclopedia of Earth Sciences Series. New York: Reinhold. Espenshade, Edward B. (ed.) 1970. *Goode's World Atlas,* 13th ed. Chicago: Rand McNally.

cast floods, to assess irrigation water supplies, to plan sewage disposal, and for engineering purposes, such as the design of dams and bridges. Stream discharge at a given location is usually inferred from the height of the stream's waters, using a previously determined relationship known as a *rating curve* (Figure 14.6).

Stream Velocity

Stream energy is closely related to stream discharge, because discharge influences the flow velocity. Flow velocity in turn determines the stream's capacity to do work in the form of erosion and sediment transport. The velocity of any segment of a stream reflects the balance between the downslope gravitational stress, which is determined by stream bed slope, and the energy lost in overcoming friction at the boundaries of the flow. The larger the flow and the smoother the channel, the less the energy lost to friction, and the greater the stream velocity.

Streams normally increase in width, depth, and discharge in the downstream direction as tributaries and groundwater flow into them. Other things being equal, this would cause streams to increase in velocity and energy in the same direction. However, as streams grow larger, their downstream slope (or rate of descent) tends to decrease. This counterbalancing effect prevents a progressive build-up of energy in the downstream direction, and produces a more uniform distribution of stream energy. Even during floods, streams seldom have a flow velocity that exceeds 3 to 5 meters per second (7 to 11 miles per hour).

FLUVIAL AND AEOLIAN LANDFORMS

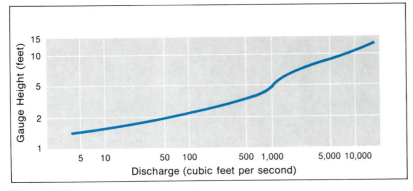

Figure 14.6

This *rating curve* for Seneca Creek, Maryland, allows one to infer the stream discharge from a simple measurement of stream height at a gauging station. The average discharge of this stream is 100 cu ft/sec, but discharges more than 50 times as great have been observed. (After *Fluvial Processes in Geomorphology* by Luna B. Leopold, M. Gordon Wolman, and John P. Miller. W. H. Freeman and Company. Copyright © 1964)

Flow Turbulence

At very low velocities water can flow as a smooth sheet, like the top card in a deck of cards that is pushed forward over the lower cards in the deck. Such flow is *laminar* (Figure 14.7). The kinetic energy of laminar flows is adequate to move only the finest particles. Therefore laminar flow is nonerosive.

If the flow velocity increases, friction within the flow and at its boundaries soon causes the flow to break into separate currents that are no longer parallel. This irregular *turbulent flow* is normal in streams. In turbulent flow, the speed and direction of motion vary continuously, with currents in every direction, including upward. Nevertheless, the average motion is in the downslope direction. The constantly moving eddies that make quiet streams so fascinating to watch are evidence of turbulent flow, as are the boiling rapids of steeply descending streams.

In turbulent flow, more than 95 percent of stream energy is consumed in overcoming the friction at the channel boundaries and between adjacent eddies and currents. Less than 5 percent of a stream's energy is available for the work of picking up rock debris and moving it through the channel. Even so, turbulent flows can be highly erosive. Rapid pressure variations near the stream bed and forceful upward-moving currents cause solid material to be lifted into turbulent flows and carried away by them. This removal of material, in both solid and dissolved form, is one of the most significant aspects of fluvial systems, as we shall now see.

Stream Load

Streams move their load of rock debris in three ways. Part of the transported material is dissolved in the water, forming the *dissolved load* (or *chemical load*) of the stream. Fine particles of clay and silt are carried in suspension within the flow of water as the *suspended load*. It is the suspended load that gives many streams their color and opaque appearance. Particles too heavy to be carried in suspension are bounced and rolled along the channel bottom as the stream's *bed load*. The bed load and suspended load together constitute a stream's *solid load*.

The effectiveness of streams as sediment carriers is one of the obvious facts of nature. China's Hwang Ho (Yellow River) earned its name

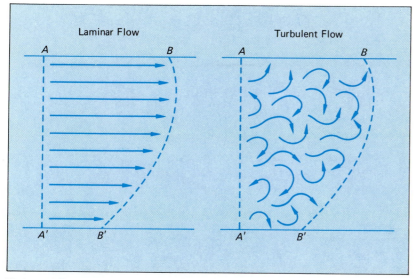

Figure 14.7
These cross sections contrast laminar and turbulent flow. Overall motion is forward from AA' to BB'. In laminar flow (left), the fluid moves like a deck of cards being deformed by internal shearing. The movement is slow and in a single direction. Turbulent flow (right) is characterized by instantaneous velocities in all directions.

because of the color of its sediment-laden waters, as did the Colorado (Red) River of the southwestern United States and the White River that drains the Badlands of South Dakota. The Mississippi River carries nearly 300 million metric tons of solid sediment to the sea each year, plus another 150 million tons of dissolved matter. To transport this much material by rail would require a daily train of 24,600 boxcars. Table 14.1 (p. 405) compares the sediment loads of a variety of large and small streams around the world.

Entrainment

A rapidly flowing stream may periodically detach particles from its bed in the process of stream bed *scour*. Specific amounts of energy are required to *entrain* (put into motion) materials of different sizes and shapes. The most easily scoured particles are those of sand size, with diameters between 0.05 and 1.0 mm. The flow velocity must be 15 to 50 cm/sec (0.3 to 1.0 miles per hour) to entrain such particles. Figure 14.8 shows that smaller clay or silt particles resist detachment more than sand. This is due to the stronger electrical bonds between the smaller particles. Particles larger than sand size, including gravel, cobbles, and boulders, also resist entrainment more than sand because of their greater weight.

As Figure 14.8 reveals, it takes more energy to put a particle in motion than it does to keep it in motion, particularly in the smaller size ranges. Although a 0.1 mm particle requires a current velocity of about 20 cm/sec to be displaced, it will subsequently continue to move at a velocity as low as 1 cm/sec. Nevertheless, a further decrease in stream velocity allows progressively finer particles to settle out of the flow and accumulate on the stream bed in the process of *fluvial deposition.*

Dissolved Load

The dissolved load of a stream is contributed largely by groundwater inflow into the stream.

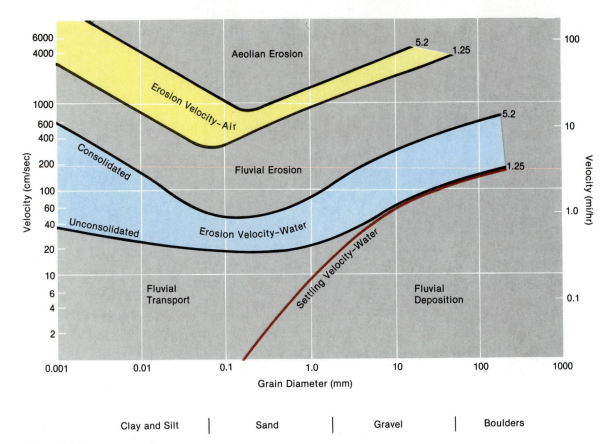

Clay and Silt | Sand | Gravel | Boulders

Figure 14.8

The ability of running water to entrain and transport particles depends on flow velocity and the size, density, and shape of the particles. This diagram relates flow velocity and particle characteristics to erosion, transportation, and deposition. The erosion velocity curve for water represents the minimum velocities at which particles having densities from 1.25 to 5.2 are entrained. The left (clay and silt) portion of the curve distinguishes between consolidated and unconsolidated material. The erosion velocity curve for air (wind) is included for comparison.

Note that once a particle is entrained at the erosion velocity, it continues to be transported at velocities well below the erosion velocity (area to the left of the settling velocity curve). Combinations of grain size, grain density, and fluid velocity to the right of the fluvial settling velocity curve (red) will cause sediment deposition on the stream bed. (Modified from A. Sundborg, 1956)

The amount of dissolved matter in groundwater varies according to the composition of the rocks and soils and the climate, weathering processes, and vegetation cover in the local area. Human pollution of streams has added greatly to their chemical load in many places, especially near industrial cities and where agriculture or mining are active. The average yearly rate of removal of dissolved matter over the entire United States is about 40 tons/sq km, or about 100 tons/sq mile. Although this is largely an invisible process, the weight of dissolved material removed from the U.S. as a whole is more than half the average annual rate of removal of solid matter (71 tons/sq km, or about 185 tons/sq mile). Thus chemical erosion of the land is a major (though largely unseen) aspect of gradation.

Solid Load

To transport its solid load, a stream must have forward and upward velocities greater than the constant downward pull of gravity on the particles carried in the flow. Material carried in suspension glides along irregular paths within the flow, while heavier material skips or rolls along the stream bed. Bouncing or skipping transport is known as *saltation,* and rolling or sliding transport is called *traction* (Figure 14.9).

As the particles of the bed load move along, they collide with one another and with the solid rock of the stream bed. Impacts against the channel floor break loose new fragments, slowly lowering the stream bed. Constant battering causes the bed load particles to become smaller and more rounded as they progress downstream—cobbles are reduced to pebbles,

and pebbles are fragmented into sand and silt. Sand, silt, and the clay washed directly into streams by overland flow are the only solid particles carried far downstream by large rivers.

Any decrease in flow velocity reduces the forces that support the suspended load and propel bed load movement. This causes some of the solid load to drop out of the flow. First to settle out are the heaviest of the particles moved by saltation and traction. These come to rest in a *channel fill* that may be either temporary or permanent. Deposition of the finest of the particles carried in suspension occurs only when the water is almost still, as in a lake or marsh.

The transporting power of a large stream is enormous. In 1933 a flash flood in California's Tehachapi Mountains caught a train crossing a trestle bridge. The locomotive and tender were

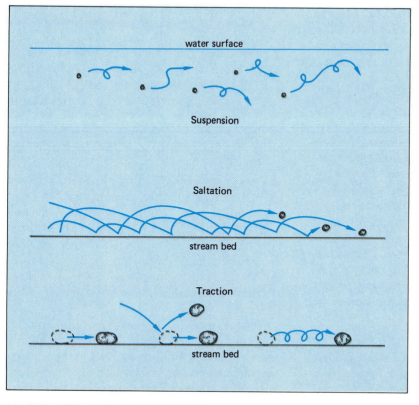

Figure 14.9
Solid material is transported in streams as suspended load and bed load. The finest particles are supported in suspension within the flow (top), while the bed load moves by both saltation and traction, in which particles are dragged or rolled by fluid shear stress, or are knocked forward by a saltating particle.

FLUVIAL AND AEOLIAN LANDFORMS

carried a kilometer downstream as part of the stream's bed load and were so thoroughly buried by gravel that a metal detector had to be used to find them.

Channel Equilibrium

A stream is a dynamic system with the ability to adjust the form of its channel in a matter of hours in response to changes in inputs of energy and material. By scouring and filling, a stream adjusts the slope of its bed and the shape of its channel so that stream energy remains in balance with the work of sediment transport.

Slope Adjustment

If a stream lacks the velocity to transport the sediment fed into it, some of the sediment is deposited in the channel. This elevates the stream bed. Any elevation of the stream bed increases the channel slope in the downstream direction, which increases the stream velocity at that point. Thus, by depositing sediment the stream increases its energy and ability to move future arrivals of sediment.

Conversely, where stream velocity exceeds that necessary for sediment transport, the stream can reduce its energy by flattening its slope. It can do this either by scouring its bed downward or by lengthening its horizontal distance per unit of vertical descent. The latter increases the stream's *sinuosity* (measured as the stream distance divided by the length of the stream valley).

In both cases negative feedback maintains equilibrium in the fluvial system. The initial disturbance triggers a reaction that either supplies or removes energy to bring the system into balance with the changed conditions.

Cross-Section Adjustment

During a flood, each section of a stream must transmit a discharge that is much larger than normal. As the discharge increases, the volume of water becomes too great for the existing channel size and shape. At the same time, the increase in flow volume is not balanced by an increase in friction, so the stream velocity increases. This enables the floodwaters to scour the bed and banks of the stream, enlarging the channel (Figure 14.10). However, the scouring process, which lowers the stream bed slightly, reduces the stream's potential energy and downstream slope, and so restrains the increase in flow velocity. If the flood should persist, and if the stream bed were easily eroded, the stream slope and channel form would eventually evolve to an equilibrium with the flood discharge.

After the flood discharge has peaked, the stream discharge begins to decrease. When this occurs the stream is once more out of equilibrium. Its channel is now larger than the discharge requires, and it has too little slope to transmit the water and sediment of the diminishing flow. As flow velocity declines, deposition of sediment begins. This elevates the stream bed, increases the stream slope, and reduces the size of the channel. These adjustments sustain the velocity of the diminishing flow so that subsequent arrivals of sediment can be carried through, thereby reestablishing stream equilibrium.

The sediment underlying the beds of most large streams is an important element in the equilibrium process. Figure 14.10 shows that at Lees Ferry about 7.5 ft of sediment is stored in the Colorado River channel at low flow. This sediment permits channel enlargement and slope adjustment during floods and the restoration of normal channel characteristics after floodwaters pass. Because flow conditions change constantly, most streams are in a state of perpetual readjustment, or "quasi-equilibrium."

Whereas stream channels formed in erodible materials can adjust to discharge changes quite rapidly, channels incised in hard rock cannot be adjusted easily. Consequently, bedrock channels may not achieve an equilibrium

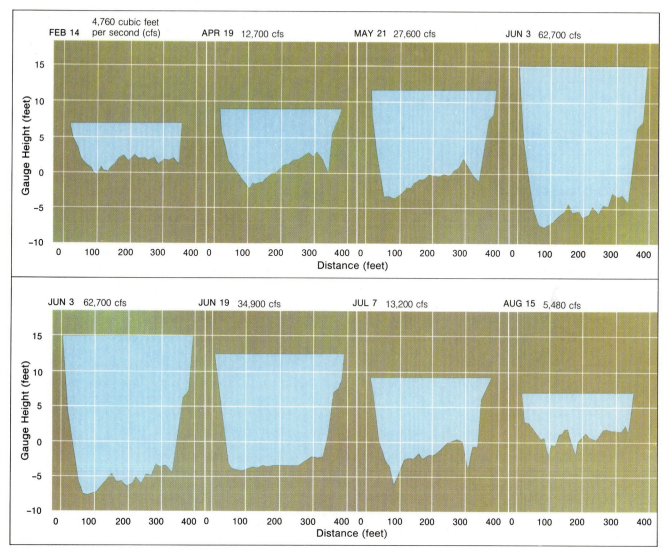

Figure 14.10
These channel cross sections for the Colorado River at Lees Ferry, Arizona, show the scouring and subsequent filling that occurred during the spring flood resulting from snow melt in the Rocky Mountains of Colorado in 1956. Discharge is in cubic feet per second (cfs). The marked enlargement of the channel by scouring was accompanied by an increase in flow velocity to accommodate the greatly increased discharge. Note that the vertical scale is magnified to 20 times the horizontal scale. (Doug Armstrong after *Fluvial Processes in Geomorphology* by Luna B. Leopold, M. Gordon Wolman, and John P. Miller. W. H. Freeman and Company. Copyright © 1964)

condition until they buffer themselves with *alluvium* (stream-deposited sediments).

Graded Channels

Any portion or "reach" of a stream that is in a quasi-equilibrium condition—neither progressively lowering its bed by scour nor raising it by sediment deposition—is said to be *graded*. A graded reach maintains a rate of descent and a channel form that give it just enough energy to transmit its fluctuating input of water and sediment. Since absolute equilibrium is impossible, a graded condition really implies that no long-term change is occurring despite repeated short-term adjustments.

Stream Gradient and the Longitudinal Profile

The rate at which a stream channel descends from higher to lower elevations is the *stream gradient*. This is measured as the vertical fall per unit of horizontal distance, expressed in like units (such as meters per meter), or as meters per kilometer or feet per mile. Gradient changes are a means of balancing stream energy to the work of water and sediment delivery. A graphic portrayal of a stream's gradient from its source to its mouth is called the stream's *longitudinal profile.*

The profiles of most streams that are hundreds to thousands of kilometers long are concave upward—steep near their sources and almost flat near their mouths, as in Figure 14.11. The initial steepness provides the en-ergy to move coarse bed loads in shallow channels. The flatter downstream gradients indicate increased discharges and larger channels, as well as a reduction in bed load particle size. Figure 14.11 also indicates that short channels commonly have nearly straight profiles. Point-to-point variations in stream profiles are a consequence of changes in discharge, sediment load, channel roughness (friction), and channel cross section (size and shape). An outstanding example is the Missouri River, which steepens its gradient from 0.8 to 1.2 feet per mile below the entry of the much smaller Platte River near Omaha, Nebraska. This large gradient increase is necessary to give the Missouri the energy required to transmit the enormous bed load of sand brought to it by the Platte. One must follow the Missouri 2,240 km (1,400 miles) upstream into its headwaters to find such a steep gradient again.

Profile Knickpoints

Stream profiles are sometimes interrupted by exceptionally steep reaches where rapids, cascades, or even waterfalls are present. Such

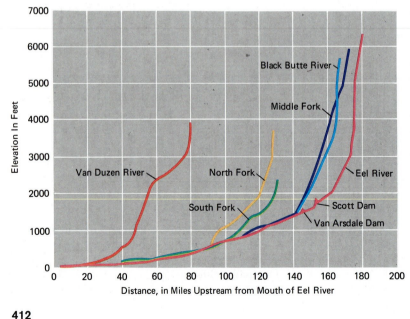

Figure 14.11
Longitudinal profiles of the Eel River and its major tributaries, which drain the Pacific Coast Ranges in northern California. Note the flattening of stream gradients in the downstream direction, resulting in concave-upward stream profiles. The longer the stream, the greater the tendency toward a concave-upward profile. (U. S. Geological Survey)

CHAPTER 14

profile interruptions are called *knickpoints*. Knickpoints in stream profiles have various origins. Some result from abrupt changes in bedrock resistance. Others may be related to disturbances of stream equilibrium by crustal movements that have caused bed scouring and channel incision to work headward (upstream) in the drainage system.

Knickpoints in stream profiles are especially common in areas that have experienced severe erosion by glaciers. The many waterfalls and cascades seen in high mountain country usually indicate that glacial erosion has deepened the major valleys, leaving smaller tributary valleys "hanging" above them, such as California's Yosemite Falls (Figure 16.16, page 498), and the many small waterfalls in New York's "Finger Lakes" region (see Chapter 16). Glacial erosion may also create steps in the main valleys themselves, each the site of a waterfall or cascade, as in the case of Nevada and Vernal falls in California's Yosemite Valley.

The historically important "Fall Line" of the Atlantic coastal region of the United States refers to rapids and cascades developed where streams pass out of the resistant crystalline rocks of the Piedmont region and into the weak sedimentary rocks of the Atlantic Coastal Plain (Figure 14.12). In colonial times the rapids in large streams at the Fall Line provided water power to turn the grindstones of flour mills. Later, the water power was used to generate electricity. The falls and rapids also necessitated the landing of ships' cargoes and the change to overland transportation. This stimulated the growth of such cities as Trenton, New Jersey; Philadelphia, Pennsylvania; Baltimore, Maryland; Richmond, Virginia; Raleigh, North Carolina; Columbia, South Carolina; and Augusta and Macon, Georgia. Their growth was assisted by the fact that the rivers were narrower and easier to cross in the hard rocks above the Fall Line.

Increased kinetic energy in the steeper section of a stream's profile gradually eliminates most knickpoints by channel scouring. However, some knickpoints wear back in an upstream direction a considerable distance before disappearing. The outstanding example is Niagara Falls (Figure 14.13), which has retreated 11 km (7 miles) since the Niagara River established its present course more than 100,000 years ago.

Knickpoints are very conspicuous in streams in seasonally wet tropical regions. The tropical forest and savanna regions of Africa and South America are noted for their spectacular cascades and waterfalls, which are present even well downstream on large rivers. This seems to be a consequence of the inability of tropical streams to cut canyons in resistant rocks. Chemical weathering in the humid tropics prevents sound rock from becoming part of stream bed loads, so that streams lack abrasive tools to wear away bed irregularities and deepen valleys.

Channel Patterns

When viewed from above, stream channels display three general patterns: meandering, braided, and straight or nonmeandering (Figure 14.14). The normal pattern is a *meandering* one, in which the stream swings back and forth in either smooth or sharp curves.

Meandering Channels

Meanders occur on streams of every size, from tiny rills on bare ground to kilometer-wide rivers. Regardless of stream size there is a relatively consistent relationship between the width of the stream and the *meander wavelength,* or distance between similar points on successive meanders (Figure 14.15). Meander wavelength is normally from 7 to 15 times the channel width.

Even streams that lack true meanders seem to show a succession of deeper *pools* and shallow-water *riffles* whose spacing is related to stream width (Figure 14.16). In fact, laboratory experiments with artificial channels in

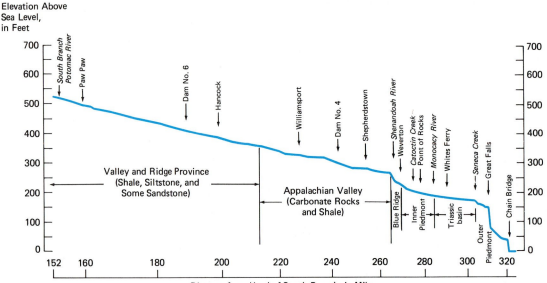

Elevation Above Sea Level, in Feet

South Branch Potomac River
Paw Paw
Dam No. 6
Hancock
Williamsport
Dam No. 4
Shepherdstown
Shenandoah River
Weverton
Catoctin Creek
Point of Rocks
Monocacy River
Whites Ferry
Seneca Creek
Great Falls
Chain Bridge

Valley and Ridge Province (Shale, Siltstone, and Some Sandstone)

Appalachian Valley (Carbonate Rocks and Shale)

Blue Ridge
Inner Piedmont
Triassic basin
Outer Piedmont

Distance from Head of South Branch, in Miles

Figure 14.12
The Great Falls of the Potomac River, just outside the city of Washington, D.C., are a striking example of the Fall Line marking the head of navigation on streams flowing from the Appalachian Highland to the Atlantic Ocean. The rivers cascade through ancient crystalline rocks of the Piedmont region, and remain ungraded probably due to gradually falling sea level during the past several million years. A second major knickpoint appears farther upstream on the Potomac River where it crosses the resistant granite rocks of the Blue Ridge Mountains. (Somerville/U. S. Geological Survey, and T. M. O.)

noncohesive sand or silt show that pool and riffle sequences in straight channels tend to evolve into sequences of meanders. When this occurs, each pool becomes the site of an out-ward-pushing channel bend, or meander. Thus the explanation of meander geometry lies in the nature of the flow that produces the initial pools and riffles.

CHAPTER 14

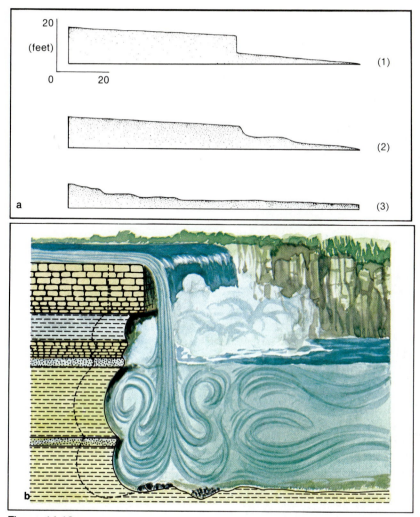

Figure 14.13

Knickpoint behavior in uniform material (a), and in layered rocks of varying resistance (b).

(a) Evolution of a knickpoint formed by an earthquake in Montana, showing the initial profile (1), the profile 3 months later (2), and the profile 3 years later (3). In uniform material, knickpoints do not retreat upstream; instead they are removed by channel scour. (M. Morisawa, *Streams: Their Dynamics and Morphology,* McGraw-Hill, 1968)

(b) Niagara Falls is formed by a dolomite caprock over weaker shales. The falls has retreated by erosion of the shale, which undermines the caprock. Retreat of the falls has left a slot-like gorge 11 km (7 miles) in length. Retreat in the last 10,000 years has been about 4 miles. (From G. K. Gilbert, in O. D. von Engeln, *Geomorphology,* Macmillan, 1942)

The stream *thalweg,* which is the line that follows the deepest part of the stream channel, swings back and forth across the channel from pool to pool, moving toward the outer bank of each successive curve. Maximum flow velocity occurs at the stream surface, above the line of the thalweg, where the stream is deepest and friction is least. Scouring is most vigorous at the outer (concave) banks where the flow is deepest and where stream velocity accelerates. Deposition of sediment is most evident at the

FLUVIAL AND AEOLIAN LANDFORMS

(a)

(b)

Figure 14.14
Meandering and braided streams.
  (a) The sinuous Animas River near Durango, Colorado, meanders over a broad floodplain produced by fluvial aggradation. (John S. Shelton)
  (b) This braided tributary of the Yukon River in northwestern Canada carries outwash from the margin of a receding glacier that is visible in the distance. (Larry W. Price)

inner (convex) banks, or "points," where flow depth and velocity are least.

The crescent-shaped deposits of sand or gravel formed on the successive points (Figures 14.16 and 14.17) are called *point bars*. Point bars commonly are composed of concentric ridges and depressions known as "bar and swale topography." The ridges are the bed load deposits accreted during the falling stages of floods, and are composed of coarse sand or gravel. Vegetation may take root in the finer material of the swales, giving the deposit a

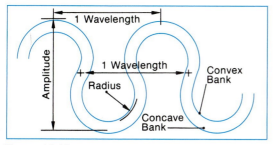

Figure 14.15
The diagram shows how the wavelength, amplitude, and radius of a meander are defined. The radius around a meander is not constant, however, because the form of a meander usually departs from a simple circular arc. The meander wavelength is 7 to 15 times the width of the channel. (After G. H. Dury, in *Water, Earth, and Man,* R. J. Chorley, ed., Methuen and Co., 1969)

"banded" appearance. The bars and swales give a clear picture of the movement of meanders through time.

The process of undercutting and collapse at the outer (concave) bank and point bar deposition directly across the channel causes the meander loops to shift outward and downstream in the process of *lateral planation.* Most scouring occurs during floods, when stream energy is at its peak. The greatest deposition of sediment occurs during the falling stages of floods, when stream banks collapse at the same time that stream energy declines. Near Needles, California, the Colorado River has moved laterally as much as 244 meters (800 feet) in a single year, and at Peru, Nebraska, the Missouri River has shifted position from 15 to 150 meters (50 to 500 feet) each year for more than 30 years.

Where stream banks are erodible, channels may migrate distances of several kilometers over a period of years. This creates significant problems: on one side of the stream property is gnawed away, while on the other side it expands; political boundaries move with the stream when it shifts slowly, but stay in place when it changes course suddenly. Mississippi's

boundaries with Arkansas and Louisiana show these effects clearly, as portions of each state have been cut off from the remainder of the state by movements of the Mississippi River.

Lateral planation increases the stream length for each unit loss in stream elevation. Therefore its effect is similar to that of vertical stream incision: both produce a flatter stream gradient, which reduces stream energy. An energetic stream that does not carry the tools (bed load) to cut vertically into bedrock can achieve equilibrium by moving laterally instead. That is why meandering channels occur where the bulk of the stream load is fine material that is carried in suspension.

Braided Channels

A stream load dominated by larger particles that move only by traction and saltation produces a different channel pattern—the *braided* type—which consists of many intertwining shallow channels, as seen in Figures 14.14 and 14.17. In shallow channels the maximum flow velocity is close to the stream bed, where it can thrust against the bed load. In comparison to meandering streams, a larger percentage of the energy of shallow streams is available to move bed load rather than being absorbed by friction within the flow itself. To develop the velocity needed to move their bed loads, such channels must compensate for their shallow depth by having a steep gradient. This is achieved initially by aggradation.

Braided streams form when bed load particles are deposited in lenticular (lens-shaped) bars in channels abundantly supplied with sand, gravel, and cobble-sized material, as in Figure 14.17. These bars divide a stream into many shallow channels and increase the downstream slope by elevating the stream bed. This provides the steeper gradient required to overcome the friction in shallow channels, permitting flow velocity to be high at the stream bed.

Braided channels develop where stream discharge fluctuates widely from day to day or

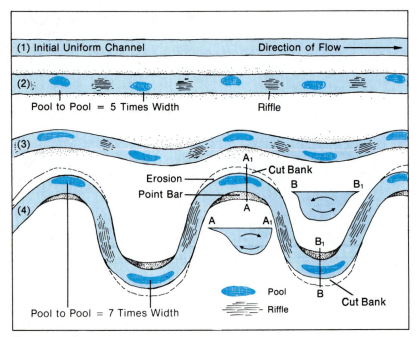

(1) Initial Uniform Channel   Direction of Flow ⟶

(2) Pool to Pool = 5 Times Width   Riffle

(3)

(4) Erosion — Point Bar
Cut Bank
A_1
B B_1
A
A A_1
B_1
B Cut Bank

Pool to Pool = 7 Times Width

Pool
Riffle

Figure 14.16
Streamflow in an erodible straight channel of uniform
cross section (1) tends to develop a sequence of
alternate deep pools and shallow bars, or *riffles* (2).
The straight channel becomes sinuous, with pools
forming toward the concave banks (3). In time, a
meandering channel may be developed (4). A
meandering stream tends to scour its outer concave
banks and to deposit sediment against its inner convex
banks where the flow velocity is least. There may also
be a lateral circulation of water, as the cross-sectional
diagrams $A-A_1$ and $B-B_1$ indicate. The lateral flow is
thought to move sediment from the cut bank to the
point bar. (Doug Armstrong after G. H. Dury from
Water, Earth, and Man, edited by R. J. Chorley, ©
1969, Methuen and Company)

season to season, as in arid and semiarid regions. Wide fluctuations lead to bank collapse,
channel widening, and sediment accumulation. The large sediment inputs required to initiate braiding occur where vegetation is
sparse, where stream banks are composed of
noncohesive sand or gravel, and where melting
glaciers feed great quantities of coarse waste to
streams.

Human activities that increase the movement of sediment into streams have caused
some meandering streams to become braided.
Such activities include certain agricultural,
mining, and logging practices, as well as urban
and suburban construction projects. By creating shallower channels, these changes generally increase the frequency and magnitude of
floods. This is an acute problem in many rainy
tropical areas, where the sediment input to
streams is enormously increased by removal of
the natural vegetation.

Straight Channels

The third channel pattern is the straight, or
nonmeandering, type. Straight stream channels are relatively uncommon and seldom extend far. Where present, they indicate that the
channel is controlled by the underlying geological structure. Streams incised in resistant
bedrock are occasionally channeled by fractures in the rock. Some meander patterns are
composed of straight segments that meet at

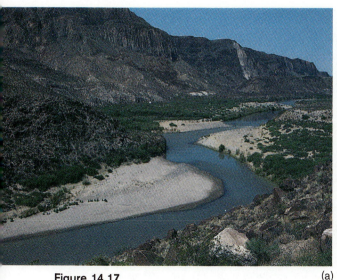

(a) (b)

Figure 14.17
Depositional features of meandering and braided channels.

(a) Point bars are well developed along this meandering reach of the Rio Grande, which forms the boundary between Texas and Mexico. These point bars consist of the sandy bed load of the Rio Grande, but elsewhere point bars may be composed of gravel or cobbles. All are built up by deposition during the falling stages of flood discharges.

(b) The White River, which issues from the Emmons Glacier on the flank of Mount Rainier in the Cascade Range, is an example of a braided stream. Here the bars separating the braids are composed of large cobbles. On other braided streams the channel bars may consist of gravel or sand. (T. M. O.)

sharp angles, indicating that joints control the channel. The channels of many river deltas are straight because cohesive clays have prevented lateral shifting of the stream beds (Figure 14.22, p. 427).

FLUVIAL LANDFORMS

Valley Development

When a stream has more kinetic energy than it requires to transport its sediment load, the stream reacts to bring its energy into balance with the work it is required to perform. For example, streams flowing down a surface recently created by motions of the earth's crust, or by the construction of a volcano, commonly have excess energy as a consequence of their steep initial gradients. Streams in such settings will cut downward, flattening their gradients until their energy is in balance with their work of sediment transport. Streams that have achieved quasi-equilibrium will begin scouring when environmental changes increase their discharge, giving them extra energy.

Stream Rejuvenation

Renewed incision to flatten stream gradients and reduce potential energy following a period of stability has been called stream "rejuvenation." The causes of stream rejuvenation are deformation of the land surface, drops in sea level, increases in stream discharge, and decreases in sediment load.

Tectonic uplift of the land surface, which increases the potential energy of streams, is probably the chief natural cause of stream rejuvenation and valley development. This reju-

Upstream and Downstream Effects of Large Dams

Dams constructed to store stream flows in artificial reservoirs are one of humanity's oldest devices for exerting control over an environmental variable—the supply of water. Most existing large dams have several purposes. They may provide flood control, be used to generate hydroelectric power, store water for agriculture and urban use, and regulate flow for river navigation. But the construction of a dam and reservoir disturbs the fluvial system, which reacts in ways that sometimes create problems.

The solid load carried into reservoirs, especially during floods, ensures that the life of any artificial reservoir will be short—no more than a few hundred years of constantly diminishing capacity. Below the reservoir, clear water reenters a channel having a slope and cross-sectional form adapted to sediment-laden natural flows. Clear water discharging through this channel has excess energy, being "hungry" for the sediment lost in the reservoir. Clear flows tend to scour their channels downward—as much as 7 meters (23 ft) in 9 years below the Glen Canyon Dam on the Colorado River, and 5 meters (16.5 ft) below Hoover and Davis dams on the same stream. Where scour is severe below a dam on a typical meandering stream, the resulting "slug" of new sediment may clog the channel farther downstream. Such clogging may also reduce the water capacity of the stream and cause overbank flooding and local change to a braided condition.

By removing peak flows, dams may also cause channels to change their patterns in the opposite manner. On the Great Plains of the central United States, many streams were naturally braided, carrying large loads of sand during their spring floods. Where such streams have been dammed, reducing their floods, their former wide beds have been invaded by vegetation, leaving single channels with meandering patterns.

venation causes valleys to deepen as stream profiles are regraded to restore equilibrium (Figure 14.18). Uplift that is continuous or more rapid than stream incision produces hilly or mountainous landscapes resulting from steady stream downcutting. Uplift in separate pulses results in stream terraces (discussed later in this chapter) and multistory valleys, in which newer valleys are cut into the floors of successively older and higher valleys.

Lowering of sea level, or uplift of the land relative to the sea, causes valley deepening near coasts. The effects of world-wide sea level changes are discussed in Chapters 16 and 17.

Increases in stream discharges have been an important cause of stream rejuvenation at various times. Increased discharges can be produced by regional changes in climate and vegetation or by local "stream piracy," in which an expanding drainage network invades and "captures" the headwaters of a neighboring stream system. The lateral planation process can also cause stream captures, as a stream meander occasionally pushes laterally into the valley of an adjacent smaller stream that flows at a slightly higher elevation. This diverts the flow of the higher channel into the lower one.

Flood discharges, during which valley deepening occurs, have been increased by such human activities as forest removal and urbanization, which covers vast areas with impermeable concrete and asphalt. A final cause of stream rejuvenation is decreased sediment load—also frequently the result of human activity, specifically dam construction, which traps both the bed load and suspended load of

If large flood discharges are not allowed to enter the channel below a dam, sediments delivered by smaller tributary channels may create obstacles in the main channel. The rapids on the Colorado River in the Grand Canyon become more treacherous each year as boulders accumulate at the mouths of tributary canyons. Before the construction of the Glen Canyon Dam many of these boulders were swept away by the annual flood on the Colorado River.

The deposits of fine sediment left by natural floods sustain the fertility of floodplain soils. The 5,000-year history of agriculture in the Nile Valley and delta of Egypt depended on the annual Nile River flood that left a veneer of new silt over the valley floor each year. Modern dams on the Nile—particularly the Aswan High Dam, which can store the entire annual flood—have destroyed the natural system of fertilization, necessitating huge imports of artificial fertilizers.

The sediment accumulations in reservoirs that receive runoff from mining areas, ore treatment facilities, or certain types of rock in arid regions may contain heavy concentrations of dangerous substances such as mercury or radioactive materials. Bottom-feeding organisms in reservoirs ingest these toxic substances, which become increasingly concentrated as they move through the food chain, eventually posing a real hazard to humans who eat fish from the reservoir. This is already a problem at Lake Powell, above the Glen Canyon Dam.

In more than one instance, the rise of water in a new reservoir has elevated the water table in adjacent fractured rocks, causing large-scale rockslides into the reservoir. The giant waves thrown up can overtop the dam itself, creating devastating floods downstream. A disaster of this type killed 2,600 people in Italy in 1963. Elsewhere, the weight of water produced by reservoir filling has caused swarms of small earthquakes that have damaged buildings and caused fissures to open in the ground.

Ultimately, even the largest reservoirs will become filled with sediment, their dams serving only as artificial waterfalls. Either the reservoir fill will have to be pumped out or the dam will have to be abandoned and its reservoir converted to cropland (possibly dependent on another reservoir for irrigation water). If the dam is not first destroyed, either artificially or by an earthquake, it will eventually be undercut and eroded away by natural fluvial processes.

streams in artificial reservoirs. Immediately downstream from large dams, streams have excess energy since they have no sediment to transport. Such streams scour downward when large volumes of water are released from the reservoirs. This is occurring now along the Colorado River below the Glen Canyon Dam (completed in 1964) and along the Egyptian Nile below the Aswan High Dam (completed in 1971). Within five years of the completion of Hoover Dam on the Colorado River in 1935, the river below the dam had incised as much as 5 meters (17 ft).

Floodplains

When streams attain a condition of equilibrium and stop deepening their valleys, the valley floors gradually become wider. Valley widening is caused both by lateral stream planation and by normal processes of erosion on valley walls. As meander planation trims back valley walls and sediment is deposited on point bars, a continuous flat-floored trough evolves. The stream swings back and forth across this open lowland (Figure 14.19). Since the stream channel develops in such a way that it can contain most of the flows through it, but not the rare large flows that occur at intervals of a year or two, this lowland is occasionally flooded and is known as a *floodplain*.

Not all floodplains are produced by stream erosion. Many of the earth's largest rivers flow across vast lowlands created by crustal movements. Examples are the Amazon River of Brazil, the Ganges River of India, the Sacramento

Figure 14.18
This image taken from an altitude of about 200 km (125 miles) shows a portion of the Colorado River's great canyon through the Colorado Plateau in northern Arizona (upper right). The river exits from its canyon at the Grand Wash Cliffs (see arrow), an escarpment produced by faulting. The area to the west has subsided, causing vertical incision of the river to a depth exceeding 1,500 meters (5,000 ft). Healthy vegetation produces the red colors on computer-generated LANDSAT images such as this. Lake Mead appears in the left (west) part of the image. (Department of the Interior, U. S. Geological Survey)

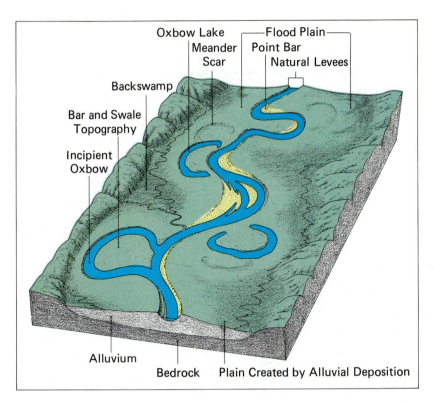

Oxbow Lake
Meander
Scar
Backswamp
Bar and Swale
Topography
Incipient
Oxbow
Flood Plain
Point Bar
Natural Levees
Alluvium
Bedrock   Plain Created by Alluvial Deposition

Figure 14.19
Floodplains may result from either fluvial aggradation or lateral stream planation that wears back valley walls. Oxbow lakes are the remnants of recently abandoned meanders. When the lake is eventually filled with sediment, a *meander scar* remains. In flood, the overflowing river deposits new sediment on its floodplain and builds *natural levees* by deposition of silt close to the river channel. (T. M. O.)

and San Joaquin rivers of California, and the lower Mississippi River. These streams carry sediment into the structural basins that they cross, which are not true valleys. Many other floodplains are products of stream aggradation that has filled deep V-shaped valleys with sediment. Such floodplains are especially common in coastal regions.

Floodplains are hazardous environments, but they have great economic value. Individual floods may leave a floodplain covered by a depth of several centimeters of fresh silt. These overbank flood deposits cover the older sand and gravel point bar deposits, and create nearly level agricultural land. This land is usually very fertile since its nutrients are renewed periodically by fresh deposition. However, where human disturbance has accelerated erosion, floods may deposit coarse sand over previously fertile areas, reducing their utility.

River floodplains can contain a variety of fluvial landforms. When overbank flooding occurs, the largest amount of deposition is just outside the stream channel where the flow velocity slows due to reduced depth and increased friction. The thickness of the overbank flood deposit diminishes with distance from the channel. This wedge-like deposition produces paired *natural levees* that slope gently away from the stream on both sides (Figure 14.19). Often there are low-lying *backswamps* between the natural levees and the valley walls that rise above the floodplain. In Louisiana, the Mississippi River's natural levees rise 5 to 6 meters (15 to 20 ft) above the backswamps. In the disastrous 1973 flood, the Mississippi River levees received as much as a meter (over 3 ft) of new sediment.

Natural levees are prime locations for agricultural settlement due to their fertile soils and good drainage characteristics. On large rivers they have become industrial sites, accessible by ocean-going ships. To protect developments on such sites, artificial levees are commonly built atop natural levees. When confined in this way, river floods of a given discharge rise much higher than under natural conditions when they could spread laterally. If such high flows break through the artificial levees, they can be even more destructive than natural flooding.

Where streams have considerable excess energy and the floodplain is composed of easily eroded sand rather than silt, meander shifting may be very active. In such places the meanders often expand into flaring "gooseneck" loops that press close against one another. Flood flows sometimes break through the narrow necks of land that separate adjacent loops, causing the meander to become cut off from the stream, forming an *oxbow lake* (Figure 14.19). Hundreds of present and past oxbow lakes are visible along the Mississippi River from Cairo, Illinois, to Baton Rouge, Louisiana. Many of these were created at the same time during major floods. Large oxbow lakes offer opportunities for recreation and water sports, and are often lined with cottages. Sometimes their creation produces political problems, as along the Rio Grande, which separates Texas from Mexico. Oxbow lakes eventually become marshes, and finally fill with silt and clay, but their traces remain clear long afterward.

Fluvial Terraces

Fluvial landforms often show that graded streams periodically adjust their profiles by either raising or lowering the stream bed. Where a stream cuts downward into a broad valley floor, the floodplain no longer acts as an overflow channel. Instead, it is left standing well above the level of the highest floodwaters to form a *fluvial terrace* (Figure 14.20).

A temporary phase of aggradation can also produce fluvial terraces. In such a case decreased stream energy or increased sediment input causes deposition that raises the stream bed, forming an *alluvial fill.* Resumption of stream incision leaves a terrace composed of the alluvial fill. Fill terraces are common where streams were temporarily aggraded by sediment washed from glaciers during the ice ages, and also where tectonic motions and major climatic fluctuations have occurred.

River terraces are important indicators of environmental change. They reveal that tectonic, climatic, or hydrologic alterations have forced streams to change their behavior to regain equilibrium. Some rivers are bordered by great flights of terraces, indicating repeated disturbances of equilibrium. The rivers of New Zealand are famed for their magnificent terraces, which reveal a history of strong tectonic activity and complex environmental changes.

Alluvial Fans

Where a stream carrying a large bed load issues from a deeply cut valley onto an open plain, it loses much of its kinetic energy. This occurs because of the spreading of waters that were previously confined to a narrow channel. Where the flow spreads, the bed load is dropped, producing a fan-like deposit issuing from the valley and spreading over the plain. This deposit, shown in Figure 14.21, is an *alluvial fan.*

Alluvial fans are most common in mountainous deserts, but can form wherever sediment loads are heavy. In deserts the lack of vegetative protection permits the infrequent heavy rains to flush large quantities of rock debris from slopes. Desert stream floods, which may last only a few hours, are highly charged with sediment. Where a stream flows out of its canyon, the coarsest part of the bed load is dropped close to the canyon mouth. In this way the apex of the alluvial fan is raised much higher than the fan's margins. Many desert alluvial fans are composed largely of the deposits of mudflows and debris flows. Since the density of these flows is many times that of water, they

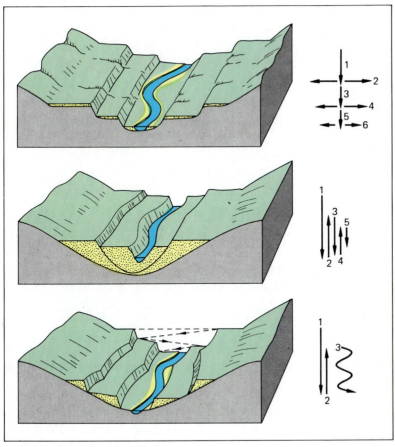

Figure 14.20
Stream terraces vary significantly in origin. To the right of each diagram is an indication of the nature of the river movements necessary to create the associated terraces: downward arrows indicate stream incision; upward arrows, aggradation; horizontal arrows, lateral planation; sinuous arrows, simultaneous incision and lateral planation.

(top) Terraces produced by periods of downward stream incision (1, 3, 5) separated by intervals of valley widening by lateral planation and slope erosion (2, 4, 6). These terraces are erosional and are said to be *paired* since those on opposite sides of the stream match in elevation.

(center) Paired fill terraces produced by two separate phases of alluvial aggradation (2, 4) followed by renewed valley excavation (3, 5).

(bottom) Valley deepening (1) is followed by aggradation (2) producing an alluvial fill. Subsequently the stream cuts downward at the same time that it is planing laterally in the fill (3), producing *unpaired* terraces. (T. M. O.)

can transport large boulders, which are often seen scattered over fan surfaces.

In mountainous deserts, neighboring fans merge to form continuous ramps of sand and gravel known as *alluvial aprons,* or *bajadas.* In California's Death Valley (Figure 14.21) these debris aprons rise as much as 600 meters (2,000 ft) over a horizontal distance of 8 km (5 miles).

There is usually abundant groundwater present at the bases of alluvial fans. In arid regions, this makes them favored locations for settlement and agricultural development. An outstanding example is Salt Lake City, which sprawls over the surface of a large fan adjacent to Utah's Wasatch Mountains. The Mormon

Figure 14.21
Single and coalescing alluvial fans.

(top) This alluvial fan is located on the east side of Death Valley, California; its size can be judged from the road crossing it. Rains are infrequent in this region, but torrential storms occur in the mountains from time to time, carrying down large quantities of coarse sediment that is deposited close to canyon mouths. White areas are salt deposits.

(bottom) Alluvial fans commonly join along the base of a desert mountain range to form an *alluvial apron,* or *bajada,* such as this example seen fringing the Panamint Mountains on the west side of California's Death Valley. (John S. Shelton)

settlers who founded Salt Lake City also established many other settlements in similar settings, including Las Vegas in Nevada and San Bernardino in California.

Deltas

When a stream flows into a lake or the sea, its velocity is checked and it loses its load-transporting ability. Its solid load is dropped at the river mouth in the form of a *delta*. The term was first used some 2,500 years ago by the Greek historian Herodotus, who noted the similarity of the depositional form to the Greek letter Δ (delta). However, delta configurations are extremely variable. The Mississippi River delta, with its bird's-foot form (Figure 14.22), is quite unlike the classic examples of the Nile and Niger river deltas in Egypt and Nigeria. Variations in delta form result from differences in the rate of sediment supply, the vigor of wave action and coastal currents, and the rate at which the alluvial deposit subsides as a result of compaction or the sinking of the sea floor under it.

When streams enter deltas, their discharge usually becomes divided among several *distributary* channels. These are created by floodwaters that spill out across low natural levees to produce new channels. The unusual form of the Mississippi delta is primarily the result of slow sinking of the sea floor, which leaves only the crests of the stream's natural levees projecting above sea level. The present bird's-foot

Figure 14.22
Varying configurations of deltas.

(a) The delta of the Nile River in Egypt, occupied by 30 million people, has a different shape from the delta of the Mississippi because the delta is not subsiding. The construction of the Aswan Dam in the 1960s has greatly reduced the supply of fluvial sediment and the delta is in danger of accelerated erosion. (NASA)

(b) The delta of the Mississippi River in Louisiana resembles a bird's foot because the delta is subsiding, leaving only the crests of its natural levees above sea level. The subsidence is caused by compaction of sediments and also by general tectonic sinking of the floor of the Gulf of Mexico. The bird's foot form was more pronounced before Garden Island Bay was partly filled by a new deposit. (Doug Armstrong after S. M. Gagliano et al., ''Hydrologic and Geologic Studies of Coastal Louisiana,'' Department of the Army, 1970)

(a)

(b)

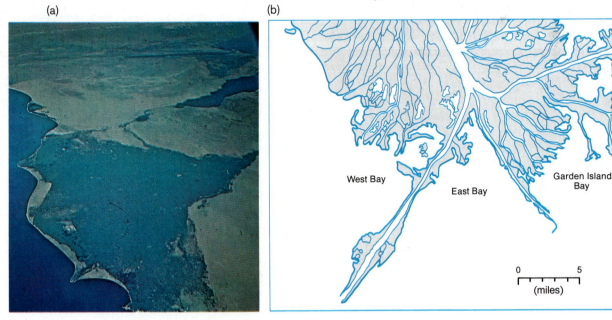

West Bay

East Bay

Garden Island Bay

0 5

(miles)

configuration developed after a change of the river's course that occurred near New Orleans only about 500 years ago.

Like floodplains, deltas are fertile and often densely populated. The Nile delta is the home of some 30 million people—two-thirds of Egypt's population. Loss of life and property can be heavy in deltas during floods. Some of the worst disasters occur when hurricanes cause flooding along low-lying deltaic coasts. This has been extremely costly to the dense populations occupying the combined deltas of the Ganges and Brahmaputra rivers of India and Bangladesh. It also is a hazard to the inhabitants of the Mississippi delta and the Gulf Coast from Florida to Texas.

AEOLIAN PROCESSES AND LANDFORMS

The greater part of this chapter has been devoted to fluvial processes and landforms because they are essential aspects of the environments in which most people live. Now we shall look briefly at the role of wind in erosion and deposition. Although the effects of wind are distinctive and significant wherever they occur, their most spectacular effects are found in certain sparsely inhabited regions of the world. But the wind can modify the land surface wherever vegetation cover is weak or absent. Human activities, by destroying the natural vegetation, can cause wind to become an effective geomorphic agent in regions where its influence would be insignificant under natural conditions.

Wind as a Geomorphic Agent

Processes related to wind action are known as *aeolian processes,* after the Greek god of the winds *Aiolis* (Latin, *Aeolus*). Aeolian erosion and deposition are natural occurrences wherever loose, unprotected sediments are present,

as along shorelines and in areas of recent glacial or fluvial deposition. The largest-scale aeolian effects are seen in the world's desert regions, where lack of moisture results in vast expanses of bare dry soils.

Erosion and Transport by Wind

A bare surface over which the wind sweeps is similar to a vast stream bed. The velocity of wind, which greatly exceeds that of flowing water, provides the energy that puts loose material in motion. Fine dust can be whirled into the air by a slight breeze. Dry sand grains 0.1 mm in diameter begin to move when the wind velocity a meter above the ground reaches 16 km (10 miles) per hour. Coarse sand (2.0 mm) is put into motion by a 50 km (30 miles) per hour wind. But, although wind can attain velocities many times those of flowing water, its density relative to the particles it affects is more than 1,000 times less, so it exerts vastly less buoyant force. Thus the wind cannot move rock particles larger than sand, and in fact rarely lifts sand more than a meter above the ground. Figure 14.8 (p. 408) compares the erosive potentials of running water and wind.

Particles put into motion by the wind behave like those moved by running water. The smallest particles are carried in suspension as windblown dust. Heavier particles move mainly by saltation (which is much more important in low-density air than in water), by traction (which is relatively less important in air), and by the impacts of saltating grains, as in the center of the traction diagram in Figure 14.9. The oblique impacts of saltating grains produces *sand creep,* in which the entire top layer of sand can be observed crawling slowly in the downwind direction.

Aeolian processes mainly affect loose particles; it is uncommon for wind action to wear away solid rock, even in deserts. Wind erosion of rock essentially means sandblasting. This requires an exceptionally strong wind to blow

(a)

(b)

Figure 14.23
Features produced by the abrasion of blowing sand and dust.

(a) This oddly shaped, 3-meter- (10-feet-) high sandstone outcrop in Egypt's Western Desert has been eroded by natural aeolian sandblast. The undercutting at the base indicates that the strongest wind blows sand from the right. (T. M. O.)

(b) This rock, about 20 cm (8 inches) in longest dimension, shows the characteristic flat faces or *facets* of a ventifact carved by blowing sand and dust. The smaller rills are a result of dust blast. (T. M. O.)

abrasive quartz particles against soft rock. Wind erosion much more often consists of the removal, or *deflation,* of fine particles provided by other gradational processes.

Aeolian deflation usually leaves no spectacular landforms. Since deflation slowly lowers large areas, there may be little left by which to gauge its magnitude. Often the principal evidence of aeolian deflation is a tree or bush left standing on a pedestal of soil, all the surrounding soil having been blown away. In a few extremely arid locations, strong winds have actually eroded parallel grooves in the silts of old lake beds, creating distinctive streamlined forms called *yardangs*. Similar features develop on a much smaller scale on more resistant rocks, where *ventifacts* ("wind-fashioned" forms) are created by natural sand and dust blasting, as in Figure 14.23.

Loess

Aeolian deflation is visible chiefly in dust storms (Figure 14.24). These are usually produced by winds related to strong pressure gradients in arid and semiarid regions. The dust,

(a)

(b)

Figure 14.24
Wind erosion and deposition of fine material.

(a) In the midst of a severe sand or dust storm, semidarkness may prevail at midday, as in this instance on the High Plains of Colorado. During the severe dust storms of the 1930s, some people died of suffocation as their breathing passages became clogged with fine particles that prevented the entry of oxygen. (Lowell Georgia/Science Source/Photo Researchers)

(b) A 6-meter (20-ft) bank of *loess* in Louisiana. The loess is typical in that its cohesiveness causes it to stand in near-vertical faces and to be gullied where vegetation protection is absent. The depth of the loess in the Mississippi Valley region varies from centimeters to as much as 15 meters (50 ft). Most Mississippi Valley loess was derived from aeolian deflation of vegetation-free glacial deposits during the late Pleistocene. (T. M. O.)

composed of silt-sized particles, may be lifted thousands of meters into the atmosphere. It settles out far from its source, to form distinctive deposits known as *loess*.

Loess, which can be centimeters to hundreds of meters thick, offers excellent agricultural opportunities. It is usually rich in calcium, and soils developed on it are young and unleached. Loess is common in the American Midwest and Mississippi Valley, and in a strip extending across Europe from northern France to the Russian Ukraine. Loess also has accumulated along the northern edge of the highlands of Central and Eastern Asia. The North American and Eurasian loess deposits originated from deflation of Pleistocene glacial deposits. Northern China has extremely thick loess deposits that reflect deflation in the Gobi Desert of Mongolia and northwest China. In northern China, loess produces a spectacular landscape that is deeply trenched by gully erosion. Since the time of *Homo erectus,* thousands of generations of humans have lived in caves carved into the soft but cohesive loess of China.

Dune Landscapes

The most impressive results of aeolian energy are the great sand seas of desert regions.

It is a mistake, however, to think of deserts in general as areas of sand dunes. Most desert terrain consists of gravelly plains, eroded rock surfaces, and stream-dissected relief. Nevertheless, deserts are distinctive in having vast expanses of dune topography as well (Figure 14.25).

The dunes most visited by large numbers of people are those associated with sea or lake beaches in climates of all types. The barrier island system fringing the majority of the Atlantic and Gulf coasts of the United States (Chapter 17) is largely a dune landscape, bordered by a beach on the ocean side and by salt marshes facing the mainland. The beaches of the Pacific Coast are likewise backed by dunes in many places. The coasts of Denmark, the Netherlands, and southwestern France are especially notable for their dune landscapes. In North America the best-known lakeshore dunes are those of Lake Michigan. Although many shore dunes carry a cover of grassy or herbaceous vegetation, the sand nevertheless flies when strong winds prevail—even vigorous sea breezes (Chapter 6).

The most recent direction of sand flow in a dune area can be determined by looking at the smallest sand accumulations—the closely spaced *sand ripples* that crawl over the dunes. The migration of sand ripples brings sand up the dune *backslope* (Figure 14.26) to the dune crest. At the crest the sand falls onto the steep *slip face* of the dune, which always has a slope angle of about 34°. Sand ripples move in different directions from day to day, often approaching the dune crest obliquely. The ripples are asymmetric, with their steep faces on the downwind side in the direction of ripple travel. Stream beds likewise commonly display sand ripples and even dunes, showing that wind and water move particles in the same way.

Dune Types

The most general distinction made among dunes is between the transverse and longitudinal types.

Transverse dunes form where sand-moving winds blow mainly from one direction. They

Figure 14.25
The sand dunes in this view are in California's Death Valley. The advance of wind-driven sand ripples, visible at the left, contributes to dune migration. These ripples change in pattern and direction of travel from day to day. (T. M. O.)

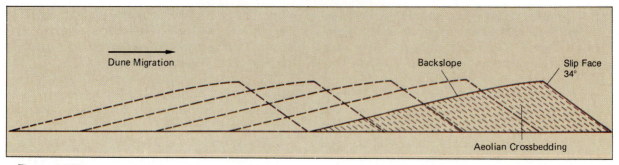

Figure 14.26
This diagram shows the migration of a dune as sand is transferred from the backslope to the slip face, causing the dune to be displaced in the direction of the slip face. The internal structure of dunes consists of steeply inclined laminations of sand produced by accretion on the slip faces. The result is *aeolian crossbedding,* which can be seen in many ancient nonmarine sandstones.

are asymmetric, with steep slip faces on the downwind side and gentler backslopes facing into the wind. The highest dune crests are at right angles to the direction of sand-carrying winds (Figure 14.27a, b, d). Over a period of years small transverse dunes may migrate hundreds of meters in the downwind direction, as in Figure 14.28.

Transverse dunes are found in both wet and dry environments and may be either simple individual features or complex forms composed of linked elements. Two simple types are especially common: one in arid regions, the other associated with beaches in humid climates.

Individual dunes with elongated tips, or *horns,* pointing downwind are known as *crescentic* or *barchan* dunes (Figures 14.27 and 14.28). This type of dune is found in dry, vegetation-free environments, and is usually seen migrating across a nonsandy surface of gravel or clay. Barchan dunes may move several tens of meters each year. Symmetrical barchans indicate a nearly constant wind direction.

Single dunes whose horns point upwind are known as *parabolic* dunes (Figure 14.27).

Often they assume a hairpinlike form. Such dunes form where sand blows into a moist area that has a vegetative cover. Vegetation or dampness in the lower part of the dune retards motion there, so the dry crest pushes ahead of the base, causing the horns to "drag" behind. Parabolic dunes are found along coastlines and often push into forests, engulfing and killing the forest trees. Similar dunes also occur where older, vegetation-covered sand accumulations become active again due to natural or artificial destruction of the vegetation.

Longitudinal dunes form where the sand-carrying winds come from more than one direction. The crests of individual longitudinal dunes may extend hundreds of kilometers and are believed to represent the sum of the vectors of the various sand-carrying winds, as shown in Figure 14.27. Many have somewhat complex topography, sometimes including branching crests. Slip faces may not be well developed, or may alternate in location with changes in wind direction. In some areas longitudinal dunes are smooth *whalebacks;* elsewhere they are knife-edged. Dunes of this general type do not migrate, but grow in length over time. They require an abundant supply of sand and are best developed in the larger deserts, such as the Sahara, Arabia, and interior Australia.

Ergs

In the deserts of North Africa and Arabia, individual sand seas called *ergs,* with dunes

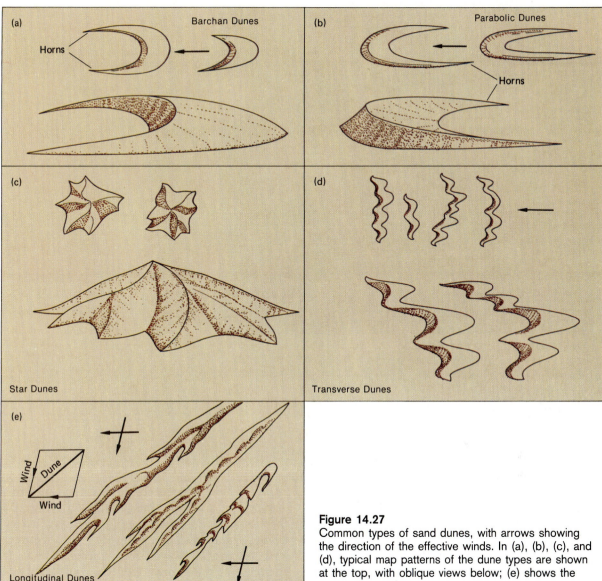

Figure 14.27
Common types of sand dunes, with arrows showing the direction of the effective winds. In (a), (b), (c), and (d), typical map patterns of the dune types are shown at the top, with oblique views below; (e) shows the map pattern only.

resembling frozen waves, cover tens of thousands of square kilometers. Somewhat less spectacular dune landscapes also cover vast areas in India, western China, and Australia, and small ergs are present in every desert region. An ancient sand sea, comparable to those of North Africa but now grass covered, occupies much of western Nebraska.

Ergs form where winds lose velocity after carrying sand hundreds of kilometers. Thus, ergs are related to large-scale wind patterns that produce aeolian sand flows involving vast areas. Once an aeolian sand deposit is initiated, it is self-enhancing. It traps arriving sand, which does not saltate as well over sand as it does over rock, pebbles, or hard-packed

Figure 14.28
Barchan dunes are crescent shaped, with the horns pointing downwind.
They form in areas where the supply of movable sediment is limited and
vegetation is sparse. Barchans migrate downwind but retain their arcuate
shape because the points of the dune migrate faster than the main body.

soil. Ergs contain dunes of many types, reflecting local variations in the sand supply and the nature of the sand-transporting winds, which may change direction seasonally. Large erg areas sometimes include regularly spaced "sand mountains" as much as 200 meters (650 ft) high; these are called *star dunes*. They seem to be fixed in position, and their formation, which is not well understood, seems to require winds from several directions. They are seen in Arabia and the Sahara.

In view of the slow rate of weathering in deserts, the immense volumes of sand stored in ergs is difficult to explain. Erg formation may require a humid period of rock disintegration and stream transport of sand out of the supply region, followed by a climatic change toward aridity, so that the wind assumes the role of the sand transporting and concentrating agent. Major ergs have probably grown in several separate stages, including climatic changes back and forth between sand generating and erg-building geomorphic systems.

Ancient ergs of vast size that are now covered with vegetation, or whose dunes rise from marshes, are graphic evidence of the oscillations in climate that occur through time. Relict ergs similar to the Nebraska Sandhills are widespread south of the Sahara in the vicinity of Timbuktu and Lake Chad, as well as on the margins of other deserts.

CHAPTER 14

Desertification

During the North American "dust bowl" years, from 1931 to 1936, wind erosion devastated cropland in Kansas, Oklahoma, and Texas. Black clouds of deflated humus and topsoil rolled eastward from the plains, darkening the sky over the Mississippi Valley and the Midwest. As the surface soil was stripped away, the sandy subsoils were exposed to the force of the wind, and dunes began to form and to move over abandoned farms. Before 1900 this landscape had been a short-grass prairie or steppe, grazed by constantly moving herds of cattle.

Dust storms and the mobilization of sand into migratory dunes is a signal of an ecological catastrophe related to drought. The dust storms of the 1970s and 1980s in the Sahel region of Africa (page 241) are recent examples, as was the enormous cloud of reddish dust that enveloped Melbourne in southern Australia in April of 1983. All were the result of droughts that caused severe economic and social problems. In inner Asia a highly visible demonstration of "desertification" is the recent reactivation of dunes in areas of sandy soils that were formerly stabilized by vegetation. The principal effect is in the short-grass steppes of Soviet Central Asia and the north and west of China, on the margins of the true deserts. Ominously, all across Africa in the Sahel region south of the Sahara there are vast areas of ancient dunes that are now inactive and covered with vegetation. Clearly, the Sahara once extended some 600 km (350 miles) farther south than it does today. The recent Sahel crisis and its counterparts elsewhere pose questions: Are climates changing? Are the Sahara and other deserts now expanding to their old limits?

In fact, most areas affected by "desertification"—especially severe wind erosion and dune activation in agricultural areas—are semiarid grasslands and dry savannas where the climate swings back and forth, pendulum-fashion, between cycles of wet and dry years. In past centuries, vegetative growth increased during wet years, benefiting livestock herders who controlled such areas. However, the shortage of water for livestock kept the numbers of animals somewhat in check. When dry cycles returned, the herds and flocks dwindled, and aeolian erosion doubtless occurred on a limited scale.

In the modern era, the once-feared pastoral nomads have been pacified, and wet years have lured agriculturalists onto the dry steppes and semidesert savannas. The soils plowed for the first time in these areas of oscillating climate are commonly used for commercial rather than subsistence crops, often encouraged by governments as a source of foreign exchange. High wheat prices after World War I brought farmers onto the North American steppes. When wheat prices tumbled in the late 1920s the land was abandoned and as the inevitable drought years struck, it began to blow. In the more tropical Sahel, a wet cycle from 1940 to 1960 drew growers of cotton and peanuts into areas of former subsistence crops, and then into the Sahelian pasturelands. Grazing pressure on the delicate vegetation of the desert margins has also intensified as the pasturelands have been squeezed in area while the water provided by drilled wells has increased the population of animals far beyond the numbers that natural water supplies could support.

Now, when the normal dry cycles occur, the land is bare as a consequence of the plow, or thinly vegetated because of overgrazing. It yields quickly to the wind. "Minideserts" soon spread outward from the most vulnerable points created by cropping and trampling. These eventually merge to form new, human-induced deserts. According to United Nations estimates, nearly 52 million sq km (20 million sq miles) of the earth's surface are either natural desert or are becoming desertified by wind and water erosion, dune formation and encroachment, vegetation change to less useful types, and salinization of the soil. Only 18 percent of this area is "natural" desert. The rest, some 40 million sq km (15 million sq miles), is becoming desertified by a lethal combination of normal drought cycles and intensified human pressure on the land.

KEY TERMS

fluvial processes
drainage network
drainage basin
catchment
drainage pattern
stream order
discharge
rating curve
laminar flow
turbulent flow
dissolved load (chemical load)
suspended load
bed load
solid load
scour
entrainment
fluvial deposition
saltation
traction
channel fill
sinuosity
alluvium
graded reach
stream gradient
longitudinal profile
knickpoint
meandering channel
meander wavelength
pool
riffle

thalweg
point bar
lateral planation
braided channel
floodplain
natural levee
backswamp
oxbow lake
fluvial terrace
alluvial fill
alluvial fan
alluvial apron (bajada)
delta
distributary channel
aeolian processes
sand creep
deflation
yardang
ventifact
loess
sand ripples
dune backslope
dune slip face
transverse dune
longitudinal dune
whaleback
crescentic dune (barchan)
parabolic dune
erg
star dune

REVIEW QUESTIONS

1. What measurable properties of streams appear to vary in a regular manner related to stream "order"?
2. What are the forces involved in stream flow?
3. What are the relationships between stream energy, stream velocity, and stream discharge?
4. In what three ways are the particles of a stream's solid load moved? Where does a stream's dissolved load come from?
5. How can a sediment-laden stream increase its own energy at a particular point? What happens when a stream has more energy than required to move its sediment load?
6. What is the meaning of "quasi-equilibrium" in reference to streams?
7. What three factors are most often responsible for knickpoints in stream profiles?
8. How is pool-and-riffle spacing related to the wavelength of stream meanders?
9. How can human activities cause a meandering stream channel to develop a braided pattern instead?

10. What is the most common landform created by fluvial processes?
11. What fluvial landforms could you expect to see on a floodplain?
12. In general, what do fluvial terraces indicate about stream behavior?
13. Where are distributary stream channels most often encountered?
14. In what specific ways are fluvial and aeolian processes similar? In what ways do they differ?

15. What chain of events is necessary to produce thick deposits of loess?
16. How does the density difference between air and water affect aeolian erosion and transport?
17. What are the most common transverse dune types seen in deserts and along shorelines?
18. Where in North America is there a very large relict erg?

APPLICATIONS

1. Most urban areas are near a permanent stream. Look at the largest stream close to your campus. How has it been modified by human activity? What were its natural characteristics in the prehistoric period?
2. Looking at the same stream, where are current areas of bed or bank erosion? of deposition? Can you make any general statement about stream behavior: (a) upstream from your city? (b) in the urban area? (c) downstream from the urban area?
3. How much have lower order streams that enter the main stream (in question 1) changed throughout historic time? Can you see evidence of recent progressive change? What explains the tendencies observed?
4. To maintain constant stream velocity (and therefore transporting power), a decrease in stream depth must be offset by a change in one of the other factors affecting flow velocity. What sorts of changes could compensate for a decrease in stream depth to keep flow velocity constant?
5. The lower Mississippi River has followed several different courses to the Gulf of Mexico. Although the old channels have been silted in, the different locations of the ancient streams are clearly evident. What would be the principal evidence of the Mississippi River's former courses?
6. The soils of arid and semiarid regions quite often contain more soluble nutrients than the underlying parent material could supply, or than could be retained against leaching over a long period of time. What is the most likely explanation?
7. What are some of the possible sources of the sand found in dune fields and ergs? What has been the *complete* history of this sand?
8. If yardangs are created by wind erosion, they should resemble aerodynamic forms. What should they look like, ideally?

FURTHER READING

Belt, C. B., Jr. "The 1973 Flood and Man's Constriction of the Mississippi River." *Science,* 189:4204 (August 1975): 681–684. This article analyzes the effects of human modification of the Mississippi River channel, which caused a flow of only moderately high magnitude to produce a flood of catastrophic proportions.

Chorley, Richard J., ed. *Water, Earth, and Man.* London: Methuen & Co. (1969), 588 pp. There are several chapters on channeled flows, river regimes, and floods in this collection of articles on water in all its forms of occurrence, written by experts on the various topics.

Cooke, Ronald U., and **Andrew Warren.** *Geomorphology in Deserts,* Berkeley and Los Angeles: University of California Press (1973), 394 pp. About 40 percent of this book concerns desert aeolian processes and associated landforms. Well illustrated with diagrams; the treatment is somewhat technical.

Goldsmith, Victor. "Coastal Dunes," in Richard A. Davis, Jr., ed., *Coastal Sedimentary Environments,* New York: Springer-Verlag (1978), 171–235. A concise but comprehensive treatment of coastal and lakeshore dunes, including an extensive bibliography.

Greeley, Ronald, and **J. D. Iversen.** *Wind as a Geological Process,* Cambridge: Cambridge University Press (1985), 333 pp. An analysis of aeolian processes and landforms on the Earth and comparisons with the widespread aeolian forms on Mars as revealed by images from orbiting spacecraft.

Gregory, K. J., and **D. E. Walling.** *Drainage Basin Form and Process.* London: Arnold (1973), 456 pp. This treatment of fluvial processes reflects English work in the field and updates Leopold, Wolman, and Miller. Most examples are from the British Isles.

Leopold, Luna B. *Water: A Primer.* San Francisco: W. H. Freeman & Co. (1974), 172 pp. A simple but technically sound introduction to the general principles of hydrology and the action of water on slopes and in channels, by the former chief hydrologist of the U.S. Geological Survey.

——, **M. Gordon Wolman,** and **John P. Miller.** *Fluvial Processes in Geomorphology.* San Francisco: W. H. Freeman & Co. (1964), 522 pp. A landmark when first published, the book presents the findings of several decades of research on fluvial processes by U.S. Geological Survey field workers. It is somewhat technical but highly readable and most informative.

McKee, Edwin D. *A Study of Global Sand Seas,* U.S. Geological Survey Professional Paper 1052 (1979), 429 pp. Detailed studies of desert dunes and ergs, including classifications and color maps of dune types in the great sand seas of the Old World deserts.

Morisawa, Marie. *Streams, Their Dynamics and Morphology.* New York: McGraw-Hill Book Co., Earth and Planetary Science Series (1968), 175 pp. This brief paperback presents most of the fundamentals of fluvial processes in nontechnical terms in a concise format.

——, ed. *Fluvial Geomorphology.* Binghamton State University of New York: Publications in Geomorphology (1974), 314 pp. This is a collection of articles on topics of current interest assembled for a 1973 symposium on fluvial geomorphology. The subject matter is diverse, including an interesting contribution on the greatest flood known, which created Washington's Channelled Scablands.

Russell, Richard J. *River Plains and Sea Coasts.* Berkeley and Los Angeles: University of California Press (1967), 173 pp. Many interesting facets of stream behavior are explored, with most examples being drawn from observation of the Mississippi River—especially good on floodplains and deltas.

Schumm, Stanley A. *River Morphology.* Stroudsburg, Pa.: Dowden, Hutchinson and Ross (1972), 429 pp. This is a collection of influential articles on fluvial processes published between 1850 and 1971. Some are technical, others are in the explanatory-descriptive vein. Each constitutes a major contribution to fluvial studies.

——. *The Fluvial System.* New York: Wiley-Interscience (1977), 338 pp. An advanced treatment of fluvial processes by a leading researcher; offers many original insights.

Wilson, I. G. "Desert Sandflow Basins and a Model for the Development of Ergs." *Geographical Journal,* 137:2 (1971), 180–199. A very readable overview and theory of the formation of desert sand seas, with emphasis on the Saharan ergs.

Winter Landscape from the Ch'ing Dynasty. (Courtesy of the Smithsonian Institution, Freer Gallery of Art, Washington, D.C.)

Towering pillars of limestone in southern China are an example of a unique landform produced by a particular erosional process and rock type. Many different geologic factors give landforms their varying characters.

15
Geologic Structure and Landforms

From the window of an airplane, one can hardly fail to be impressed by the geometrical patterns made by landforms in many regions. Seen from a cloud-free flight across the United States, the linear mountains and basins of Nevada, the layered tablelands and sinuous lines of cliffs of central Utah, and the seemingly endless winding ridges of Pennsylvania stand out as highly distinctive forms. The scenic character of such areas results from their peculiar geologic structures. Resistant rocks project boldly above their surroundings, and weaker rocks have been worn to low relief. Jointing often controls drainage patterns and relief forms. Faults and folds produce particular types of landforms that are different from those developed in undisturbed rocks, and sedimentary rocks that are flat-lying display landforms unlike those seen where rock strata have been deformed by tectonic activity.

The distinctive "personalities" of many local landforms, as well as of entire regional landscapes, are largely results of the types and arrangement of the rock masses present. The varying architectural elements of the earth's crust were introduced in Chapter 12. Now we must explore their specific effects on landforms.

DIFFERENTIAL WEATHERING AND EROSION

Figure 15.1 illustrates the complexity of the terrain in a small portion of the eastern United

Figure 15.1

This *physiographic diagram* by the master cartographer Erwin Raisz reveals how geologic structure affects the scenery of the lower Hudson River region. North is at the top. In the upstream area, the cross section shows that gently dipping resistant sandstones form the Catskill Mountains, with metamorphic rocks underlying the Hudson River Valley. In the cross section farther south, the highlands are shown to be produced by granite rocks (darkest symbol), with basaltic lavas making ridges in the lowland west of the river. (From the Report of the International Geological Congress, Washington, D.C., 1933)

States. The variety of relief features seen in this region results from *differential weathering and erosion,* which refers to the varying responses of different rock types to the weathering processes that fragment rocks and the erosional processes that remove the fragmented material. In the lower Hudson River Valley portrayed in Figure 15.1, differential weathering and erosion are possible because various disturbances of the earth's crust have created complex geological structures, allowing many different rock types to be exposed. Figure 13.17 (page 381) shows differential weathering and erosion on a smaller scale. In arid regions, even minor variations in rock resistance are revealed, for the processes of weathering and erosion are even more selective where there is no cover of vegetation or soil.

The layering of sedimentary rocks—principally shale, sandstone, conglomerate, and limestone—provides the best opportunity for differential weathering and erosion, as each rock stratum has its own physical and chemical characteristics that may be quite different from those of adjacent strata.

Shale

Shale is mechanically weak, tending to fragment into thin flakes. When chemically weathered, it produces clay soils that are slow to accept water. The resulting high runoff during rainstorms produces an above average number of stream channels per unit of area. Shale is, in fact, the most erodible of all rock types. Because of this, it generally evolves into lowlands or dissected slopes fringing higher hills that are composed of other material (Figure 15.2). In dry regions, the erosion of weakly cemented shales often creates a maze of closely spaced ridges and ravines known as *badland* topography. An excellent example of such a landscape has been set aside as South Dakota's Badlands National Park, but similar topography occurs in many places in the arid West of the Rocky Mountains.

Sandstone

Sandstone can be physically weak or strong depending on the substance that cements its grains together. It is often porous, permitting water to soak into it or into the sandy soils that form on it. This reduces runoff and erosion. Since it erodes slowly, sandstone frequently forms ledges or cliffs (Figure 15.2) or hills rising above areas of shale or limestone. Conglomerate, which is composed of cemented gravel or cobbles, is the least common of the clastic sedimentary rocks, but when well-cemented is even more resistant than sand-

stone and almost always gives rise to bold ledges.

Differential weathering and erosion in sedimentary rocks often produce a series of cliffs and slopes. These vary in form depending upon the tectonically produced tilt, or *dip*, of the strata, as shown in Figures 15.3 and 15.4. Where the dip is gentle, a *cuesta* is formed. This is a plateau capped by a thick layer of resistant sedimentary rock that gradually dips below the land surface. Every cuesta has a steep erosional *scarp* and a gently dipping *backslope* capped by the resistant layer of rock. The cuesta scarp retreats as a consequence of erosion of the weak rock (usually shale) under the caprock; this leads to collapse of the edge of the caprock, due to lack of support.

Flat-topped *mesas* (Figure 15.3) may be detached from the cuesta as its scarp retreats in the direction of bedrock dip. Subsequent erosion eventually reduces mesas to smaller *buttes* that no longer preserve extensive flat summits. Where the rock strata dip steeply, cuestas are replaced by *hogback ridges* (Figure 15.4). The steep slopes that meet at the sharp crest of a hogback are similar in angle, but one slope cuts across the edges of rock layers while the other is the top of the dipping bed that creates the ridge.

Figure 15.2
Where shale is interbedded with limestone, sandstone, or conglomerate, the shale erodes to a gentler slope below cliffs of the more resistant rock type. Erosion of the shale slope undermines the resistant caprock, causing its edge to retreat by repeated small-scale collapses. This sandstone promontory, with underlying shale, is seen on the west flank of the San Rafael Swell in eastern Utah (see Figure 15.21, p. 462). (T. M. O.)

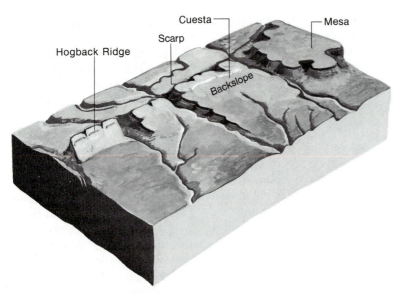

Cuesta

Scarp

Mesa

Hogback Ridge

Backslope

Figure 15.3
Mesas, cuestas, and *hogback ridges* are produced by differential erosion where a resistant caprock overlies less resistant strata. A flat-topped mesa is formed where the rock strata are horizontal. A cuesta is formed where the strata are slightly inclined, so that a cuesta possesses a steep face, or *scarp,* and a gradually descending *backslope.* A hogback ridge is formed where the strata are steeply inclined. One slope consists of the dipping caprock, and the other is the erosional scarp that cuts across the caprock and the underlying layers. (John Dawson)

Figure 15.4
In arid environments, where a protective cover of soil and forest vegetation are absent, hogback ridges assume a serrate form. This sandstone hogback north of the San Gabriel Mountains in southern California resembles a row of shark's teeth. (T. M. O.)

The arid Colorado Plateau region of the western United States, including parts of Colorado, Utah, Arizona, and New Mexico, is famous for its sharply defined cuestas, mesas, buttes, and hogbacks of naked rock. Forms that are similar, but softened by a forest cover, are present in the eastern United States. In Kentucky and Tennessee, cuestas, buttes, and mesas are important landscape elements, and in large areas of Pennsylvania, Virginia, and Arkansas, hogbacks dominate the scenery. Any cuesta or linear ridge formed by a dipping layer of rock is known as a *homoclinal* (inclined in one direction) structure.

Limestone

Limestone is peculiar in that its dominant mineral, calcite ($CaCO_3$), dissolves in slightly acidic water. Soil moisture contains enough dissolved CO_2 from the atmosphere and from plant decay to become a weak carbonic acid solution, and it also absorbs organic acids released by plants. Moving ground water in the zone of saturation is also slightly acid and therefore corrosive to limestone. The solution (dissolving) of limestone is concentrated along joints and bedding planes, where water can circulate freely and be replenished before it becomes saturated with calcium bicarbonate, the product of limestone solution.

Dolines

Where joints intersect, the solution process is especially rapid, resulting in underground voids of some size. These may enlarge upward to the surface, or the surface rocks may collapse into them. In either case, a surface depression known as a *sinkhole,* or merely a "sink," results. In the scientific literature these are termed *dolines,* a word borrowed from the Serbo-Croatian language, reflecting the importance of limestone landforms throughout Yugoslavia. All landforms produced by solution of limestone are known as *karst* landforms. The name is derived from the Karst region of northern Yugoslavia, where solution features were first described in detail in 1893.

Some areas of Indiana have as many as 300 sinkholes (dolines) per sq km (Figure 15.5). Portions of Kentucky and Florida are likewise thoroughly pitted with dolines, which often divert small surface streams to underground routes. Dolines may hold small ponds or marshes, but where the development of dolines is well advanced, surface water can be quite scarce, especially if a dry season occurs.

Limestone Caverns

The most unusual feature created by carbonate solution is the *limestone cavern.* Vast subsurface voids can form wherever a thick mass of limestone is present at or below the water table in an area of high local relief. Where there is significant topographic relief the water table slopes downward toward the major streams, resulting in active lateral flow of groundwater. This prevents saturation of groundwater with calcium bicarbonate and is necessary for large-scale solution to continue. Accessible limestone caverns are formed in two stages: a period of major solution at or below the water table, and a second period after the water table is lowered, allowing air to enter the cavern.

Of the 48 conterminous states, all but Rhode Island, Vermont, New Hampshire, and Louisiana have accessible limestone caverns, and some states have hundreds. Some caverns have more than 100 km (60 miles) of passages, including rooms tens of meters high and thousands of square meters in area. Many limestone caverns occupy several levels vertically and are beautifully decorated with hanging icicle-like *stalactites* and more massive upthrusting *stalagmites,* both of which are composed of calcium carbonate (Figure 15.6). These and a variety of other types of cave deposits, known collectively as *speleothems,* are formed where downward percolating water, enriched in both carbon dioxide and calcium bicarbonate, seeps into the cavern. Some of the carbon dioxide gas escapes from the water, and this decreases the water's ability to retain calcium bicarbonate, which is precipitated as either calcite or aragonite, two slightly different crystalline forms.

Effect of Climate

Because limestone varies in strength and massiveness (degree of jointing and bedding) it can be either a resistant or a weak rock. In humid areas it may dissolve to form lowlands surrounded by higher land on other rock types. Hard limestones can create cliffs and even mountain peaks, such as those in the Alps and Himalayas. In arid regions there is little water available to dissolve limestone, which then acts as a resistant rock, forming cuestas,

turn now to this aspect of structure as a control of landform development.

VOLCANIC LANDFORMS

The relationship of volcanism to lithospheric plates was outlined in Chapter 12. There we saw that two dissimilar types of volcanic activity are easily distinguished—nonexplosive basaltic eruptions where plates of the lithosphere are pulling apart, and explosive eruptions of silicic tephra (volcanic ash) and lava along lines of plate convergence and subduction. Different landforms are associated with these two contrasting eruptive styles. These landforms indicate the degree of volcanic hazard in the locations where they occur.

Basaltic Volcanism

Where lithospheric plates are being pulled apart, basic magma from the earth's mantle rises to fill the opening between them. This erupts at the surface as basaltic lava. The rising magma creates the oceanic ridges, which occasionally push above the sea surface, as in Iceland. Basic magma also erupts through sea-floor fractures, where portions of oceanic plates are sliding past one another. The ocean floor is studded with volcanoes that are now inactive. Many of these appear to have been carried away from the ocean ridge systems by sea-floor spreading. A peculiar form known as the *guyot* is discussed in Chapter 17.

Oceanic Hot Spots

Not all active oceanic volcanoes are associated with plate boundaries. Some lie at the ends of linear chains of submarine mountains that can be followed for a thousand kilometers or more. Such volcanoes mark the position of so-called "hot spots" in the earth's mantle, where rising plumes of magma seem to remain fixed in place while lithospheric plates move over them. The extinct volcanoes that form the majority of such mountain chains were built above the hot spot and then moved away from it so that the trend of the submarine mountain chain records the direction of plate motion. The Hawaiian chain is the best-known example. Its volcanic rocks increase in age with distance from the hot spot presently situated under the southeastern portion of the island of Hawaii itself (Figure 15.10).

Shield Volcanoes

Volcanic eruptions on the sea floors build basaltic *shield volcanoes,* which have gentle slopes and broad rounded summits (Figure 15.11). Hawaii's Mauna Kea and Mauna Loa are the world's largest mountains, rising some 9,000 meters (30,000 ft) above the sea floor. Their slopes rarely exceed an angle of 10 degrees. The Hawaiian shield volcanoes clearly display two contrasting types of lava (Figure 12.12, p. 337; 15.12). The hottest flows issue in the form of very fluid *pahoehoe* (pāh-ho-ay-hó-ay) lava, which solidifies with a satin-smooth skin that is often conspicuously wrinkled by the continued flow of the lava underneath. Somewhat cooler basalt issues as slag-like rough-surfaced *aa* (ah-ah) lava. Pahoehoe lava is made fluid by the gases still dissolved in the lava. When these gases are lost by excessive agitation before eruption or within a flow, pahoehoe alters to aa lava.

Scoria Cones

In addition to producing enormous volumes of free-flowing basaltic lava, nonviolent eruptions of the type seen in Hawaii and Iceland are still forceful enough to hurl solid particles hundreds of meters into the air. These rain down to form deposits of *pyroclastic* ("fire-broken") debris. This includes solidified clots of lava and fragments of rock torn loose during the eruption. The larger particles, or *scoria* (Figure 15.13), fall close to the vent and build up a *scoria cone* or *cinder cone.* Sometimes the

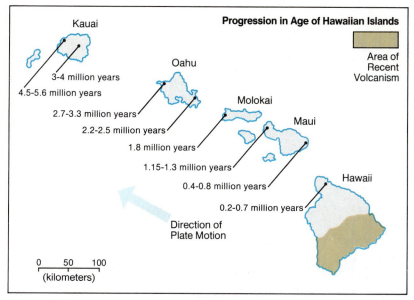

Figure 15.10
Radiometric dating reveals that the lavas of the Hawaiian Islands are progressively younger southeastward along the chain. This reflects the northwestward movement of the Pacific plate over the location of a "hot spot" deeper in the earth's mantle. The hot spot causes repeated volcanic eruptions on the ocean floor, eventually building up islands of basalt. (Andy Lucas after data by Ian McDougall, *Nature,* Vol. 231, 1971)

magma is so gaseous that it rises to the surface as a froth, like the foam on beer. This solidifies as *pumice,* an extremely porous rock that is light enough to float on water.

Flood Basalts

Basaltic eruptions similar to the oceanic type also occur on the continents. Several land

Figure 15.11
Basaltic *shield volcanoes* such as Hawaii's Kilauea (foreground) and Mauna Loa (horizon) have very gentle slopes, making their true sizes hard to appreciate. Mauna Loa rises almost 3,000 meters (9,800 ft) above the summit of Kilauea. In the foreground is the summit crater of Kilauea volcano. (T. M. O.)

GEOLOGIC STRUCTURE AND LANDFORMS

(a)

(b)

Figure 15.12
Contrasting types of basaltic lava, resulting from varying contents of dissolved gases.

(a) Its content of dissolved gases makes *pahoehoe* lava, shown here at Kilauea crater in Hawaii, much more fluid than *aa* lava even though the chemical composition is the same. Note the thin layer of solidified crust on the molten rock. When the lava solidifies, the surface is left in the form of smooth billows or ropes. (Frank Sojka © Island Pictures, Honolulu, Hawaii)

(b) This advancing *aa* lava flow on Hawaii is characterized by chunky, angular blocks mixed with still-molten lava. When an *aa* flow solidifies, an almost impassable field of jagged lava is formed.

Figure 15.13
A *scoria cone,* produced by the fall of pyroclastic materials that have been hurled into the air by the force of escaping magmatic gases and steam during a volcanic eruption. The example here is Paricutín volcano, which rose from a Mexican cornfield in 1943 and erupted sporadically until 1952. (R. Segerstrom/U.S. Geological Survey)

areas have been buried by enormous outpourings of very fluid basaltic lava, which has accumulated, flow upon flow, to thicknesses exceeding a thousand meters. Individual flows of these continental *flood basalts* have spread

more than 160 km (100 miles) from their source, covering thousands of square kilometers to depths of tens of meters. The volumes of such flows measure in the tens to hundreds of cubic kilometers. A lava flood of this type occurred in Iceland in 1783, producing some 12 cu km (3 cu miles) of new basaltic rock.

Areas of continental flood basalts include the Columbia Plateau region of Washington and Oregon and the Snake River Plain of Idaho, as well as portions of India, Brazil, Patagonia, South Africa, and Antarctica. Some of the lava floods on the land may have occurred during the breakup of continents as new oceans were being born. The lava flows of Triassic age that form the Hudson River Palisades and the Watchung Mountains of New Jersey probably originated in this way when North America began to separate from West Africa to initiate the present Atlantic Ocean, some 200 million years ago.

All areas of flood basalts on the land have been somewhat dissected by stream erosion. The resulting canyon walls expose the separate flows clearly, one atop the other, as can be seen along portions of the Columbia and Snake rivers in Washington and Idaho (Figure 15.14). Smaller basaltic lava fields with cinder cones

Figure 15.14
The Columbia Plateau in the state of Washington consists of layers of basaltic lava having an aggregate depth of about 1,000 meters (3,300 ft). The distinctive columnar jointing of the lava is evident wherever it is trenched by an erosional canyon, as in this view at Grand Coulee. Such *flood basalts* occur in areas of crustal tension and rifting. (T. M. O.)

and shield volcanoes are widely scattered in areas of recent tectonic activity, marking localities where fractures have penetrated through the earth's crust to the mantle beneath.

Andesitic Volcanism

Despite the enormous volumes of lava generated by basaltic volcanism on the continents and ocean floors, the eruptions are seldom violent or dangerous. But where crustal plates are converging and subduction zones are present, the volcanism is quite different. Subduction-related volcanism generates much smaller amounts of lava, but releases pent-up gases so explosively that vast areas are denuded, incinerated, or suffocated with tephra (ash). The volcanoes themselves are occasionally decapitated by the eruptive events.

The force of some volcanic eruptions is incredible. When Krakatau, an island volcano between Java and Sumatra, exploded in 1883, the sound carried 3,000 km (1,900 miles) to Australia. Krakatau itself vanished. Ash falls from a similar explosion at Santorin in the Mediterranean Sea may have destroyed the ancient Minoan civilization of Crete about 3,500 years ago. The explosive eruption of Mount St. Helens in the state of Washington in 1980 leveled 400 square km (150 sq miles) of forest, and removed 2 cu km (.75 cu miles) of the volcano itself. A similar event at El Chichón in Mexico in 1982 ejected so much tephra into the upper atmosphere that the climate of the northern hemisphere seemed to have been cooled measurably. Similar eruptions of other volcanoes in the past have lowered the average temperature of the northern hemisphere by about 0.3°C, with the effect lasting three to five years.

Andesitic volcanism occurs along continental edges where subduction is occurring, as in the Andes Mountains running the length of western South America; in areas of continental collision, as in the Mediterranean region and eastern Turkey; and in island arcs where one sea-floor plate is being subducted beneath another, as in the western Pacific and Indonesia.

Strato-Volcanoes

The first products of volcanic eruptions on the land are blankets of tephra and scoria cones like Paricutín, born in a Mexican cornfield in 1943 and built to a height of 400 meters (1,300 ft) in eight months (Figure 15.13). Long-continued eruptions veneer scoria cones with lava, causing them to become *composite cones*. Flows of andesitic lava usually consist of a stiff pasty interior covered by a rubble of hardened blocks, unlike either aa or pahoehoe lava. As lava coats the cone, it is also injected into cracks in the volcano, where it solidifies as lava *dikes*. These reinforce the cone's structure. As the cone grows higher, lava may issue from fissures opening in its sides rather than from the summit crater. Over hundreds of thousands of years giant steep-sided *strato-volcanoes,* composed of both scoria and lava, are constructed, such as Japan's Fujiyama, Washington's Mt. Rainier, and California's Mt. Shasta. Such volcanoes often rise 3,000 meters (10,000 ft) above the surrounding countryside (Figure 15.15a). All the earth's large intact strato-volcanoes have been built within the last 1 million years—some within the last 10,000 years—and every one is potentially dangerous.

Plug Domes

A peculiar type of volcano is the so-called *plug dome*. These are seldom of large size, but their eruptions are extremely hazardous. Plug domes are formed when magma that is more silicic than andesite pushes to the surface. Such lava (dacite or rhyolite) is stiff and pasty and congeals as it rises, jamming the surface vent so that enormous pressure builds beneath the volcano. This is released in catastrophic explosions that may hurl large blocks several kilometers. Plug domes can be recognized by

(a)

(b)

(c)

Figure 15.15
Volcanic landforms produced by crustal subduction.

(a) *Strato-volcanoes,* such as Mount Hood in Oregon, are among the most impressive of all constructional landforms. Cones of this type are composed mainly of andesitic lava and pyroclastic material resulting from explosive eruptions.

(b) This volcanic plug dome in eastern California is composed of silicic dacite and rhyolite emplaced in association with violent volcanic eruptions that often include searing nuée ardente blasts lethal to all life in their paths. In the foreground is a "coulee" of the same highly viscous lava that forms the dome in the background.

(c) Oregon's Crater Lake is a caldera formed by the collapse of a large strato-volcano during an immense eruption that showered the Pacific Northwest with ash only 6,600 years ago. The resulting crater filled with a lake that is presently about 600 meters (2,000 ft) deep and 10 km (6 miles) across. Portions of the crater rim, such as the high point visible in the distance, rise 600 meters (2,000 ft) above the lake surface. Subsequent volcanic activity has built three new cones in the caldera, one of which protrudes above the lake surface as Wizard Island, visible at the right. (T. M. O.)

Associated with the plug dome is the *nuée ardente* (glowing cloud) type of eruption. A nuée ardente is an avalanche of glowing volcanic ash that moves down the flank of a volcano at speeds of more than 160 km (100 miles) per hour. Everything in the path of a nuée ardente is baked to a cinder or set afire instantly, and nearby life that is not incinerated may be asphyxiated by lack of oxygen or mummified by heat. In 1902 a nuée ardente issuing from Mt. Pelée, on the Caribbean island of Martinique, obliterated the town of St. Pierre, snuffing out almost 30,000 lives. Reports of the number of survivors range from two to four— one of whom was in an underground dungeon.

Calderas

Many giant strato-volcanoes have satellite plug domes, or have at times behaved like plug domes. They have erupted so violently that their summits have collapsed, producing vast craters known as *calderas.* This occurred in

the stubby masses of hardened lava that often ring their summits (Figure 15.15b).

GEOLOGIC STRUCTURE AND LANDFORMS

recent eruptions of both Mount St. Helens and El Chichón. If new plug domes do not refill them, these calderas usually hold circular lakes. Crater Lake in Oregon is a particularly beautiful example (Figure 15.15c). It partly fills a caldera 10 km (6 miles) wide and initially 1,200 meters (4,000 ft) deep. Mt. Vesuvius, near Naples, was partially destroyed in its great eruption of A.D. 79, which buried the Roman city of Pompeii in volcanic ash. In the last 1,900 years Vesuvius has rebuilt itself within the caldera created by that eruption.

Dormant Volcanoes

It is difficult to know whether a volcano is active, dormant (potentially active), or extinct. For example, Vesuvius had been dormant for centuries prior to its eruption in A.D. 79. Oregon's Crater Lake was created by a violent eruption about 6,600 years ago, but subsequently much of the crater was filled by further volcanic activity. In fact, Wizard Island, at one side of the lake, is the summit of a volcanic cone built within the caldera. In the volcano-studded Cascade Range of Washington, Oregon, and northern California, only Mount St. Helens in Washington and Lassen Peak in California have produced violent eruptions in the last 200 years. Nevertheless, all of the major Cascade volcanoes have displayed some eruptive activity during this period. Thus all must be regarded as dormant rather than extinct.

The Cascade Range is part of the "Pacific Ring of Fire"—a nearly continuous chain of andesitic volcanoes circling from South and Central America through Alaska, Japan, the Philippines, and New Zealand (Figure 12.17, page 347). This gives clear evidence of crustal subduction around nearly the entire rim of the Pacific Ocean. The only interruptions in the volcanic ring are north and south of the Cascade Range. To the north in Canada, and to the south in California and northern Mexico, crustal plates seem to be sliding past one another rather than converging and being subducted.

Extinct Volcanoes

All volcanoes eventually become inactive. When they do, erosion wears them away. All that is left in some areas are landforms resulting from the differential erosion of the lavas and ash composing the volcano. The most impressive of these forms are *volcanic necks* that rise as pinnacles composed of the lava left in the "throat" of a volcano that has been eroded away; *dike ridges* that form free-standing rock walls, often radiating from a volcanic neck; and *table mountains* consisting of winding, flat-topped ridges capped by lava that originally was channeled through a valley. Because the lava is more permeable and more resistant to erosion than the former valley walls, it is left, millions of years later, as a ridge. This is an example of *relief inversion* by erosion (Figure 15.16).

Volcanism and Humanity

Although individual volcanic eruptions endanger the living organisms in the vicinity, the various phenomena related to volcanism have affected environments in ways that can be quite beneficial to later human inhabitants.

Volcanic Hazards

Large volcanoes pose many kinds of hazards. Violent eruptions cause death by baking and suffocation. The force of the 1980 blast at Mount St. Helens stripped forested hillsides bare, removing even the soil to a depth of over a meter. Volcanic ash can be toxic to humans, animals, and plants, and its weight on roofs can collapse houses. It damages machinery and engines, including those of aircraft flying in the vicinity of eruptions. Lava flows have been destructive to towns as well as to agricultural fields. Equally damaging are volcanic mud-

(a)

(b)

Figure 15.16
Results of differential erosion in areas of past volcanic activity.

(a) Flat-topped *table mountains,* such as this one in southern California, form when lava enters valleys cut in erodible rocks. The lava itself erodes very slowly since its great permeability allows rainwater to soak into it rather than running off over its surface. Erosion of the surrounding area results in *relief inversion.* (T. M. O.)

(b) Shiprock, in northwest New Mexico, rises 430 meters (1,400 ft) and is a striking example of a *volcanic neck.* The original volcano has been completely removed by erosion, and all that remains is the lava that solidified in the subsurface conduit to the former cone. The volcano itself was active about 30 million years ago. Note the volcanic dikes radiating from Shiprock. (John S. Shelton)

GEOLOGIC STRUCTURE AND LANDFORMS

flows, known as *lahars* (Figure 15.17). These result from a number of causes: rainfall on fresh volcanic ash; lava eruptions under glaciers or snowfields; and slope failure on volcano summits weakened by eruptions, earthquakes, or acidic fumes vented from subsurface magma. Lahar deposits in the valleys surrounding some Cascade Range volcanoes are as much as 300 meters (900 ft) deep. Minor volcanic eruptions can trigger major lahars that engulf villages, often with great loss of human lives. The major eruption of Nevado del Ruiz in Colombia in 1985 created lahars that killed some 20,000 people in one of the greatest volcanic disasters in history.

Effect on Soils

There are also positive aspects to volcanism. Fresh mineral-rich volcanic ash and decomposed basaltic lava produce excellent agricultural soils. This is especially important in the wet tropics where leaching of nutrients leaves older soils infertile. The productivity of volcanic soils in Japan, the Philippines, Indonesia, and South America have, in fact, lured dense populations onto volcanoes, close to the hazards of explosive blasts, lava flows, ash falls, and lahars.

Ore Formation

Another important aspect of volcanism is its role in the formation of metallic ores, such as those of iron, copper, lead, zinc, tin, tungsten, nickel, and chromium, as well as gold and silver. Metallic ores are deposited where hot, richly mineralized solutions move upward from magma bodies through fissures in overlying older rock. Ore formation does not occur automatically, however; it depends on chemical reactions between the local rock and the invading magmatic solutions. Recent discoveries (Chapter 17) show that the volcanic vents at sea-floor spreading centers include veritable "sludges" of metallic minerals. It is assumed that these are incorporated into the sea-floor

Figure 15.17
This lahar deposit, generated by the 1980 eruption of Mount St. Helens (background) in the Cascade Mountains of Washington, is as much as 200 meters (600 ft) in depth. The torrent of mud that created it was boiling hot and vaporized the water of the Toutle River, whose valley it invaded, causing steam eruptions to build pseudo-volcanoes of mud on the surface of the lahar deposit. (T. M. O.)

lithosphere and later remobilized into the new magma generated by melting in subduction zones. Thus metallic ores originating on the sea floors are eventually emplaced in the continental crust.

Geothermal Energy

Pockets of subsurface magma can also be made to serve humanity. In a few areas of active, dormant, or extinct volcanism there are such natural phenomena as steam vents, geysers that periodically erupt boiling water, and sulphur springs with their smell of rotten eggs (hydrogen sulfide fumes). They all indicate that there is a subsurface *geothermal energy* source that humans can utilize (Figure 15.18).

To bring geothermal energy directly from the zone of fluid magma that is present everywhere beneath the earth's crust would require wells many tens of kilometers deep. These are costly and hazardous to drill. But the hot springs and steam jets in volcanic regions bring heat energy to the surface naturally. Hot springs occur where groundwater rises to the surface after being in contact with hot rock or steam vented from magma. The steam itself may remain trapped in the subsurface within reach of wells, or it may vent to the surface through rock fissures.

Steam from subsurface sources is already used to heat water and generate electricity in Italy, Iceland, New Zealand, and California. However, many technological problems have been encountered in developing and maintaining heat and pressure in these geothermal fields. Numerous attempts to develop new fields in promising locations have failed or been disappointing for reasons having to do with the nature of the aquifers and fracture systems, and inadequate subsurface water or water pressure. Therefore, great expansion of geothermal energy extraction and use, especially in nonvolcanic areas, awaits future technological advances.

CRUSTAL DEFORMATON AND LANDFORMS

Volcanism is a spectacular aspect of tectonic activity that creates impressive landforms on a variety of scales. However, less visible, slow movements that continue over millions of years have had much wider effects on the continental surfaces. These movements create the structures that control the evolution of landscapes in many parts of the world. In Chapter 12 we outlined the major geologic structures

Figure 15.18
The hot water issuing from Castle Geyser in Yellowstone National Park, Wyoming, is driven upward by steam generated from groundwater in fissured hot rock deep below the surface. Seismic studies and continuing measurable uplift in the Yellowstone area suggest a large body of magma unusually close to the surface. (David Miller)

produced by crustal tension, compression, and shearing due to lithospheric plate motions. The faults, folds, and broad warps of the lithosphere that result from plate motions are very important in the appearance of landscapes. It is now time to examine the consequences of these slow motions of the earth's crust.

Broad Crustal Warping

A geologic map of the eastern United States (Figure 15.19) shows many circular or arcuate patterns. These are outcrops of sedimentary rocks whose varying ages are indicated by the map colors. Strata of different ages outcrop at the surface because the strata are not entirely flat, but are somewhat warped by gentle deformation so that the land surface occasionally exposes their edges.

Taking a closer look at one of the arc patterns, the peninsula of Michigan may be thought of as the center of a saucer, surrounded by outfacing edges in the form of low cuestas. The Paris Basin in France, crossed by the Seine River, is a similar structural basin surrounded by outfacing cuestas such as those of the Champagne region to the east.

To the south of the Michigan Basin we find the opposite configuration—two structural domes surrounded by *infacing* cuestas. These are the Nashville Basin in Tennessee and the Bluegrass Basin (or Lexington Plain) in Kentucky. They are topographic lowlands because the sedimentary rocks exposed in their centers are more erodible than those forming their rims, as in Figure 15.20b. Note that the Michigan Basin is a structural basin, but the Nashville Basin is an erosional basin in a structural dome!

All of these humid-region cuestas are unspectacular forested slopes or well-dissected hill masses of low elevation. But farther westward in the semiarid and arid portions of the United States we encounter more sharply defined uplifts, with mountainous cores surrounded by infacing cliff-like cuestas, or even

Figure 15.19
This portion of the geologic map of the United States displays the outcrop patterns of gently deformed sedimentary rocks from Michigan in the north to Tennessee in the south. The colors indicate the ages of the rocks exposed at the surface. Concentric circular patterns indicate structural domes and basins surrounded by cuestas.

jagged hogback ridges (Figure 15.21). The classic example is the Black Hills Uplift of South Dakota—a partially unroofed granitic dome surrounded by a series of concentric homoclinal structures as in Figure 15.20a.

Mountain Building

Although volcanism has created some of our planet's most majestic individual mountains,

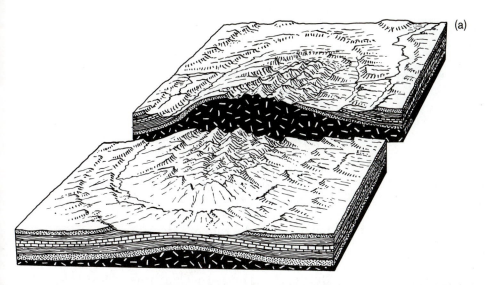

Figure 15.20
These drawings are diagrammatic representations of erosional forms developed in structural domes.

(a) Erosion of a structural dome exposing a core of resistant granitic rock. Encircling sedimentary rocks form hogbacks or cuestas with scarps facing the center of the uplift. Erodible sedimentary rocks form circular valleys between the homoclinal ridges. This diagram is representative of the Black Hills of South Dakota or the Adirondack Mountains of New York.

(b) Erosion of a dome structure in which only sedimentary rocks appear at the surface. As the rocks exposed in the center of this dome are relatively weak, a relief inversion results, with the structural high expressed in the landscape as a topographic low. The basin is surrounded by in-facing cuestas of resistant rock and encircling valleys carved in erodible rock. The Nashville Basin of Tennessee and the Blue Grass (Lexington) Basin of Kentucky are similar to this model. (T. M. Oberlander)

all sizable mountain ranges, even those capped by volcanoes, have been formed by larger-scale motions of the earth's crust. Nearly all mountain systems began as enormous accumulations of sediment on the sea floors, close to the margins of continents. In most instances, thicknesses of more than 10,000 meters (over 33,000 ft) of sediments accumulated along the edges of the continents before being crumpled into the complex structures seen in major mountain ranges.

GEOLOGIC STRUCTURE AND LANDFORMS

Figure 15.21
This LANDSAT false-color image taken from space depicts the structural dome in eastern Utah known as the San Rafael Swell. Clearly visible are the hogbacks and cuesta scarps that encircle the uplift, which exposes only sedimentary rock. The red areas at the left are irrigated cropland. (NASA and the Earth Satellite Corporation)

Geosynclines and Subduction

Until the 1960s it was not agreed how or why these great sediment traps, called *geosynclines,* were formed, or why their deposits were eventually crushed together in the process that transformed them into mountains. During the 1960s, however, earth scientists came to understand the subduction process. It is now accepted that the descent of crustal plates in subduction zones first produces oceanic trenches that fill with sediment, and then crushes the sediment mass collected in the trenches and adjacent continental rise areas (Chapter 12). The present ocean trenches and continental rises are, in fact, living geosynclines—the

birthplaces of future mountains. The blankets of sediment from which mountains are made also include the underwater continental shelves, though there the sediments are a thinner veneer over older rocks of the continent itself. Although every coast has a continental shelf and a continental rise, future mountain ranges of sedimentary rock will be built only where an ocean trench system is also present and lithospheric plate convergence occurs.

The crushing of geosynclinal sediments by horizontal movements is not the entire explanation for the building of mountain systems. Later, when the rate of compression slackens, strong vertical uplift of the rock that was squeezed in the geosyncline seems to occur. During the slow rise of California's Sierra Nevada range, between 15 and 20 km (9 to 12 miles) of sedimentary rock has been removed by erosion, causing plutonic rocks to be widely exposed at the surface. The prior phase of subduction, folding, and plutonic injection ended some 70 million years ago.

Even greater uplift occurred in portions of the Rocky Mountains, which are exceptional in their mid-continent location, far from known lithospheric plate boundaries. The Middle and Southern Rockies, from Wyoming to New Mexico, appear to have formed almost entirely by vertical uplift, without the prior existence of a deep-water geosyncline.

The most recent period of world-wide mountain building began about 20 million years ago, with the first compression and uplift of the Alps, Himalayas, and coastal ranges of western North America. The geologic structures of the Rocky Mountains and Sierra Nevada are somewhat older, but the present ranges were lifted by the latest mountain-building movements. The structures of the Appalachian Mountains of the eastern United States were formed earlier still, about 250 million years ago, when seafloor spreading drove the African continent against the edge of North America, eliminating an ancient Atlantic Ocean. The present Atlantic began reopening about 200 million years ago, accelerating subduction around the margins of the Pacific Ocean.

It is these large-scale tectonic motions that produce the individual geologic structures we can see expressed in landforms in regions with variable relief. We shall now turn our attention to these local structures and the landforms they create.

Fault-Controlled Landforms

The Great Basin, which is centered in Nevada, has experienced a long and complex geologic history, involving early phases of compression and massive overthrusting (Chapter 12). But it stands now as the world's greatest example of a fault-controlled landscape produced by crustal extension and collapse. The extension began some 30 million years ago and continues to the present.

Tensional Faulting

The Great Basin is bordered on its east and west sides by two great tilted blocks, the Wasatch Range of Utah and California's Sierra Nevada. Each of these presents a bold wall, or *fault scarp,* facing the center of the Great Basin. Between the two are a host of *tilted blocks* with fault scarps evident on only one side, upthrust *horsts* with fault scarps on both sides, and down-dropped *grabens* bordered by horsts or tilted blocks (Figure 15.22). The range-bounding faults are normal dip-slip faults in which the block resting on the inclined fault plane has slipped downward. This type of motion not only creates vertical relief, but also causes horizontal extension of the crust, as indicated in Figure 15.23.

Fault scarps are recognized by the abrupt rise of mountain slopes from a linear base line, as in Figure 15.24. The base line of a fault block often cuts cleanly across both resistant and erodible rock. Where the base line is irregular, with spurs projecting different distances, no active fault is present, which is to say that there has been no movement along the fault in many thousands of years. Well-preserved fault scarps are dissected into *triangular facets* (Figure 15.24) that sometimes reveal polished or grooved *slickensided* surfaces caused by the friction of one fault block rubbing against another. Since faults extend downward many kilometers into the earth, they are often marked by hot springs and even volcanic cinder cones. Springs are common along all types of faults due to disruption of groundwater flow.

Compressional Faulting

Thrust faults (Chapter 12) occur in areas of strong compression, and are often associated

Figure 15.22
Structures developed by normal faulting in a zone of crustal extension. In the Great Basin, centered in Nevada, the relative movement of fault blocks is made evident by the tilting of formerly horizontal lavas and volcanic tuffs erupted just prior to crustal rupture. Depressions created by the subsidence of blocks between bounding faults are *grabens*; high-standing blocks bounded by faults on both sides are *horsts*; *tilted blocks* have rotated, sinking on one side and rising on the other to form fault scarps. Often the bounding fault systems are complex, with several separate fault planes present, as illustrated here. (T. M. O.)

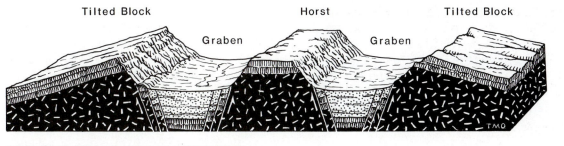

Tilted Block   Horst   Tilted Block

Graben   Graben

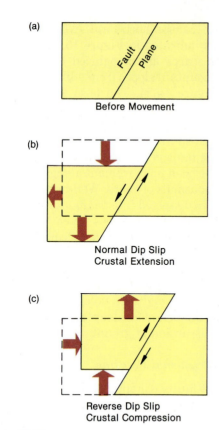

(a)

Fault Plane

Before Movement

(b)

Normal Dip Slip
Crustal Extension

(c)

Reverse Dip Slip
Crustal Compression

Figure 15.23
Horizontal and vertical effects of normal and reverse
fault slip. Black arrows show relative directions of slip
on the fault plane. Red arrows indicate the horizontal
and vertical components of fault slip.

with folds. Where *thrust faults* (also known as *reverse faults*) occur, the block resting on the inclined fault plane moves upward and over the lower block, causing crustal shortening (Figure 15.23). Thrust faults can form linear mountain ranges, such as those in eastern Idaho, but the scarps of thrust faults never retain the angle of the fault plane itself—to do so they would have to be overhanging.

Overthrusts, in which enormous slabs of rock are pushed horizontally up and over other rocks (Chapter 12), affect scenery mainly through the type of rock they carry into an area. Most have been folded after being displaced, so their landforms are those developed in folded rocks. Being associated with high mountains, their topography has often been retouched by glacial erosion (Chapter 16).

Strike-Slip Faulting

On both the continents and the ocean floors portions of the earth's crust slide past one an-

Figure 15.24
The east front of California's Sierra Nevada is a fault scarp that has been dissected by erosion. Note the straight and regular appearance of the mountain range and its abrupt rise from Owens Valley, a graben. The low hill area in front of the Sierran scarp is a second, much lower fault block. (U. S. Geological Survey)

The diagram at the right shows the evolution of an escarpment produced by faulting. Two periods of uplift are indicated by arrows. The forms are successively older toward the right. Note how the growth of valleys cutting back from the fault scarp gradually converts the scarp into a series of spurs ending in blunt *triangular facets*. These become smaller and less apparent with time. Alluvial fans issue from the valleys, becoming larger as the valleys reach farther into the uplifted block. Most existing fault scarps resemble the forms on the far right. (T. M. Oberlander)

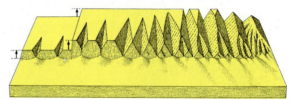

Prediction of Earthquakes and Volcanic Eruptions

How possible is it to predict massively destructive geological events? Prediction of such hazards itself entails danger. A false alarm will surely cause many people to deride subsequent better-founded warnings. Thus geologists are faced with a difficult question: When are the signs of impending disaster so indisputable that the public must be informed?

This problem arose in California between 1980 and 1983 when the Mammoth Lakes area, a major ski resort and summer recreation hub, began to experience earthquake activity—the strongest there in 100 years. The area lies in the ancient Long Valley caldera, which produced a colossal volcanic eruption about 700,000 years ago. The renewed activity in the early 1980s included thousands of small seismic shocks, an ominous horizontal stretching of the earth's crust accompanied by uplift of almost half a meter, increases in the activity of hot springs and heat flow from the earth, changes in the local magnetic field, and emissions of radon gas. Combined with detailed seismic surveys, these phenomena suggested the slow rise of a dike of molten magma that was fracturing solid rock as it moved upward, perhaps pushing water ahead of it. The possibility of an explosive steam eruption seemed high. Volcanic activity had occurred nearby as recently as 200 years ago. With the 1980 Mt. St. Helens disaster fresh in all minds, the U.S. Geological Survey issued an official notice of a "potential volcanic hazard" in the Long Valley area in 1982. Since 1983 geologic activity in the area has slackened.

With the lessons learned at Mt. St. Helens there is little fear that a violent volcanic eruption can take scientists by surprise. Even the most famous destructive eruptions of the past followed weeks or months of preliminary activity in the form of spectacular steam clouds, ash eruptions, and lahars. Mt. St. Helens was active for two months before its paroxismal blast. Vesuvius, Krakatau, and Mt. Pelée all gave ample warning before the catastrophes they unleashed on Pompeii, Southeast Asia, and Saint Pierre. Still, the actual point of breakout, the direction of the principal blast, and the exact type of eruption cannot be predicted. It is impossible to save structures from a volcano's fury, but loss of life should be avoidable.

Major earthquakes are a more difficult problem. Although scientists have identified many phenomena that precede individual earthquakes, and though these are monitored closely in quake-prone areas, "sneak earthquakes" occur with discouraging frequency, while predicted earthquakes either do not happen or are insignificant in scale. Unlike volcanic violence, an earthquake's first punch is its hardest. The precursory phenomena are subtle: small changes in ground tilt or elevation, increased emissions of certain gases (radon, H_2, SO_2), and the sudden fall of water in wells as rock begins to fracture before rupturing massively.

In some locations earthquakes produced by fault motion seem to have some periodicity, which may be detectable in studies of sediments and soil phenomena. By this crude gauge, several locations in western North America are overdue for a large seismic event. Two examples are the San Andreas fault north of Los Angeles and Utah's Wasatch fault from Ogden through Salt Lake City to Provo. Likewise there are "seismic gaps"—the last unruptured places along lines of weakness that have already released most of their pent-up energy. The Mexico City earthquake of 1985 had been forecast on the basis of a seismic gap there.

Finally, Chinese scientists, who are very experienced in earthquake prediction, put considerable emphasis on animal behavior as an earthquake precursor. Will North Americans heed an earthquake warning based on unusual actions of dogs, cats, pigs, and cows? The likelihood seems small.

Figure 15.25
The San Andreas fault (arrows) is visible crossing this vertical aerial photograph from the upper left to the lower right near the base of the hills. Note the offset ridges and valleys that mark the line of this right-lateral strike-slip fault, some 3 km (2 miles) of which can be seen here. This view is in the Carrizo Plain and Temblor ("earthquake") Range of central California. (U. S. Department of Agriculture)

other laterally rather than vertically, the motion being parallel to the fault strike rather than to the fault dip. The relation of strike-slip faulting to sea-floor spreading was noted in Chapter 12. Strike-slip faults are vertical and either straight or gently curving for distances of hundreds of kilometers. On the land they create a distinctive set of forms known as *rift topography,* which is very well displayed along California's San Andreas fault system (see Figure 12.24, p. 354). Three features are most evident: linear valleys eroded in crushed rock; elongated hills that are either compressed "welts" of pulverized rock or "slivers" of solid rock dragged along in the fault zone; and closed depressions created by shifts in the ground. Small ponds form in the latter. When these so-called *sag ponds* can be seen forming a definite line, a strike-slip fault is clearly indicated.

One of the interesting features of strike-slip faulting is the horizontal offset of ridge lines and stream beds that cross the fault (Figure 15.25). Fences, orchard rows, streets, sidewalks, and buildings all may be similarly torn apart and offset. Some of this separation occurs in the slow process of *fault creep,* in which offsets of 1 or 2 cm a year may occur without any earthquake activity.

From the air, or on a topographic map, active strike-slip faults are relatively easy to detect despite the fact that they do not normally create major relief effects. Rift topography and the abrupt change in the pattern of stream dissection due to the juxtaposition of different rock types along the fault—all forming a conspicuously straight line—are unmistakable indicators of a zone of high earthquake risk. Unfortunately, many such areas were settled and developed long before the seismic hazard was fully appreciated, as in the case of many of California's coastal cities.

Fold Structures and Landforms

The terrain that develops in a region of folded sedimentary rock depends on many factors. These include the size and shape of the folds, the local stratigraphy (number, thicknesses, and relative resistances of rock strata present), the rate of tectonic uplift, and the cli-

Figure 15.26
The folded mountains of southwestern Iran have formed so recently that the uppermost resistant layer is only beginning to be attacked by erosion. This young limestone fold is crossed by a transverse stream that creates the gap in the center of the anticlinal arch. (Aerofilms and Aero Pictorial, Ltd.)

mate. The landforms that develop at a particular point in such an area depend on the rock type, its thickness, and its dip at that location. Thus a great diversity of form is possible.

In real landscapes, anticlinal structures (Chapter 12) rarely produce simple ridges like those in Figure 15.26, and synclines are seldom seen as simple troughs. Generally, the anticlines have been "unroofed" by erosion, revealing layer upon layer of rock—some resistant, some weak (Figure 15.27). The scenery is dominated by edges of inclined rock strata that are resistant to erosion. These edges form *homoclinal ridges,* meaning that they are produced by rock layers that dip in one direction (rather than in two directions as along the axis, or center line, of an anticline or syncline). Often there are erosional basins in the centers of the anticlines, indicating that weak rock has been exposed there.

Relief Inversion

Synclines frequently form elongated plateaus or mountain peaks standing above anticlinal valleys. This is a second type of *relief inversion;* the structural highs become topographic lows, and the structural lows create topographic highs, as on the right side of Fig-

ure 15.27. Rock strata that have been folded are stretched and weakened by jointing at anticlinal crests, but are compressed in synclines. This makes the rock at the crests of anticlines less resistant to weathering and erosion than the same rock in synclines, thus facilitating relief inversion.

Most anticlines are elongated domes that die out eventually. Homoclinal ridges wrap around the plunging "noses" of folds, as shown in Figure 15.28. Where folds press together, as in Figure 15.28, the homoclinal ridges loop back and forth, bending one way at the axis of an anticline and the other way at the axis of the neighboring syncline.

Drainage Pattern

In most regions of fold structure there are two types of drainage. The smaller streams are controlled by geological structure and have a "trellis" pattern. Their longer segments are parallel and lie in synclinal, anticlinal, and homoclinal valleys, with short connecting links across the grain of the topography. More often than not the main streams appear to disregard the geological structure. They cut through homoclines, anticlines, and synclines impartially, often crossing resistant layers in

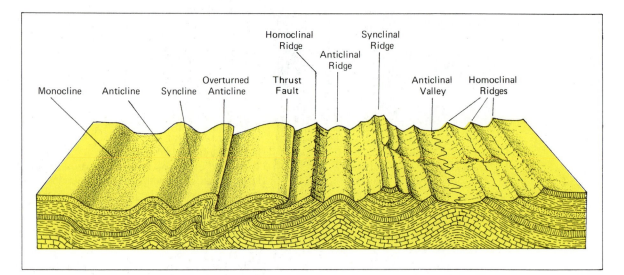

Figure 15.27
The left side of this diagram illustrates the geometry of folded rock strata and the terms used to describe it. Fold structures are rarely seen intact, as in this half of the drawing.

The right side of this diagram illustrates the variety of landforms that can be produced by erosion of folded rock strata. The thinner layers are assumed to be more resistant than the thicker layers. If a portion of the top layer of resistant rock over an anticline has been worn away, *homoclinal ridges* are formed by the exposed edges of rock strata, with an *anticlinal valley* between them. Exposure of a deeper stratum of hard rock may resurrect an *anticlinal ridge* between the facing homoclinal ridges. A *synclinal ridge* is formed where a remnant of resistant rock in a syncline acts as caprock to retard the erosion of the underlying layers. (T. M. Oberlander)

Figure 15.28
Folded strata truncated by erosion form a distinctive ridge and valley landscape in southeastern Pennsylvania. This LANDSAT image spans a distance of approximately 50 miles. The ridges are resistant sandstone, and the valleys are eroded into shale or limestone. Note the nested hairpin ridges that reverse their direction at the axes of anticlines and synclines. The course of the Susquehanna River, which cuts across many anticlinal axes, has never been fully explained. (GEOPIC by Earth Satellite Corp.)

CHAPTER 15

deep gorges. Near Harrisburg, Pennsylvania, the Susquehanna River cuts through five successive massive ridges of resistant sandstone (Figure 15.29). The Indus and Ganges rivers of Pakistan and India cleave through the folded overthrusts of the Himalaya Range in valleys more than 5 km (3 miles) deep.

Several hypotheses have been offered in explanation of streams that are transverse to geological structure. These include the possibilities that: (1) the streams were in place before the structures were formed and are therefore *antecedent streams;* (2) the streams developed on a blanket of younger unfolded sediments that buried older fold structures and are thus *superimposed streams;* and (3) the streams extended headward through points of structural weakness by a process of *progressive stream piracy.* Seldom is there any clear indication how transverse streams actually were established. In fact they constitute the principal mystery associated with landforms controlled by geological structure.

KEY TERMS

physiographic diagram
differential weathering and erosion
cuesta
scarp
backslope
mesa
butte
hogback ridge
homoclinal structure
sinkhole
doline
karst
limestone caverns
stalactite
stalagmite
speleothem
cockpit karst
cone karst
tower karst
corestone
tor
guyot
shield volcano
pahoehoe lava
aa lava
pyroclastic
scoria (cinder) cone
pumice
flood basalts

composite cone (strato-volcano)
dike
plug dome
nuée ardente
caldera
volcanic neck
dike ridge
table mountain
relief inversion
lahar
geothermal energy
geosyncline
fault scarp
tilted block
horst
graben
triangular facet
slickensides
reverse (thrust) fault
rift topography
sag pond
fault creep
anticline
syncline
homoclinal ridge
relief inversion
antecedent stream
superimposed stream
progressive stream piracy

REVIEW QUESTIONS

1. Would you expect differential weathering and erosion to be more important in areas of igneous rocks or areas of sedimentary rocks? Why?
2. What is the relationship between buttes, mesas, and cuestas?
3. What sequence of events is necessary to produce a limestone cavern system that is decorated with stalactites and stalagmites?
4. What are some of the hydrological problems encountered in areas of karst topography?
5. Are any distinctive landforms associated with granitic rocks?
6. In what two types of locations do we find the majority of active oceanic volcanoes?
7. What phenomenon is exemplified by the Columbia Plateau of the state of Washington?
8. In what ways do andesitic volcanoes differ from basaltic volcanoes?
9. How are volcanic calderas formed?
10. What landforms are produced by long-continued differential erosion in volcanic regions after volcanism ceases to be active?
11. In what ways is volcanism important to humanity?
12. What is the modern conception of a "geosyncline"?
13. Describe the geological structure and related landforms of the state of Nevada.
14. What distinctive landforms are encountered in areas of strike-slip faulting?
15. What is the dominant landform in areas of anticlines and synclines? Describe the drainage pattern in such areas.

APPLICATIONS

1. Draw north to south and east to west geological cross sections across your state, using information from state geological surveys, published articles, and maps.
2. How does geological structure affect the landscape in your area? If no specific structures can be seen nearby, where is the closest clear expression of a structural influence on the landscape?
3. In the library, consult the periodical *Geotimes* and look at the section titled "Geologic Events" for a period of twelve consecutive months. Make a catalogue of all of the geological events noted over this period.
4. What is the earthquake history of your area? How often have earthquakes occurred there, and what magnitudes have been experienced? What geological feature has caused the earthquakes?
5. Volcanism seems to occur both where crustal plates are pulling apart and where they are pushing together. Explain this apparent contradiction.
6. How do the individual mountains in regions of folded strata differ from those produced by faulting?
7. Obtain a copy of Erwin Raisz's physiographic map *Landforms of the United States* and divide the nation into physiographic regions of fairly uniform characteristics. Write an essay on the problems of such an undertaking.

FURTHER READING

Green, Jack, and **Nicholas M. Short,** eds. *Volcanic Landforms and Surface Features*. New York: Springer-Verlag (1971), 519 pp. This is a large collection of spectacular high-quality photographs covering every aspect of volcanism.

Harris, Stephen L. *Fire and Ice*. Seattle: The Mountaineers and Pacific Search Press (1980). This fascinating paperback presents the eruptive history of each of the volcanoes of the Cascade Range. Written with authority and enthusiasm.

Hunt, Charles B. *Natural Regions of the United States and Canada*. San Francisco: W. H. Freeman (1974), 725 pp. In Hunt's scheme, natural regions are primarily differentiated by geologic structure. His book presents good general descriptions of the physiography of each of the natural regions of North America, including information on the climate, vegetation, and human use of each region.

Iacopi, Robert. *Earthquake Country*. Menlo Park, Calif.: Lane Books (1964), 192 pp. Well-illustrated popular account of California's earthquake-prone fault landscapes.

Lobeck, A. K. *Geomorphology*. New York: McGraw-Hill (1939), 731 pp. Lobeck's textbook was constructed on the premise that a photograph or diagram is worth a thousand words. This book includes about 200 pages on folding, faulting, and volcanism.

Macdonald, G. A., A. T. Abbot, and **F. L. Peterson.** *Volcanoes in the Sea: The Geology of Hawaii,* 2nd ed. Honolulu: University of Hawaii Press (1983), 517 pp. The geology, geomorphology, and eruptive histories of the Hawaiian volcanoes. Comprehensive, readable, and very well illustrated.

Shelton, John S. *Geology Illustrated*. San Francisco: W. H. Freeman (1966), 434 pp. Nowhere is there published a better collection of photographs illustrating structurally controlled landforms. The text is very readable and original in approach.

Shimer, John A. *Field Guide to Landforms in the United States*. New York: Macmillan (1971), 272 pp. This handbook outlines the nature of the various structurally determined physiographic regions of the United States, with portions of Erwin Raisz's detailed physiographic diagram of the United States used as index maps of each region. Following this is a catalogue of individual landforms, each illustrated by an excellent line drawing.

Simkin, T., L. Siebert, L. McClelland, D. Bridge, C. Newhall, and **J. Latter.** *Volcanoes of the World: A Regional Directory, Gazetteer and Chronology of Volcanism During the Last 10,000 Years*. Stroudsburg, Pa.: Hutchinson Ross (1981), 232 pp. The most complete record of volcanic activity available. Fascinating details and statistics on 1,343 volcanoes and the 5,564 known volcanic eruptions since the Ice Ages.

Sweeting, Marjorie M. *Karst Landforms*. New York: Columbia University Press (1972), 362 pp. This text by an English geomorphologist deals exclusively with the distinctive landforms that develop where limestone is the bedrock.

Twidale, C. R. *Analysis of Landforms*. New York: Wiley (1976), 572 pp. Twidale's massive geomorphology text contains many excellent illustrations of structurally controlled landforms, a particular interest of this author.

Wilcoxson, Kent H. *Chains of Fire*. New York: Chilton (1966), 235 pp. This stimulating book on volcanism, written for the general public, includes many eye-witness accounts of volcanic eruptions of various types. Excellent reading.

Williams, Howell, and **Alexander R. McBirney.** *Volcanology*. San Francisco: W. H. Freeman (1979), 397 pp. This is one of the best of a large number of authoritative texts on volcanic processes and resulting landforms.

land surface. This is an exceptional condition that has occurred only occasionally throughout geologic time. The world's existing glaciers hold an amount of water equal to 5,000 years' flow of the earth's greatest river (the Amazon), or 60 years of rain and snow over the whole planet. By comparison, the storage of water in lakes, rivers, swamps, and artificial reservoirs is trivial.

Most of this frozen water resides in the immense Antarctic and Greenland ice sheets—85 percent in the Antarctic ice sheet alone (Figure 16.1). During much of the preceding 2 million years, two additional continent-sized ice sheets existed, covering all of northern Europe and Canada, as well as 2.6 million sq km (1 million sq miles) of the United States. Smaller ice fields blanketed portions of the world's great mountain ranges. Strangely enough, most of Siberia, widely regarded as almost glacial today, was too dry to sustain large glaciers even at the peak of the Ice Age.

Glaciers of the past are of interest because they have significantly altered the environ-

ment we inhabit. Over vast areas in the higher latitudes, as well as at high altitudes, slowly moving currents of ice have scraped away all the soil and much of the weathered mantle that formerly covered the land surface. Elsewhere the melting of enormous masses of debris-laden ice has completely submerged the previous landscape under a blanket of boulders, gravel, sand, silt, and clay. This has created rolling or flat plains where once there were hills and valleys produced by fluvial erosion. In some areas the effects of glaciers were detrimental to later human use of the land; in oth-

Figure 16.1
Ice caps and outlet glaciers.

(a) The eastern edge of the Greenland ice cap, showing the upper portion of an outlet glacier (right). The ice cap submerges an area of about 1.7 million sq. km (650,000 sq. miles). The featureless expanse of ice in the far distance is as much as 3 km (2 miles) thick. (T. M. O.)

(b) Outlet glaciers from an unconfined highland ice cap on Baffin Island in the Canadian Arctic. (Air Photo Division, Energy, Mines & Resources © Canadian Government)

(a)

(b)

CHAPTER 16

ers the effects were very beneficial. This chapter explores these effects and the processes responsible for them.

GLACIATION PRESENT AND PAST

Glaciers form in the cool climates of high latitudes and lofty elevations where more snow falls each winter than can melt or evaporate during the succeeding warm season. In such locations the snow accumulates from year to year, and the basal layers are gradually recrystallized into solid ice. Its increasing weight and low internal strength cause the ice to deform. A glacier is a mass of ice that is "flowing" under gravitational stress.

Glacier Types

Glaciers exist in several different forms (Figures 16.1 and 16.2). The broadest distinction is between confined and unconfined glaciers, which produce quite different erosional and depositional effects. Dwarfing all other types are the great *continental ice sheets*. These are unconfined blankets of glacial ice that submerge the land surface over areas of millions of square kilometers. Two exist today—the Antarctic ice sheet, which covers 12.5 million sq km (4.8 million sq miles), and the Greenland ice sheet, which has an area of about 1.7 million sq km (650,000 sq miles).

Somewhat similar, but far smaller, are unconfined *highland ice caps* that blanket hundreds to thousands of square kilometers of mountainous terrain in the higher latitudes. Highland ice caps are conspicuous in Iceland, the Canadian Arctic Islands, and the Canadian Rockies. Often they find outlets to lower elevations through valley systems, resulting in confined *outlet glaciers* (Figures 16.1 and 16.2)

More common in high mountains are still smaller *alpine glaciers*. This term includes both confined ice fields, which occupy depressions below high mountain crest lines, and streams of ice that are hemmed in by valley walls as they drain from mountain crest lines to lower elevations. Alpine glaciers originate in high altitude rock-walled ice reservoirs called *cirques* (French, from Latin *circus,* or circle). Cirques are steep-walled rock basins having a distinctive appearance resulting from frost action and glacial erosion. In many mountain areas the only glaciers present are *cirque glaciers*—masses of ice that are restricted to cirques and do not enter valleys. Cirque glaciers range in area from less than 1 to 5 or more sq km (2 sq miles).

Where glacial ice does spill from cirques into the valleys below, it moves slowly in channeled streams called *valley glaciers*. These may have lengths of several tens of kilometers and depths of 1,000 meters or more (over 3,000 ft). Both valley glaciers and the outlet glaciers draining highland ice caps occasionally flow into open areas where they spread out in unconfined pools of ice known as *piedmont glaciers* (Figure 16.2). Thus alpine glaciers can take three forms: cirque glaciers, valley glaciers, and piedmont glaciers. Alpine glaciers are best developed in Alaska; the Canadian Rockies; the European Alps; the Caucasus of the Soviet Union; the Himalaya, Karakorum, and Tien Shan ranges of Asia; the Andes of South America; and the Southern Alps of New Zealand.

More than 1,000 glaciers exist in the conterminous United States, most of them in Washington's Cascade Range. Nearly all are small, having a combined area of barely 500 sq km (200 sq miles). Nevertheless, they are important water sources, providing about 2.1 billion cu meters (1.7 million acre ft) of meltwater each year. Melting of glacial ice in the summer provides water for irrigation just when the demand is highest and when streams fed by rainfall and groundwater inflow are lowest. North America's greatest glaciers are in Alaska, where they cover more than 50,000 sq

Cirque Glacier

Highland Ice Cap

Outlet Glacier

Valley Glacier

Piedmont Glacier

(a)

(b)

(c)

Figure 16.2 (opposite)
Confined and unconfined glaciers.

(a) Glaciers may be unconfined, as *highland ice caps*, which submerge the older erosional topography in mountain regions, or confined, as *cirque* and *valley glaciers*. The valley glacier descending from the unconfined ice cap is an *outlet glacier*. In some regions, the ice from one or more valley glaciers spreads out over a plain and forms a *piedmont glacier*. (John Dawson)

(b) The surface of this valley glacier in the French Alps is broken by numerous crevasses. Cirques are visible in the far distance. (Anita Kolenkow)

(c) The Malaspina Glacier in southern Alaska forms an extensive piedmont ice sheet that terminates at the coastline of the Gulf of Alaska. The ice flow is from right to left in this picture. The dark bands are deformed layers of rock debris carried within the ice. (Austin Post/U.S. Geological Survey)

km (20,000 sq miles). The Malaspina glacier system of Alaska's St. Elias Range is itself more than 4,000 sq km (1,500 sq miles) in area (Figure 16.2c).

Pleistocene Glaciation

We live in the *Holocene* epoch of the Cenozoic era (see Figure 1.8, p. 16)—the "post-glacial" period initiated with the rapid shrinkage of the great northern hemisphere ice sheets, beginning about 14,000 years ago. The last remnants of these ice sheets melted away some 6,000 years ago. The preceding *Pleistocene* epoch—the time of alternating cold (glacial) and warm (interglacial) periods—began about 1.8 million years ago, and is probably not truly ended.

The Holocene epoch seems to be the latest in a series of interglacial interludes preceding glaciations yet to come. Most scientists attribute climatic cooling to astronomical phenomena such as the tilt of the earth's axis, the elongation of the earth's orbit, and the locations in the orbit at which the equinoxes and solstices occur. On this basis it has been predicted that the earth should enter another Ice Age in approximately 23,000 years.

In the past 800,000 years there have been seven or more major episodes of glaciation, each lasting tens of thousands of years, separated by warmer interglacial periods when only Antarctica and Greenland maintained ice covers. There have been other Ice Ages in the earth's history, but they occurred hundreds of millions of years ago and have left no imprint on present landscapes.

At their maximum, Pleistocene glaciers sprawled over some 44 million sq km (17 million sq miles)—almost one-third of the earth's present land area. The North American continental ice sheets, known as the Laurentide ice sheets, spread outward from the vicinity of Hudson Bay and at their maximum terminated close to the present line of the Missouri and Ohio rivers (Figure 16.3). Europe was invaded by the Fennoscandian ice sheet, which originated in northern Scandinavia and spread westward to an ice cap covering the British Isles, and southward to the uplands of Central Europe. The advances and retreats of these glaciers left clear evidence in the form of ice-scoured bedrock and vast blankets of glacially transported rock debris. All around the world Pleistocene glacial erosion completely transformed the scenery of the high mountains, destroying their previously rounded summits and giving them their picturesque "alpine" appearance.

The Pleistocene ice sheets and alpine glaciers left a magnificent legacy of landscapes especially suited for human recreation and enjoyment. The pinnacle of the Matterhorn in Switzerland, the fjords of Norway, the ski slopes of Tuckerman Ravine in New Hampshire, the deep gouge of California's Yosemite Valley, the sandy arm of Cape Cod in Massachusetts, and the ten thousand (and more) lakes of Minnesota—all were fashioned by Pleistocene glaciers.

Figure 16.4

Cryergic landscape. Where temperatures remain below freezing the larger part of the year, landscapes take on a highly distinctive appearance. Below the top meter or so of the land surface, the water in all pore spaces in rock and soil remains frozen solid throughout the year. Permanently frozen rock and soil is known as *permafrost*. Permafrost reaches to depths exceeding 600 meters in northern Alaska and Siberia.

Above the permafrost, the surficial blanket of soil thaws each summer, becoming saturated with water that cannot escape downward due to the frozen substrate. The water-saturated soil moves downslope in lobes as the inset (bottom right) indicates. This type of movement is known as *gelifluction*. Gelifluction lobes are a meter or so in height. Studies show that despite the appearance of great activity produced by these gelifluction lobes, most are inactive, movement occurring only in a few places during each thaw. However, considered over a long period of time, the entire soil cover is draining downslope at a far more rapid pace than that produced in warmer regions by the creep process.

The portion of the soil that freezes and thaws annually, expanding, contracting, and moving downslope, is known as the *active layer*. The long-term effect of the gelifluction process is to smooth the landscape, filling preexisting depressions and peeling down projections. Small valleys are infilled with *colluvium* (slope deposits), forming flat-floored *dells* with poorly developed watercourses. Water leaks through the colluvial valley fills.

Bare rock within the active layer or projecting above it is subject to intense frost weathering during the long winter season. Projecting solid rock masses are rapidly reduced to rubble, which becomes incorporated into the gelifluction lobes. High areas are therefore lowered effectively. Broad summits are covered with angular rock rubble produced by the intensity of the freezing process in exposed sites. The result is a "sea of rocks," or *felsenmeer*. At the edges of rock exposures, large talus accumulations reflect the downslope movement of frost-riven blocks.

The annual freeze and thaw process produces a host of unusual minor landforms in addition to gelifluction lobes. Repeated

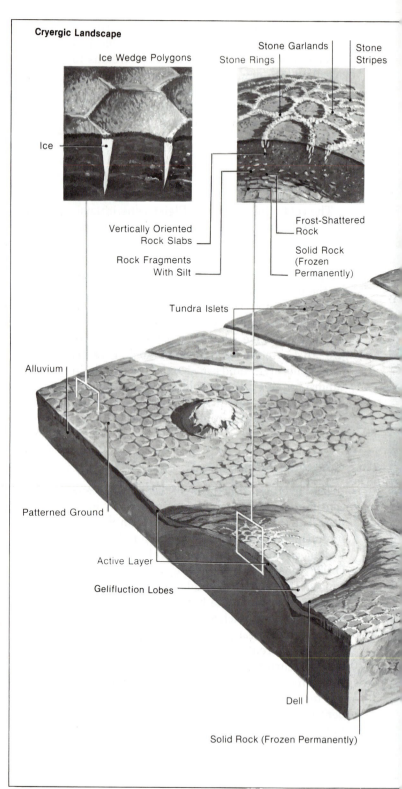

Cryergic Landscape

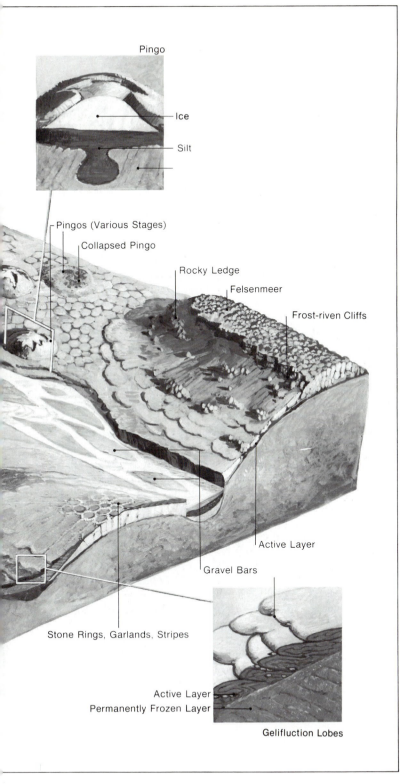

Pingo

Ice

Silt

Pingos (Various Stages)

Collapsed Pingo

Rocky Ledge

Felsenmeer

Frost-riven Cliffs

Active Layer

Gravel Bars

Stone Rings, Garlands, Stripes

Active Layer

Permanently Frozen Layer

Gelifluction Lobes

volume changes in the weathered mantle have the effect of sorting out coarse and fine material. Coarse debris becomes shunted away from the fines. The result is the stone rings shown in the inset (top center). On slopes these rings are drawn out downhill into "garlands" and "stripes" by movements of the active layer.

In lowlands composed of silty alluvial deposits, another type of *patterned ground* is developed. During the freezing process, water-soaked silt first expands, but at prolonged low temperatures eventually begins to contract, cracking into a network of polygonal fissures. Ice forms in these and eventually produces wedges, often a meter across and 5 meters deep (see inset at top left). These are *ice-wedge polygons.* Once formed, they are permanent, continuing to grow very slowly.

Clear ice also forms in the ground in lenslike masses that heave up the overlying sod (see inset at top right). This phenomenon is known as a *pingo.* Pingos develop in old lake beds or marshes that are filling with sediment and vegetation. If a pingo is destroyed by thawing, it leaves a depression ringed by an earth rampart. A large pingo may be 100 meters across and 30 meters high.

Other oddities of landscapes formed by *cryergic* (low-temperature) processes are overhanging (frozen) riverbanks and unusually smooth stream profiles due to the abundance of fresh rock waste, which provides streams with abrasive tools. (John Dawson after T. M. Oberlander)

The Frozen Record

Existing glacial ice is the repository for at least a million years of the earth's history. Frozen into the world's large glaciers is a record of climatic oscillations, the chemical and particulate fallout from the atmosphere, and even the arrival of extraterrestrial debris. The thickest masses of ice, and those preserving the longest record, are the polar ice caps of Greenland and Antarctica, where the annual snowfall and rate of formation of glacier ice are exceedingly low. Here ice cores a few hundred meters long can include a record extending over thousands of years.

The principal use of Greenland and Antarctic ice cores has been to develop an accurate record of climatic oscillations over the past several thousands of years. This is achieved by measuring the ratio of heavy to light isotopes of oxygen in the ice. During colder phases, snowfall is slightly enriched in ^{16}O relative to ^{18}O. There is also a summer-to-winter change in the oxygen isotope ratio. This can be used to identify annual layers in the ice of the last several thousands of years, resulting in very precise age determinations. Ice cores up to 2,000 meters (6,500 ft) in length have been extracted from the Greenland and Antarctic ice caps. These provide a record extending back about 100,000 years in both the Arctic and Antarctic. Analysis of these cores verifies that major climatic changes in the northern and southern hemispheres have been simultaneous.

Studies of ice cores have revealed periodicities in glacial activity that seem related to variations in solar activity. This helps to explain the "Little Ice Age," from about A.D. 1580 to 1900, when all glaciers, large and small, advanced. To prove that variations in solar activity have been responsible for glacier fluctuations, the nitrate content of the ice has been analyzed. Nitrates are formed high above the earth's surface during periods of the aurora, when energetic particles from the sun bombard the earth's magnetosphere. Unusual concentrations of nitrates in glacial ice are interpreted as a signal of increased solar output. The nitrate record in the ice mirrors the oxygen isotope record, supporting the solar explanation of climatic change. When nitrates are reduced (during cold periods) the sodium content of the ice rises, suggesting a change in the atmospheric circulation, with more salt-bearing marine air brought over the ice caps.

Human effects appear in the younger ice. Rising atmospheric pollution is well recorded. Lead concentrations in the ice increase 200-fold over the past 3,000 years. There is an abrupt rise in the methane content of air bubbles in the ice about 400 years ago, after nearly 30,000 years of constant methane levels. As methane is second only to carbon dioxide as a "greenhouse gas" that causes atmospheric warming, this finding, and its interpretation, are matters of serious concern.

Among the most interesting of the phenomena preserved in polar ice are large numbers of meteorites that have impacted on the Antarctic ice sheet. These have subsequently been conveyed by ice flow to barriers where flow stagnates, ablation continues, and the meteorites become concentrated. The ages of Antarctic meteorites range upward to 1.5 million years, and the variety of meteorite types is astonishing. The highly unusual chemical compositions of some suggest that they were produced as "splash" from catastrophic meteorite impacts on both Mars and the earth's moon. The largest iron meteorite yet discovered, weighing 136 kg (300 lb) was found on the Antarctic ice. The cushion of the glacier and the absence of chemical weathering and organic activity in the frozen environment account for the excellent preservation of the individual specimens. They are sealed against contamination and, quite appropriately, are kept refrigerated for their voyages to the scientific laboratories of the world.

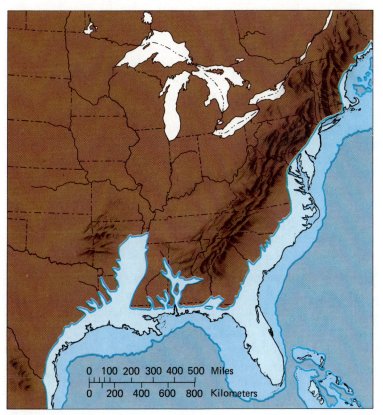

Figure 16.5
This map of the Atlantic and Gulf coasts of the United States shows the degree to which the storage of water as glacial ice affects the position of coastlines. The darkest blue tone represents ocean areas deeper than 130 meters (430 ft), which remained submerged throughout the Pleistocene. The middle blue tone shows presently submerged areas that were exposed as land during low stands of the sea that accompanied maximum advances of Pleistocene ice sheets. The lightest blue tone indicates land areas that would be submerged if the present continental ice sheets in Greenland and Antarctica were to melt, raising sea level by about 65 meters (210 ft).

understood. An ice sheet is in a delicate balance between growth and retreat, and relatively small changes in climate can cause the balance to shift either way. Some of the hypotheses advanced to account for the Ice Ages include changes in the positions of the continents and in ocean circulation patterns as a result of sea-floor spreading; increased altitude of the land masses after a period of geologic upheaval; and variations in the amount of solar radiation the earth receives. The last could be caused by changes in the sun's energy output, in the earth's relationship to the sun, or in the atmosphere's content of carbon dioxide and volcanic dust.

Recent analyses seem to support the long-argued contention that cyclic changes in the inclination of the earth's axis, in the elongation of the earth's orbit around the sun, and in the orbital positions of the equinoxes and solstices account for the periodicity of warm and cold climates during the Pleistocene epoch. These astronomical cycles of varying lengths periodically reinforce and negate one another, resulting in intervals of above or below average solar energy input to the earth. Changes in oxygen isotope ratios in ocean sediments composed of microscopic marine organisms also reveal a periodicity. The isotope ratios are a reflection of land ice volumes that affect the chemistry of seawater. The oxygen isotope ratios have been found to fluctuate in rhythm with astronomical cycles we have just discussed, making it appear that the latter have controlled the pulse of glaciation over the past 2 million years, as first proposed in 1912 by the Serbian scientist Milutin Milankovitch. Quite probably the movement of continents into arctic latitudes in late Cenozoic times was the major factor making continental ice sheets possible, with the astronomical phenomena subsequently producing cycles of ice sheet growth and decay.

ANATOMY AND DYNAMICS OF GLACIERS

Glaciers are formed above the annual *snowline,* the elevation at which some winter snow survives the high temperatures of the following summer. Snowline elevation is thus a function of both the amount of winter snowfall and the temperature of the warm season.

In polar areas the snowline is at sea level. With distance from the poles, the snowline rises progressively higher and reaches a maximum of 6,000 meters (20,000 ft) or more in the dry zone, some 20 degrees north and south of the equator. Closer to the equator, cloudiness and precipitation increase, permitting the snowline to descend almost 1,000 meters (3,300 ft) below its level in the subtropics.

In the Colorado Rockies and California's Sierra Nevada the snowline on slopes exposed to direct solar radiation would be at about 4,500 meters (15,000 ft), just above the highest peaks of both ranges. But on the shaded north- and east-facing slopes the snowline descends to between 3,600 and 4,000 meters (12,000 to 13,000 ft). Thus snow can accumulate, forming glaciers in cool north- and east-facing depressions in the higher portions of both ranges. Farther north, the gradual descent of the snowline with increasing latitude permits even the south- and west-facing slopes of several peaks in Washington's Cascade Range to be glacier-clad, although their summits are lower than ice-free peaks in both the Rockies and Sierra Nevada.

Formation of Glacial Ice

Snowfall is the raw material for glaciers, but glaciers themselves are solid ice. Snow usually arrives at the surface in the form of lacy hexagonal ice crystals. Eventually, compaction, vaporization, and local melting and refreezing convert the initial delicate snowflakes to rounded granules of ice. When the snow is wet and heavy, the transformation of snow crystals to ice granules requires only a few days. But in the extreme cold of Antarctica the snow remains fluffy for years, and the transformation to ice is extremely slow.

Newly fallen snow contains much air space and has a specific gravity (weight relative to water) between 0.05 and 0.15. With time, and under the pressure of overlying snow, the ice granules formed from snowflakes gradually pack closer together. In midlatitude areas, the granular material begins to coalesce after one summer, and attains a specific gravity of about 0.55. Such material is called *firn*. In tens or hundreds of years, depending on summer temperatures, the pore spaces in the lower portions of the firn gradually disappear until the basal firn has become solid ice with a specific gravity of about 0.85. Then, as air bubbles gradually disappear, the specific gravity of the ice increases to about 0.9. The result is true glacier ice. Such ice usually forms at a depth of some 50 meters (150 ft) beneath the firn at the surface.

Glacier Mass Balance

Glaciers are dynamic ice flow systems—expanding or contracting, thickening in some sections and thinning in others, but always feeding ice forward to replace ice lost by melting or evaporation below the snowline. Whether a glacier advances or recedes depends on the glacier's *mass balance*—the balance between the input of snow above the snowline, and the loss, or *ablation*, of glacial ice by melting or evaporation below the snowline.

All glaciers consist of two portions, an upper *accumulation zone,* in which the annual input of snow exceeds the loss to ablation, and a lower *ablation zone,* in which the annual ablation loss exceeds the direct input by snowfall. Glacial flow transports the excess mass from the accumulation zone to the area of deficit in the ablation zone (Figure 16.6). The *equilibrium line* separates the zones of net gain and net loss, and more or less coincides with the *firn line,* which is the annual snowline on the glacier itself. During the winter, snow accumulation exceeds ablation on a glacier. In the summer, when snowfalls are less frequent or absent altogether, ablation is dominant (Figure 16.7). Whether a glacier experiences visible growth or contraction depends on the an-

Figure 16.6
The dynamics of a cirque glacier. Bold arrows show relative gains and losses due to accumulation and ablation. The *firn line* marks the boundary between the new accumulation of firn and older glacial ice; the glacier's surface is white upslope from the firn line and bluish on the downslope side. Cirque glaciers exhibit a deep crevasse, called a *bergschrund,* at their heads. The bergschrund is a clear indication of the downslope motion of the glacier.

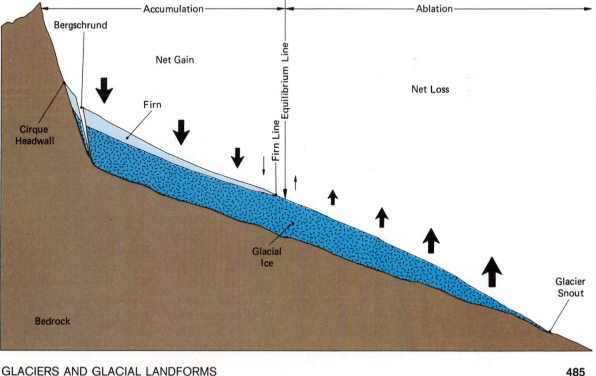

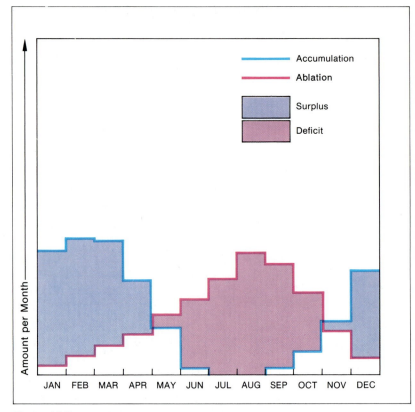

Figure 16.7
This diagram illustrates a typical glacier mass balance in the northern hemisphere. During the winter months, the accumulation of new snow on the glacier exceeds the loss of ice and snow by melting and evaporation. During the summer months, ablation exceeds accumulation, and the glacier experiences a deficit. The mass of the glacier increases during periods of surplus and decreases during periods of deficit. If the net annual surplus exceeds the net annual deficit over a period of years, the glacier grows and advances.

nual mass balance between accumulation and ablation.

Ice is always moving forward to the front of an active glacier; this is true whether the glacier margin is advancing, retreating, or fixed in position. Where the forward ice margin (or "snout" in the case of a valley glacier) is stable in position from year to year, the volume of ice flowing to the margin is exactly balanced by that removed by melting and direct evaporation (sublimation). An increase in the rate of ice arrival or a decrease in the ablation rate will cause the glacier snout to advance. Such a glacier would have a *positive mass balance*. If the rate of ablation exceeds the rate of ice replacement, the snout retreats, and the glacier has a *negative mass balance*.

Glaciers in the Canadian Rockies, Alaska, and Scandinavia lose a depth of about 12 meters (40 ft) of ice to ablation each year. If they are to hold their positions, this loss must be replaced by inflow from above the equilibrium line. Most of these glaciers have been retreating since about 1900, indicating that negative mass balances have prevailed throughout the present century.

Movement of Glaciers

Because of its much greater resistance to deformation, ice flows in a very different manner from water. When a mass of ice attains a thickness of about 50 meters (150 ft), it begins to spread and flow outward as a consequence of its own weight. It can move in two ways: by internal deformation and by slipping over its bed. Internal deformation is very gradual, amounting to a few centimeters per day, and depends on the fact that ice melts under pressure. The flow is generally laminar (Figure 16.8), with little internal mixing, and is the result of repeated partial melting and recrystallization of ice that is subjected to various stresses within the moving ice mass.

Glaciers may also move by *basal slip,* in which the ice mass as a whole slides over its bed on a film of water. This water is produced by the melting of ice that is pressed against obstacles on the glacier bed (called *pressure melting*). Basal slip permits more rapid flow than does internal deformation alone. However, the total forward movement of ice in a midlatitude alpine glacier is rarely more than a meter a day.

Temperate and Cold Glaciers

The nature and rate of movement within a glacier is closely related to the temperature of the glacial ice. In *temperate glaciers,* meltwater is produced by pressure melting both within and at the base of the ice and by summer melting on the glacier surface. When the meltwater refreezes within the glacier, it liberates the la-

Figure 16.8
This diagram, in which the vertical scale has been exaggerated, shows the ice velocity at various points in the Saskatchewan Glacier in the Canadian Rockies. The movement of a temperate glacier is accomplished partly by basal slip over its bed (black arrow) and partly by internal deformation of the ice (red arrows). Internal deformation causes the surface ice to move more rapidly than deeper portions of the glacier. The gray *velocity-depth profiles* show forward displacement at various levels within the ice. Streamlines (blue) show paths of points within the glacier and indicate laminar flow of the ice.

Only velocities at the surface of a glacier can be measured directly; the internal velocities indicated in the diagram are inferred. The measured velocity of this glacier's surface in its thicker upper portion is about 400 meters (1,300 ft) per year. Near the snout, the velocity is approximately 100 meters (330 ft) per year. Near the snout of the glacier, most of the movement is by water-lubricated basal slip. (Doug Armstrong after M. F. Meier, 1960)

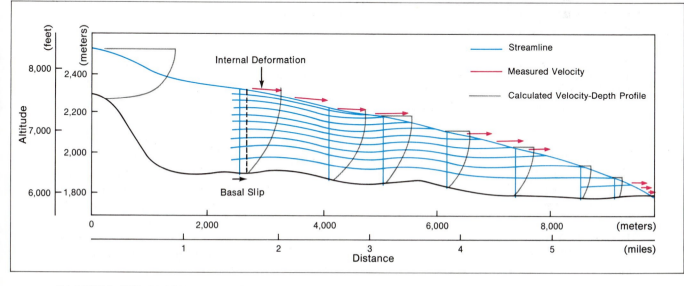

tent heat of crystallization, which keeps the glacier "warm"—near 0°C (32°F), so that it can continue to melt when pressed against obstacles on its bed. Under such conditions, water-lubricated basal slip is a major component of ice flow.

In polar regions, however, summer melting is minimal and temperatures remain far below the freezing point throughout the depth of the ice. The result is *cold glaciers,* or *polar glaciers,* such as those in Antarctica. Cold glaciers are frozen to their beds, at least on their accessible margins. Because there is no basal slip and internal deformation is greatly reduced, glaciers of this type are extremely slow-moving. We shall see shortly that the distinction between cold and temperate glaciers is important in the development of landforms by glacial erosion.

Glacier Structures

In confined valley glaciers, the fastest ice currents are at the center of the ice surface, with the margins of the ice stream retarded by friction with the valley walls. The shearing that results within the ice stream causes its surface to split open in a series of parallel fissures called *crevasses* (Figures 16.2b and 16.9). Where steepening of the glacial bed causes a valley glacier to accelerate in speed, crevasses arc across the breadth of the ice surface. In some places glacial beds steepen so abruptly that ice streams collapse and break up in

Figure 16.9
These ice streams descending steeply from the summit of Mont Blanc in the French Alps are broken by crevasses. As the glaciers pass over convexities in their beds, they produce small icefalls. (T. M. O.)

Glacier Bursts

The snouts of glaciers tend to be inhospitable places—devoid of life, somehow ominous. In fact, there is a human hazard associated with large, confined glaciers. Occasionally, enormous sudden floods of water discharge from the glacier snout, often carrying masses of ice along. These abrupt floods, known as glacier bursts, have taken a toll of life and property in many valleys downstream from large glaciers.

Glacier bursts have different causes. Where glaciers cover the summits of volcanoes, the danger is obvious. Even small volcanic eruptions will melt the ice around the volcanic vent, producing a sudden flood, often laden with debris eroded from the volcano's slopes. The English term "glacier burst" is a direct translation of the older Icelandic term, *jokulhlaup*. Subglacial volcanic eruptions have been a constant hazard for the people of Iceland, where ice caps cover the largest and most active of the shield volcanoes that compose the entire island. Past jokulhlaups have caused much devastation here, sometimes burying entire valley floors—their farms, people, and livestock—under volcanic mud.

The term *jokulhlaup* is now widely used to refer to any flood produced from a glacier, regardless of its cause, while the term *lahar* indicates a debris flood from a volcano. An especially catastrophic jokulhlaup-lahar event occurred in 1985 near Bogota, Colombia, when a violent eruption of the 5,400-meter (17,716-ft) volcano, Nevado del Ruiz, melted its ice cap and sent torrents of mud into surrounding valleys. Some 20,000 people were killed by mudflows that obliterated one sizable town.

A second type of glacier burst occurs where a valley glacier blocks the drainage of water from an unglaciated tributary valley. The result is a lake backed up into the tributary valley. When the lake attains a critical depth, its hydrostatic pressure is sufficient to buoy up the blocking ice, allowing the lake waters to escape beneath the glacier, flooding out at its snout. Some ice-dammed lakes empty according to a regular schedule, so that the flood hazard is well known.

Floods from ice-dammed lakes are common in the Alps, Andes, Himalayas, Canadian Rockies, Alaskan ranges, and Norway. The largest flood on record resulted from the collapse of an ice dam in late Pleistocene time. The ice was a lobe that pushed south from the Canadian Rockies to block westward drainage from Montana, creating Glacial Lake Missoula, a vast body of deep water. When the ice dam failed, the resulting westward discharge of water exceeded the maximum recorded flow of the Amazon River by about eight times. It scoured the basaltic plateau of eastern Washington into the unique landscape now known as the Channeled Scabland.

Jokulhlaups are also associated with glacier surges. Holes bored through one surging glacier in Alaska reveal that during the surge there is an increase in water pressure at the base of the ice, which buoys up the ice mass. In this stage a bulge of ice moves through the glacier as an internal wave. The glacier studied thickened by about 100 meters (330 ft) as the heavily crevassed "surge front" moved past at a velocity of about 80 meters (260 ft) per day. About 95 percent of the forward motion in the thickened section was by basal slip facilitated by a cushion of pressurized water under the ice. Slowdowns in the rate of ice flow coincided with large-scale flooding from the glacier snout. The surge stopped when muddy water gushed massively from the crevasses in the surge front, dissipating the subglacial water cushion. The conclusion is that when subglacial water is unable to drain normally through open tunnels under the ice, the water increases in pressure, spreads laterally, and lifts the ice, triggering the surge. The surge itself hinders the escape of water by disrupting subglacial drainage paths. When the water finally finds an escape route, it floods out as a jokulhlaup and the surging phase concludes.

With some knowledge of the processes affecting glacial surges, jokulhlaup prediction based on monitoring of subglacial water conditions should help to minimize loss of life to this particular natural hazard.

icefalls (Figure 16.9). Such areas are dangerous, since icefalls include enormous unstable blocks that frequently topple with crushing force.

Glacier Surges

Occasionally an alpine glacier moving by basal slip exhibits behavior called a *surge,* in which the ice front advances as much as 20 meters (65 ft) a day. Glacial surges may be caused in various ways. Some occur when large pockets of water form under the ice, or when lake water in tributary valleys blocked by the ice seeps out under the glacier. Other surges occur after periods of unusually high snow accumulation, which cause the glacier to thicken in the accumulation area. This produces a "bulge" or wave of ice that subsequently moves through the glacier at a speed that exceeds the normal ice velocity. When the bulge reaches the glacier snout, the ice front begins to push forward. This merely signifies that, for a time, much more ice is arriving than is melting away.

GLACIAL MODIFICATION OF LANDSCAPES

Flowing ice is a far more powerful agent of landscape change than running water. A 1,000-meter-thick alpine glacier can excavate rock material from both the floor and walls of a valley, whereas a stream's direct action is confined to the valley floor. Furthermore, the special properties of glacial ice enable it to remove and transport debris by processes not available to running water. Like streams, glaciers have their greatest erosional potential when their depths and velocities are greatest, but unlike streams, they do not drop their loads of debris wherever their velocity decreases. In fact, large-scale glacial deposition occurs only where the ice terminates or where great masses of ice stagnate and melt with no forward motion at all.

Glacial Erosion

Glacial ice by itself has little destructive ability. Ice cannot generate enough pressure to break away unweathered rock, since slowly moving ice yields by melting as it is pressed against any solid obstacle. However, a glacier that is frozen onto its bed will tear out rock and soil material as it is pushed forward by the ice behind it. Ice that freezes onto rocks and pulls them loose soon becomes armored with rock debris. It is similar to coarse sandpaper, with the ice acting as the glue that holds the abrasive particles in place. A rock-studded glacial bottom, or *sole,* is an effective file. The faster the glacier moves, the greater the number of abrasive tools scraped over a given surface and the greater the potential for erosion.

Erosional Processes

All glacial erosion can be attributed to the processes of tearing out, known as *plucking* or *quarrying;* filing down, known as *abrasion;* and *crushing* of rock projections in the glacier's path. Plucking is best accomplished by cold glaciers in which the ice is frozen to the glacier bed. Abrasion occurs under temperate glaciers where basal slip is caused by water at the base of the ice. Crushing occurs where there is basal slip and the subglacial surface is highly irregular.

In general, plucking roughens the surface under the ice and provides the glacier with tools for abrasion. Glaciers probably remove far more material by plucking than by abrasion, particularly where the ice flow encounters unconsolidated (though frozen) soil and where well-developed jointing divides rock into blocks of movable size. Abrasion tends to smooth rock surfaces and to gouge out grooves parallel to the direction of ice flow (Figure 16.10a).

(a)

(b)

Figure 16.10

Effects of abrasion by debris-laden glacial ice.

(a) These bedrock outcrops on Vancouver Island, British Columbia, have been smoothed and streamlined by the abrasion of glacial ice, which moved from left to right at this location. (T. M. O.)

(b) Crescentic fractures (or ''chatter marks'') on glacially polished granitic rock. Scale is provided by a 9-cm-long jackknife in the center of the view. The direction of the ice flow is indicated by the *dip* of the fracture, which is to the right here. (Daniel O. Holmes)

The effects of glacial erosion can be seen on many scales, from an individual rock outcrop to an entire countryside. The first alteration a glacier produces when it invades a region is the removal of the soil and weathered mantle. This demonstrates that glacial erosion is more potent than the gradational processes that preceded it. The resulting exposure of the underlying bedrock enables the glacier to begin to remove unweathered rock, which would remain relatively untouched by normal processes of erosion.

Erosional Forms

The abrasion process causes bedrock to be scratched, chipped, gouged, and filed down (Figure 16.10b). Crescent-shaped fractures and gouges called *chatter marks* or *friction cracks,* produced by the pressure of rock upon rock, are characteristic small-scale effects of glacial abrasion on individual rock outcrops. Sometimes large expanses of rock are sculptured into smoothly contoured furrows and concavities. Smoothly sculptured forms may be created by either flowing meltwater under a glacier or a subglacial slush of sandy mud mixed with ice crystals.

The difference between abrasion and plucking can be seen on single rock outcrops, which often display *stoss and lee topography.* As Figure 16.11 indicates, abrasion tends to smooth and streamline the upglacier (stoss) face of rock knobs, while plucking and crushing steepens and roughens the downglacier (lee) side.

Magnitude of Erosion

Severe glacial erosion is most evident where it has quarried out numerous rock basins that subsequently fill with water, as in Figure 16.12. Some rock troughs have been deepened more than 600 meters (2,000 ft) by glacial erosion. Examples are Norway's Sogne and Hardanger fjords and California's Yosemite Valley. Glacial erosion excavated some 400 meters (1,350 ft) of bedrock to form the basin now occupied by Lake Superior. The ice-scoured areas of Canada, New England, and Minnesota contain countless large and small lakes in basins produced by glacial quarrying.

Available data indicate that glacial erosion by abrasion alone is 20 to 50 times more rapid than the rate of erosion of nearby unglaciated areas. Glacial abrasion of marks chiselled into bedrock, or of objects artificially emplaced in the basal ice of glaciers themselves, indicates removal rates of 1 to 40 mm (0.04 to 1.5 in.) per year, the variation being related to differences in rock resistance to abrasion.

Figure 16.11
In this diagram illustrating the characteristic action of glaciers on outcrops of jointed rock, the direction of glacier flow is from left to right. The upglacier *(stoss)* side of the rock is smoothed and polished by abrasion, and the downglacier *(lee)* side is roughened as the glacier crushes projections and plucks large rocks from the outcrop. (T. M. O.)

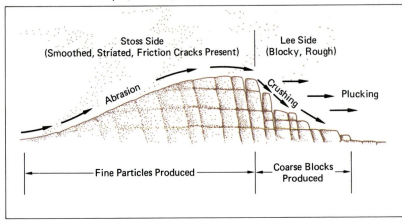

Stoss Side
(Smoothed, Striated, Friction Cracks Present)

Lee Side
(Blocky, Rough)

Abrasion

Crushing

Plucking

Fine Particles Produced

Coarse Blocks Produced

CHAPTER 16

Figure 16.12
Ice scouring in crystalline rocks produces multitudes of basins that later fill with water, as seen in this view of the Laurentian Shield north of Ottawa, in the Canadian province of Quebec. (Rolf and Mari Wesche)

Glacial Deposition

In some ways glaciers resemble conveyor belts that steadily feed fragmented rock forward to the glacial margin. Most of the debris carried by glaciers is concentrated in the lowest few meters of the ice, and in the case of alpine glaciers, along the ice margins. Because flow in glaciers is laminar, there is little upward or lateral transport of debris into the body of a glacier.

At the glacier terminus, or snout, transported debris is melted out. This material is overridden by advancing ice. Along a stationary ice front it accumulates in a ridge-like glacial dump. Even when the ice front is retreating, the "conveyor belt" remains in action, continuing to transport debris forward to the glacier snout. Forward transport stops only when the ice thins to the point where flow cannot be maintained over obstacles in the glacial bed. When this occurs the ice mass decays by melting in place.

Glacial Drift

The general term for all types of material deposited directly or indirectly from glaciers is *glacial drift*. The term "drift" dates from about 150 years ago, when it was supposed that the blanket of rock debris covering vast areas in Europe and North America was left by the Biblical flood associated with Noah.

Glacial drift is quite variable in character, for it can be produced in many different ways. When ice advances it may become so laden with soil and rock detritus that it begins to release some of its load, which is plastered over the subglacial land surface in the form of *glacial till*. Till is a distinctive mixture of coarse and fine material, including fragments of all sizes, from boulders to clay; indeed, it was once called "boulder-clay." Till is most easily distinguished by the absence of any form of sorting or bedding (Figure 16.13a). Usually some boulders within the till have facets that have been ground smooth by glacial abrasion. Till can be a problem for engineers because excavations in it are often delayed by unexpected encounters with large boulders or extremely tough clays.

When the ice front has pushed to its limit and finally begins to retreat, its marginal portions sometimes stagnate, melting in place, with great volumes of rock debris collecting amid the irregular topography of the decaying ice. When the ice finally disappears, a rather disorganized mass of hills and depressions is

Figure 16.14
The steep-walled basins in this photograph of the
Wind River Mountains, Wyoming, are *cirques* formed
by the quarrying action of glacial ice. Small lakes, or
tarns, are present in the cirques and appear in the
glacially scoured area in the foreground. This part of
the Wind River Range preserves large areas of the
rolling preglacial topography between the separate
cirque basins. (Austin Post/U.S. Geological Survey)

cirques into the valleys below. The results are
nearly as impressive as glacial modification of
uplands. Valley glaciers deepen fluvial valleys

and steepen their walls, converting them into
glacial troughs (Figure 16.17). The interlocking spurs created by the irregular paths of
deeply incised mountain streams are trimmed
back. Valleys are expanded and simplified into
open grooves as they are deepened. The result
is a semicircular or U-shaped valley cross section.

As a consequence of the greater volume of
channeled ice, glacial deepening of major valleys is more rapid than deepening of smaller
tributary valleys. Retreat of the ice leaves the

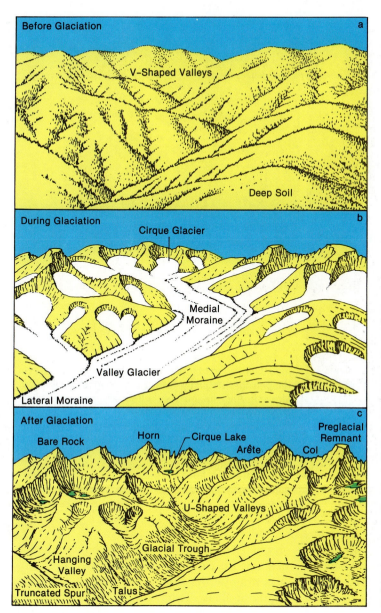

Figure 16.15
This series of diagrams shows the transformation of a fluvially dissected mountain region (a) into a glacially eroded landscape (c). At the highest levels, snowfields develop into *firn* basins that gradually produce glacial ice (b). Cirque growth by glacial scouring and frost action creates sharp crested *arêtes* that come together in *horns* (c). At lower elevations, ice streams convert sinuous fluvial valleys into open *glacial troughs*. Valleys are deepened and expanded, and their walls are made much steeper by glacial erosion. The greater modification of trunk valleys by large ice streams leaves smaller tributary valleys hanging along their margins, producing *hanging valley* waterfalls and cascades. Uneven glacial scouring of jointed bedrock excavates basins that later fill with *tarn lakes*. The result is alpine scenery.
(T. M. Oberlander)

tributary valleys hanging along the margins of the major glacial troughs. Streams occupying these valleys descend to the floor of the trough by picturesque cascades, or by spectacular *hanging valley waterfalls* (Figure 16.16b).

Because glaciers do not excavate weathered or jointed rock uniformly, erosion by valley glaciers produces irregular longitudinal valley profiles. The result may be a *glacial stairway* with high steps and waterfalls in the center of the trough itself. Lakes known as *tarns* occupy basins quarried out by the ice. Strings of tarns, such as those visible in Figure 16.14, are quite common in alpine topography, being called *pa-*

(a)

(c)

(b)

Figure 16.16
Erosional forms resulting from Alpine glaciation.

(a) Severe frost weathering may reduce the surfaces between adjacent cirques to the narrow, ragged rock ridges, or *arêtes,* seen here in the French Alps. (T. M. O.)

(b) The deepening of the principal stream valleys in mountain regions by glacial erosion is more rapid than deepening of tributary valleys that channel smaller ice streams. After the glaciers retreat, the tributary valleys may open into the main glacial trough high above its floor. The waterfall in the photograph is a stream falling from such a *hanging valley* in Yosemite National Park, California. (T. M. O.)

(c) The Matterhorn in Switzerland is an extreme example of a *horn peak* formed by erosion where the headwalls of three or more cirques intersect. (John E. Kesseli)

ternoster lakes because of their resemblance to the beads on a rosary.

Fjords

Downward erosion by valley glaciers is best displayed in the *fjord* landscapes of Greenland, Norway, eastern and western Canada, Alaska, Chile, and New Zealand (Figure 16.17b). Fjords are glacial troughs that have been cut far below present sea level, resulting in deeply penetrating arms of the sea hemmed in by rock walls 1,000 or more meters (over 3,300 feet) high. The glacial origin of fjords is demonstrated by their longitudinal profiles, which show deep basins that could not have been excavated by fluvial processes. The rock floors of some fjords are more than 1,000 meters below sea level. This is because a glacier with a den-

sity of 0.9 entering the sea, which has a density of about 1.0, continues to erode downward until it is nine-tenths submerged. Only then will it begin to float. Thus a 1,000-meter-thick ice stream in a fjord still has the potential to erode its bed in water that is more than 800 meters (2,700 ft) deep.

Depositional Forms

The depositional landforms produced by alpine glaciation are no match in scenic grandeur for alpine erosional forms, but they are

(a)

(b)

Figure 16.17
Erosional effects of valley glaciers.

(a) Valley glaciers follow preexisting fluvial valleys as they flow toward lower elevations. The glacier scours the valley floor and walls, removing irregularities and forming a smooth *glacial trough* such as this example in California's Sierra Nevada. (T. M. O.)

(b) If a valley glacier reaches the sea, it can continue to erode its channel below sea level until the ice is nine-tenths submerged, when it will finally float. When the glacier retreats, it leaves a deep rock-walled ocean inlet, or *fjord*, such as Milford Sound, New Zealand. (Alvin Lynch)

nevertheless quite conspicuous. The largest depositional features produced by valley glaciers are *lateral moraines* built along the sides of the ice streams. These are ridges of till banked against valley walls or issuing from mountain canyons. They frequently rise more than 300 meters (1,000 ft) above surrounding lowlands. The waxing and waning of glaciers during the Pleistocene created many smaller end moraines that commonly make a succession of arcs across valleys, impounding lakes between the morainic crests (Figure 16.18). Many mountain campgrounds in the Sierra Nevada and Rocky Mountains are situated on these bouldery end moraines.

Climatic cooling since the postglacial temperature maximum about 6,000 years ago pro-

duced glacial readvances in all the world's high mountains. The advances created Neoglacial (Holocene) moraines, visible mainly in cirques. These moraines are usually ridges or tongues of coarse rock rubble. However, present glaciers have retreated well behind their various Neoglacial moraines as a result of the warming trend of the present century.

Valley glaciers are also sources of glacial outwash, which aggrades stream valleys beyond the glacial margin (Figure 14.14b, p. 416). The deposits consist of sand and gravel, changing to silt at greater distances from the sediment source. In intervals between major glaciations, such as the present period, the decrease in stream loads increases available stream energy, causing the streams to incise the outwash deposits, converting them into terraces.

LANDFORMS RESULTING FROM CONTINENTAL GLACIATION

The dual effects of continental glaciation are clearly illustrated in three dissimilar North

KEY TERMS

continental ice sheet
highland ice cap
glacier
 outlet
 alpine
 cirque
 valley
 piedmont
Holocene
Pleistocene
cryergic
eustatic
snowline
firn
bergschrund
accumulation zone
ablation zone
equilibrium line
firn line
mass balance
 positive
 negative
basal slip
pressure melting
glacier
 temperate
 cold
 polar
crevasse
icefall
glacial surge
plucking
quarrying
abrasion
crushing

chatter marks
friction cracks
stoss and lee topography
glacial drift
glacial till
ice stagnation topography
glacial outwash
glacial erratic
moraine
 end
 terminal
 recessional
 ground
till plain
drift plain
outwash plain
valley train
arête
glacial horn
glacial trough
glacial stairway
hanging valley
tarn
paternoster lakes
fjord
lateral moraine
fluted drift
drumlin
washboard moraine
kame
kettle
kame moraine
pitted outwash plain
esker

REVIEW QUESTIONS

1. Approximately how large a temperature decrease, averaged globally, would be necessary to cause the return of an Ice Age?
2. Indicate three different ways of classifying glaciers. Name the glacier types within each classification.
3. What is the Holocene epoch? How does it relate to the Pleistocene epoch? What are the approximate beginning and ending dates for the Wisconsinan stage of the Pleistocene epoch?
4. How many major Pleistocene glaciations can be distinguished, based on oxygen isotope records from the deep seas?

5. What were the effects of Ice Age climates in unglaciated areas?
6. How is the "snowline" relevant to glacier formation? How does the snowline vary with latitude from pole to pole?
7. Define glacier "mass balance." How is the glacier mass balance reflected at the glacier "snout"?
8. How do "temperate" and "cold" glaciers differ? How are the differences reflected in glacial movement and related glacial erosional processes?
9. What is "stoss and lee" topography?
10. Under what conditions does glacier ice stagnation occur?
11. What types of deposits are included in the term "glacial drift"?
12. What landforms are created by the enlargement of adjacent cirques?
13. How could you distinguish between a valley that has been glaciated and one that has not been glaciated?
14. How do the moraines produced by continental and alpine glaciation differ?
15. What are the dominant landforms of the Laurentian Shield? of the areas of glacial deposition in the Midwest of the United States and in New England?
16. In what two ways are New York's Finger Lakes similar to fjords?
17. What are the beneficial results of glaciation in North America?

APPLICATIONS

1. It is usually possible to determine the limit of the major late Pleistocene glacial advance by merely looking at a map of moderate scale. Why?
2. If both drumlins and eskers are seen in the same area, which must have formed first?
3. How might the stones found in glacial till differ in appearance from those deposited by stream or wave action? If your campus is in a glaciated region, collect or photograph the most distinctively formed till stone you can find (often in or washed out of a road cut through a moraine). Compare your find with those of your classmates.
4. Obtain a topographic map of the closest area of unglaciated mountains. On such a map, relief is shown by topographic contours. Redraw the bold index contours to show the area after transformation by alpine glaciation. Now redraw the contours again to represent the area after continental glaciation.
5. If your campus is in (or close to) an area of alpine or continental glaciation, make a catalogue of all the glacial landforms you can identify, indicating the specific location of each.
6. Obtain the *Atlas of American Agriculture* from your campus library. The atlas graphically displays agricultural statistics for the nation, using counties as data units. On how many of these economic maps can you identify the effects of continental glaciation? Wherever the glaciated areas stand out, attempt to explain the specific effect of glaciation on the particular economic phenomenon.

FURTHER READING

Andrews, John T. *Glacial Systems: An Approach to Glaciers and Their Environments.* North Scituate, Mass.: Duxbury Press (1975), 191 pp. This treatment of glaciers, glacial landforms,

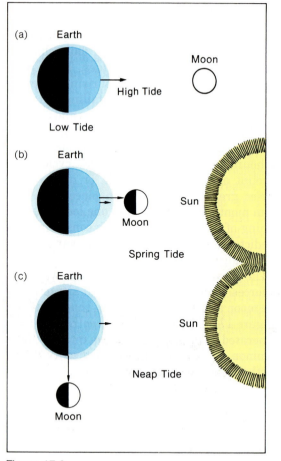

Figure 17.3
Effect of the moon and sun on ocean tides.
(a) Tidal bulges, here greatly exaggerated in height, are produced by the gravitational effects of the moon and sun (arrows) and the centrifugal force resulting from the rotation of the earth-moon system.
(b) The highest high tides and the lowest low tides, or *spring* tides, occur during new moon (shown here) and full moon, when the earth, moon, and sun are aligned.
(c) The lowest high tides, or *neap* tides, occur during the moon's first and last quarters, when the gravitational pulls of the moon and sun are at right angles, as shown. (Tom O'Mary)

wheel with an off-center axis of rotation independent of its own rotational axis. As with a spinning off-center wheel, there is a strong component of outward centrifugal force on the "heavy" side, tending to fling it still farther outward. This centrifugal force directed *away* from the moon forces ocean water outward in a second tidal bulge on the side of the earth opposite the tidal bulge created by simple lunar gravity.

Tidal Regimes

Since the two tidal bulges draw water from the areas between them, at 90° from the moon, low tides tend to occur when the moon is lowest in the sky. Theoretically, the earth should rotate through both watery bulges every 24 hours and 50 minutes, or one every 12 hours and 25 minutes. This is seldom the actual case due to various complicating effects, including impediments to the movement of water imposed by the land masses themselves.

The sun too has an effect on the tides. But the sun's great distance from the earth limits its gravitational force to about half that of the much smaller but closer moon. The greatest tidal range in any region, known as a *spring tide,* occurs when the sun aligns with the earth and moon. This increases either the gravitational pull in the direction of the moon (at the time of new moon, when the moon is between the sun and earth), or the centrifugal effect (at the time of full moon, when the earth is between the sun and moon). Conversely, the tidal range is at a minimum (*neap tide*) when the positions of the sun and moon form a 90° angle with the earth, so that the sun's gravity draws off some of the water of both tidal bulges (Figure 17.3c)

The oceanic response to these tide-raising forces is affected by many factors, including ocean currents, variations in water density and atmospheric pressure, and the shapes of land masses. As a result, there are many local peculiarities in tidal characteristics. The tides at

of the earth on the side facing the moon (Figure 17.4). Thus the earth component of the earth/moon system resembles an eccentric

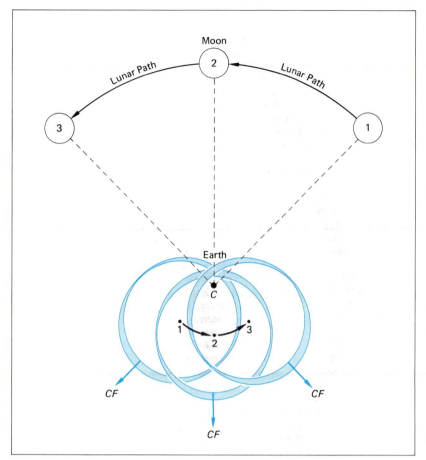

Figure 17.4
Three successive positions of the earth and moon are shown, numbered 1, 2, 3, with arrows indicating the path followed by the center of each. The earth and moon rotate together about the center of gravity of the earth-moon system at *C*, which lies just below the earth's surface on the side closest to the moon. The centrifugal force *(CF)* created by the earth's rotation about this off-center axis forces water toward the region on the earth that is farthest from the moon. At the same time, the gravitational attraction of the moon creates an opposing tidal bulge on the side of the earth nearest to the moon.

Dover, England (Figure 17.5), exhibit the usual pattern of twice-daily high tides. The heights of the daily tides vary through the period because the earth, moon, and sun change their relative positions. At Victoria, British Columbia, there is one high and one low tide per day. Once-daily tides can occur where certain of the twice-daily tides are unusually small or delayed because of special land configurations. If tides arrive at a location by way of two channels, with a relative delay of 6 hours between the separate flows, the twice-daily tides may be canceled. The tides at many locations show a mixture of twice-daily and once-daily tides.

Tidal ranges are greatest where the flood tide is channeled into narrowing estuaries; there the mass of water is crowded together so that it surges strongly upward. Where the tidal bulge pushes into the English Channel, its range increases from less than one-quarter of a meter in the open sea to 7 meters (23 ft) at the narrowest part of the channel. The Bay of Fundy in Nova Scotia, Canada, has the world's highest tidal range, the average spring tide range being 15.4 meters (50.5 ft). This is a consequence of rocky walls that confine the water in a channel that splits abruptly into two narrow arms that "shoal" (become shallow) rapidly.

Tidal regimes are extremely important in marine navigation. Certain harbors can be entered only at high or flood tides. In some places

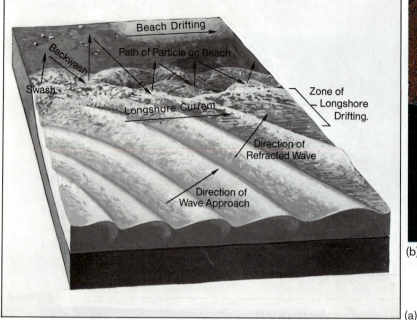

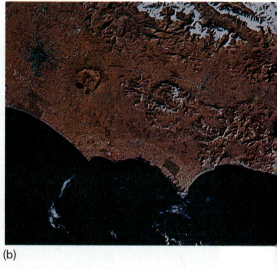

(b)

(a)

Figure 17.13

Beach processes and coastal configuration.

(a) Waves that strike a beach obliquely cause sediment to be shunted along by *swash* and *backwash,* as wave break causes water to wash up the beach obliquely and then return downslope. This produces the movement of particles known as *beach drifting.* The longshore current established by the oblique waves also transports material parallel to the beach in the process of *longshore drifting.* (Tom O'Mary)

(b) This LANDSAT false-color image of the western coast of Italy near Rome (gray spot at left) illustrates repeated "half-heart" curvature of bays of varying sizes. This form is created by longshore drifting of sediment, and reveals the refraction of swell by projecting headlands and deltas. The direction of longshore drifting is from left to right, with the dominant swell from the lower left. (NASA)

drifting (Figure 17.13a), results in a flow of sand along the shore that is necessary to maintain beaches and associated depositional landforms.

Longshore Currents

A second process is equally important in the flow of coastal sediment. In the offshore zone of wave break, the water is kept in an agitated condition and is often clouded with fine suspended sediment that is swirled this way and that. The sediment may be suspended only momentarily after each wave impact, but while it is in suspension it moves with the water. Because waves generally meet coastlines at a slight angle, they push water against the shore obliquely. This continual input of water cannot pile up vertically, so much of it drains off laterally, parallel to the coast in the general direction of wave advance. The result is a *longshore current.*

Longshore Drifting

The oblique onshore piling of water by the arrival of one breaker after another causes the sediment-laden water in the zone of breaking waves to migrate slowly along the shore in the direction of the longshore current, resulting in *longshore drifting* of sediment. The direction may reverse from day to day, but most coasts have a preferred direction of longshore drifting during the season of maximum storminess and most vigorous wave action. This is often visible in the configuration of the coast, as illustrated in Figure 17.13b.

The combined processes of beach and longshore drifting produce a constant flux of sediment along smooth coastlines. This flow waxes and wanes and even changes direction from time to time, depending on the state of the sea, but it creates and sustains all depositional landforms present in coastal regions.

Coastal Deposition

Sediment transport under marine conditions follows the same principles as sediment transport by running water and wind. Transport occurs by suspension, saltation, and surface creep. It is easy to see (and hear) sand and gravel moving in all these ways in the swash zone. Deposition occurs where wave energy decreases, permitting sediment to come to rest.

Every marine depositional form has a characteristic sediment budget in which gains must balance losses if the form is to persist. Many marine depositional forms periodically grow and shrink as a result of shifts in the balance between sediment input and outgo. Although waves strike coasts from varying directions from day to day, every coast has an average "wave climate" and an average direction of sediment flow.

Artificial structures that interfere with the flow of sediment along coasts give us a measure of the amount of material passing by any point over a period of time in the same way that fluvial sediment trapped in reservoirs records the load carried by stream systems. As an example, the breakwater built in 1928 to protect the marina at Santa Barbara, California, traps an average of 215,000 cubic meters of sand per year. This has caused erosion of beaches for a distance of nearly 20 km (32 miles) in the normal direction of sand flow, accelerating the retreat of sea cliffs they formerly protected.

Beaches

Beaches are surfaces of sand or disk-shaped cobbles (known as "shingle") deposited by wave action between the low-tide line and the highest levels reached by storm waves (Figures 17.14 and 17.15). On the landward side of a beach is material of a different type or a surface of different character from the beach, such as sand dunes, permanent vegetation, or a sea cliff.

Beaches form along both irregular and straight coastlines. On irregular coasts, wave refraction causes sand or shingle to collect in the embayments between projecting headlands, creating *bay head beaches* or smaller *pocket beaches*. Beaches also develop in continuous fringes along straight coasts where unresistant geological materials are present. Figure 17.15 indicates the main features of beach morphology.

Beaches are among the most sensitive indicators of changes in geomorphic systems. We usually enjoy beaches in the summer, when the relative absence of destructive storm waves allows them to be well filled with sand (Figure 17.16). The sediment may originate from wave erosion of headlands or from rivers carrying sediment from the land. Along a coast that has a prevailing longshore drift, sand eroded from one beach may be passed along to the next beach. On the southern beaches of Long Island, New York, sand produced by wave erosion of glacial drift moves from east to west. At the present time, loss seems to be outrunning accumulation, and some beaches are retreating 1 or 2 meters (3 to 6 ft) per year. Certain beaches elsewhere have been shrinking in recent years because the construction of dams on streams has reduced the amount of sand supplied to the coast, or because residential construction on coastal dunes has cut off part of the sand supply for the beach system.

One way to prevent beach erosion is to intentionally interrupt longshore drifting with a wood, concrete, or rubble dam, or *groin,* that

Figure 17.14
In many locations beaches are composed of "shingle" composed of pebbles or cobbles. This is often true where waves are eroding stony glacial drift or periglacial gelifluction deposits, or along steep coasts where there are no rivers to transport sand to the sea. This shingle beach is on Anacapa Island, off the coast of southern California. (T. M. O.)

Figure 17.15
The diagram shows the principal geomorphic divisions of a beach. The *berm* is the nearly flat portion at the top of a beach; it is covered with material deposited by waves and constitutes the *backshore*. The *foreshore* extends from the edge of the berm to the low-tide line. Within this zone is the *beach face,* which is the area subject to swash and backwash. The *offshore,* which is permanently under water, contains *bars* and *troughs.* This is the zone of wave break and surf action. (After Francis P. Shepard, *The Earth Beneath the Sea,* 2nd ed., 1967, Johns Hopkins Press)

extends into the water at right angles to the shoreline (Figure 17.17). The longshore current is partially blocked at the groin and drops part of its sediment—just as a stream drops part of its sediment when its velocity decreases. Inevitably, the deposition caused by the groin robs beaches farther down the coast of incoming sediment. Those beaches erode all the more rapidly, to the anguish of resort owners. Miami Beach must import sand to balance erosional losses, despite efforts of individual hotels to maintain their beaches by groin construction.

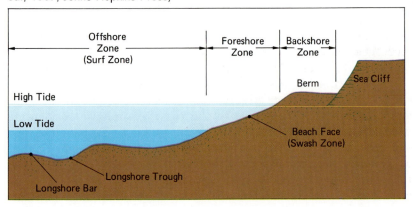

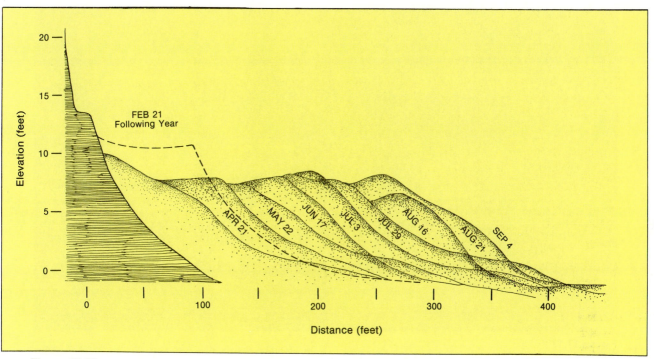

Figure 17.16
The growth of the berm on the beach at Carmel, California, is shown in this series of measured profiles. By February of the following year, most of the berm had been cut away again. The vertical scale is exaggerated 10 times. (After "Beaches" by Willard Bascom, *Scientific American*. Copyright © 1960 by Scientific American, Inc. All rights reserved)

Changes in the size of breakers, or a change from long wavelength waves of low height to steeper short wavelength waves of greater height, greatly accelerate the removal of the particles composing the beach. Large steep waves can peel a beach downward 2 meters or more in a day. In areas affected by seasonal storminess, beaches tend to be stripped downward and steepened by the erosive action of forced waves during the storm season— usually the winter in the middle latitudes. Some beaches periodically lose all their sand, retaining only large cobbles that are too heavy to be entrained even by the largest storm surges. During the season of calms, the sand stored offshore is slowly moved back toward the beaches by low energy swell, eventually refilling the beaches to their former level.

Spits

Beach drifting and longshore drifting frequently produce deposits of sediment that extend outward from initial shorelines (Figure 17.18). Linear sediment accumulations that are attached to the land at one or both ends are termed *spits*. The term *bar* is reserved for submerged depositional forms.

Some spits form where sediment is moving along a coastline that changes direction sharply, causing wave refraction and creating a low-energy environment in which sediment can accumulate. Where a coastline turns abruptly inland, as in a bay, the waves pivot and diverge, decreasing wave energy and sediment transporting ability. Sediment moved by beach and longshore drifting comes to rest where the

MARINE PROCESSES AND COASTAL LANDFORMS

Figure 17.17
The sandy beach at Dover, England, is maintained by
groins, seen projecting into the water at the left.
These barriers trap sand that moves by beach drifting
and longshore currents. (T. M. O.)

incoming waves begin to pivot around the corner. The resulting deposit becomes the new shoreline, which causes the point of wave pivoting to be shifted laterally. This, in turn, produces more deposition. In time a linear spit takes shape, usually with a curved end. At Cape Cod, Massachusetts, a large spit has been extended westward from the end of a wave-eroded glacial moraine.

The formation of spits from one or both sides of small bays may close off the bays, transforming them into *lagoons.* The lagoons eventually fill with fine sediment and are colonized by vegetation. This is one of the most important processes in the straightening of irregular coasts by marine action. Sand spits are common at river mouths, but river currents and floods normally maintain openings through such spits, preventing lagoon formation.

Sediment deposition also occurs in the low-energy wave environments in the lee of coastal islets or large sea stacks. Here wave refraction sweeps sediment behind the obstacle from two sides, producing a type of spit known as a *tombolo,* which ties the island to the land (Figure 17.18).

Spits are especially vulnerable to damage during hurricanes and storm surges associated with midlatitude cyclones. The topography of many spits reveals that they are compound features, having been rebuilt many times after partial destruction by storm waves.

Barrier Islands

Portions of the world's coastlines are paralleled offshore by narrow strips of sand dunes, beaches, and saltwater marshes, known as *barrier islands,* (Figures 17.18 and 17.19). Low, grassy dunes form the axes of the islands, with beaches facing the sea and salt marshes on the landward side. Barrier islands are more regu-

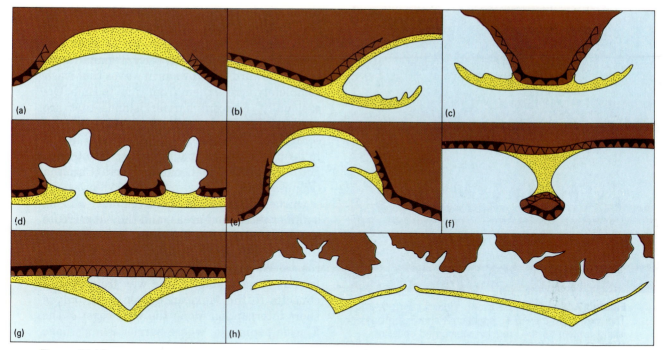

Figure 17.18

These diagrams illustrate the most common depositional forms produced by wave action. The dark sawtooth symbol indicates an active sea cliff; the open sawtooth symbol shows an inactive sea cliff. Land areas are dark. Depositional forms are yellow.

(a) *Bay head beach* with sediment deposition resulting from a low-energy wave environment.

(b) *Recurved spit* formed by sediment transport to the right, prolonging the previous line of the coast.

(c) *Winged headland* with sediment moved both ways from the eroding cliff as a consequence of changing directions of wave approach.

(d) *Bay mouth* or *barrier spits* straightening an initially irregular coast of submergence. Changes in wave approach produce spits extending from both sides of the bays, eventually closing them off and converting them into lagoons.

(e) *Mid-bay spits* with sediments accumulating before reaching the head of the bay.

(f) *Tombolo* formed by the deposition of sediment to the lee of an island and eventually linking the island to the mainland.

(g) *Cuspate spit* developed by reversals in the direction of longshore drifting along a straight coast.

(h) *Barrier island* developed along a coastline that is low-lying but irregular, suggesting recent submergence. The points on such barrier islands are called *cuspate forelands*. Cape Hatteras, North Carolina, is an outstanding example of such a form.

MARINE PROCESSES AND COASTAL LANDFORMS

that have *prograded* (advanced seaward) by deposition of deltas, sand spits, and lagoonal fills; *retrograded* (retreated) by erosion; or developed their characteristics as a consequence of *submergence* or *emergence* of the land relative to the sea. Coastal emergence, or rise of the land relative to the sea, clearly is required for the creation of marine terraces, which are common on tectonically active coasts. Where tectonic activity is weak, emerged areas of sea floor are low-lying coastal plains. Along the Atlantic and Gulf coasts of the United States,

there has been a long-term tendency toward emergence, with Pleistocene sea-level fluctuations superimposed on this general trend.

Holocene Submergence

Two types of landforms have resulted from the world-wide coastal submergence caused by the melting of the Late Pleistocene northern hemisphere ice sheets. Very apparent on a map of eastern North America are deeply penetrat-

(a)

Figure 17.22
Coral reefs in the South Pacific.
  (a) This island in the Belau group has both a fringing reef (left) and a barrier reef (right), separated by a lagoon. The presence of the two reef types suggest island subsidence, causing the barrier reef to grow upward from a submerged fringing reef, followed by tectonic stability, permitting a new fringing reef to grow outward from a stable shoreline. (Douglas Faulkner/Photo Researchers)
  (b) These atolls photographed from the Apollo spacecraft are in the Tuamotu Archipelago. The view spans a distance of about 500 km (300 miles). Note that the atolls are irregular rather than circular in plan. The white flecks are cumulus clouds. (NASA)

(b)

CHAPTER 17

Glossary

aa lava lava having rough, irregular surface, formed when sufficient gas escapes from lava to cause it to lose its fluidity.

ablation processes of melting and evaporation that result in loss of ice or snow from glacier.

ablation zone downstream portion of glacier in which removal of snow and ice by ablation processes exceeds annual input of new snow and ice.

abrasion erosional removal of rock due to friction with other rock particles moved by running water, waves, wind, glacial ice, and gravity.

abrasion platform, *see* wave-cut platform

absolute humidity weight of water vapor present in given volume of air, usually expressed in grams of water per cubic meter of air.

absorption process in which energy of electromagnetic radiation is taken up by a molecule and converted to different form of energy.

abyssal plains flat portions of deep ocean floors that are blanketed by marine sediments covering older volcanic rocks.

accumulation zone upstream portion of glacier in which annual additions of new snow and ice exceed annual removals by ablation processes.

acid any substance containing hydrogen ions that are available to take part in chemical reactions.

acidity concentration of hydrogen ions.

active layer portion of tundra soil that thaws each summer, underlain by permanently frozen ground.

actual evapotranspiration (AE) estimated depth of liquid water actually removed from the land surface by evaporation and plant transpiration in a unit of time.

adiabatic cooling drop in temperature occurring in parcel of rising air that is forced to expand without receiving or losing energy.

adiabatic heating rise in temperature occurring in parcel of descending air that is forced to contract without receiving or losing energy.

adiabatic lapse rate rate of change of temperature with altitude occurring in parcel of air rising freely through atmosphere without significant gain or loss of energy.

advection horizontal movement of air across earth's surface.

advection fog fog formed when warm, moist air moves over a cool surface and is chilled to its dew point.

aeolian related to wind.

aeolian crossbedding fine laminations seen in sand dunes and aeolian sandstones, consisting of deposits accreted on dune slip faces.

aggradation, fluvial prolonged deposition of sediment by streams, causing elevation of stream bed.

A horizon soil layer in which humus and other organic materials are mixed with mineral particles; also zone from which eluviation has removed certain fine particles and soluble substances.

air mass large, nearly uniform body of air that moves as unit, usually in association with secondary high- or low-pressure systems.

albedo portion of incident short-wave radiation reflected by a surface.

alfisols soils with yellowish-brown *A* horizons that have been partially leached of bases, causing upper soil to be colored by iron and aluminum compounds. Better alfisols are agriculturally productive.

algae simple photosynthetic organisms, mostly aquatic, and often microscopic in size.

alkaline having fewer hydrogen ions than does neutral water; rich in sodium and potassium.

alluvial apron extensive, continuous ramp of sediment deposited by streams issuing from an upland; formed by coalescence of adjacent alluvial fans; bajada.

alluvial fan fan-shaped deposit of sediment laid down by stream issuing from mountain canyon.

alluvial fill thick deposit of alluvium produced by stream aggradation.

alluvial plain plain covered with sediment deposited by one or more streams.

alluvial soils soils developed on sediment deposited by streams.

alluvial terrace previous valley floor standing above stream channel due to stream incision.

alluvium sediment deposited by streams.

alpine glacier stream of ice confined by valley walls as it drains from higher to lower elevations in a mountain environment.

alpine topography serrate (jagged) high mountain topography produced by frost weathering and erosion by alpine glaciers.

altitude vertical angle between 0° and 90°, starting from level horizon; also elevation above given reference level, such as average sea level.

altitude tinting use of color tints on maps to indicate ranges of altitudes; also called hypsometric tinting.

altocumulus clouds clouds of medium height that form extensive area of fluffy white patches called mackerel sky; commonly associated with good weather.

altostratus clouds clouds that blanket sky in gray layer; commonly associated with light rain or snowfall.

andesite igneous rock intermediate in chemical composition between the most siliceous and most basic types; the common rock of stratovolcanoes.

Andesite Line the nearly continuous ring of andesitic volcanoes encircling the Pacific Ocean; related to subduction around the Pacific margin.

angle of repose steepest angle that can be maintained by loose fragments on slope.

anion negatively charged ion having surplus electrons.

annual plants plants that die off during periods of temperature or moisture stress, but leave behind crop of seeds to germinate during next favorable period for growth.

Antarctic Circle latitude at which the sun remains above the horizon 24 hours at the southern-hemisphere summer solstice, and below the horizon 24 hours at the southern-hemisphere winter solstice; latitude 66½° south.

antecedent streams   streams presumed to have been in place longer than the geological structures they cut across.

anticlinal axis   surface trace of plane passing through points of maximum curvature of each stratum in anticlinal fold.

anticlinal ridge   topographic ridge that coincides with axis of anticlinal fold; normally composed of resistant rock.

anticlinal valley   valley eroded in unresistant rocks exposed along axis of anticline.

anticline   structural arch or ridge in folded rock strata.

anticyclones (highs)   diverging flows of air around high-pressure regions in atmosphere. Flow is clockwise in northern hemisphere; counterclockwise in southern hemisphere.

aphelion   point in the earth's orbit when earth is farthest from the sun in July.

aquifer   underground layer of permeable material that can store, transmit, and supply water.

Arctic Circle   latitude at which the sun remains above the horizon 24 hours at the northern-hemisphere summer solstice, and below the horizon 24 hours at the northern-hemisphere winter solstice; latitude 66½° north.

Arctic outbreaks   extremely cold air masses that penetrate into the southern United States, causing severe frost-damage to subtropical crops.

arête   sharp alpine ridge created where two glacial cirques intersect.

aridisols   desert soils that receive shallow and infrequent penetration of water, characterized by minimum of organic matter and maximum stoniness.

artesian well   flowing well that penetrates confined aquifer containing water under hydrostatic pressure; no pumping is necessary.

asteroids   astronomical bodies smaller than planets or moons; satellites of sun; they orbit primarily between orbits of Mars and Jupiter.

asthenosphere   weak zone of mobile material in upper part of earth's mantle, permitting movement of lithospheric plates.

atmosphere   envelope of gases that surround solid portion of planet.

atmospheric pressure   force per unit of area exerted by weight of gases in atmosphere.

atmospheric window   wavelength range in which atmosphere does not strongly absorb radiation; principally from 8.5 to 11 micrometers.

atoll   coral reef in form of ring or partial ring, which encloses lagoon.

atom   smallest particle of element that can take part in chemical reaction; has nucleus composed of protons and neutrons, surrounded by shells of orbiting electrons.

autotrophs   organisms capable of subsisting on inorganic materials and energy; primary producers.

autumnal equinox   in northern hemisphere, time when midday sun is directly overhead at equator; about September 21.

average lapse rate   average upward rate of temperature change in atmosphere.

avulsion   sudden change in stream channel location occurring during flood.

axis, fold   surface trace of plane of maximum curvature of strata forming ridges or troughs (anticlines and synclines) in folded rocks.

azimuth   angular distance from north, usually measured clockwise and expressed in degrees as angle between 0° and 360°.

azimuthal projection   projection of portion of globe onto tangent plane.

azonal soils   soils lacking profile development, such as those formed on recent deposits of alluvium or volcanic ash.

b

backshore   upper part of shore, beyond the reach of ordinary waves and tides.

backslope   in cuestas or fault blocks, the slope that descends gradually in the opposite direction from the steeper scarp slope.

backswamp   swamp trapped between valley wall and natural levee of a floodplain.

backwash   water that drains back to sea after wave break pushes swash up beach.

bajada   (pronounced bah-HAH-dah), *see* alluvial apron

bar   (geomorphology) ridge of sand or gravel deposited offshore or in a river; (meteorology) unit of pressure.

barchan   (pronounced bahr-KAHN)   sand dune shaped like crescent, with horns pointed downwind.

barometer   device used to measure atmospheric pressure.

barrier island   marine sediment embankment completely separated from land, forming linear island bordering a coast of low relief.

barrier reef   coral reef that is separated from island or landmass by lagoon.

basal slip   form of glacial motion in which glacier as a whole moves by slipping over its bed.

basalt   extrusive igneous rock (lava) of dark gray color and chemically basic composition; volcanic material forming sea floors.

basaltic volcanism   generally non-explosive volcanism, producing large volumes of basaltic lava with minor tephra.

base   (chemistry) cation; substance that tends to gain proton from donor; substance that dissolves in water to liberate OH^- ions; reacts with acids to form salts.

base flow   portion of stream's discharge that is maintained by almost steady groundwater inflow.

base level of erosion   lowest point to which stream can erode its bed; also lowest level to which land can be eroded (normally sea level).

base line   east-west reference line for U.S. Bureau of Land Management land survey; in surveying, line of known length and position from which survey is extended.

base saturation   degree to which soil is saturated with exchangeable cations other than hydrogen.

batholith   mass of intrusive igneous rock composed of many separate plutons, having surface area of hundreds to thousands of square kilometers.

bauxite   end product of process of laterization; residue of aluminum oxides and hydroxides; principal ore of aluminum.

bayhead beach   beach on irregular coasts where wave refraction has caused sand to collect in embayments between projecting headlands.

baymouth spit   spit that projects across mouth of bay from adjacent headland.

beach   accumulation of loose sediment maintained by marine depositional processes; zone of transition between land and sea.

beach cycles cyclic removal and replacement of beach material related to seasonal changes in wave energy.

beach drifting zigzag movement of marine sediment parallel to shoreline by swash and backwash.

beach face area of foreshore subject to wave uprush, swash, and backwash.

bed (stratum) layer of sedimentary rock, representing period of sediment deposition.

bedding planes separations between strata of sedimentary rock that signify periods of nondeposition.

bed load solid material transported along bed of stream by saltation and traction.

bedrock in-place rock underlying an area, as opposed to transported material.

Bergeron ice crystal model model of precipitation in which early stage is growth of ice crystals from water vapor in atmosphere containing supercooled water droplets.

bergschrund (pronounced BERK-schroont) crevasse between head of glacier and rock wall; scene of active erosion of rock by snow and ice.

berm relatively flat area at backshore of beach, composed of material deposited by wave action.

B horizon illuvial soil layer that receives solid and dissolved material eluviated from A and E horizons.

Big Bang theory hypothesis that energy and matter in universe were originally concentrated together at a point, rose to high temperature, then expanded rapidly outward.

biogeographical realms geographic divisions reflecting the evolutionary centers and patterns of dispersal of plants and animals.

biogeography study of the processes affecting the distributions of various life forms on the earth.

biomass weight of dry organic matter per unit of ground area.

biomass pyramid schematic diagram that represents deceasing biomass in each stage of food chain.

biomes nine global groupings of interacting plant and animal communities.

biosphere the realm of living organisms; one of the earth's major systems along with the atmosphere, hydrosphere, and lithosphere.

black body most efficient emitter of thermal radiation at given temperature.

block diagram perspective diagram for representing landforms and their underlying geologic structures.

block faulting break-up of continental crust by tension, producing horsts, grabens, and tilted blocks.

block lava lava too depleted of gas to produce flow; moves as rubble of hardened blocks over pasty liquid interior.

blowout dunes dunes formerly stabilized by vegetation, but reactivated by disruption of plant cover; convex downwind as crest moves forward.

boreal forest, northern, see coniferous forests, northern

bornhardt isolated rock dome, usually granite, rising above savanna plain.

braided stream stream composed of intertwining channels separated by sand and gravel bars.

breaker, wave failure of wave form because of steepness, causing sliding or toppling of crest.

budget accounting of energy or material that enters, leaves, is stored, or is balanced in a system.

bulk density refers to mass per unit of volume, including pore space.

butte erosional remnant of much larger rock mass; smaller than a mesa, and less flat on top.

Buys Ballot's law states that an observer standing with the geostrophic wind at his back will have high pressure to his right and low pressure to his left in northern hemisphere, but high pressure to his left and low pressure to his right in southern hemisphere.

C

calcic horizon subsidiary horizon formed in subsoil by deposit of calcium carbonate.

calcification pedogenic regime that produces calcium carbonate enrichment in B or C horizon of soil.

calcite common rock-forming mineral, $CaCO_3$; the main constituent of limestone as well as stalactites and stalagmites.

calcrete rocklike lime horizon within soil, formed by slow accumulation of calcium carbonate; known as caliche in southwestern United States.

caldera circular, steep-walled basin formed by collapse of volcanic cone or larger volcanic area.

caliche, see calcrete

calorie amount of energy required to heat 1 gram of water from 14.5°C to 15.5°C.

capillary action upward movement of liquids in confined spaces, caused by attractive forces between liquid and confining solids.

capillary fringe zone in soil in which water moves toward surface from permanently saturated zone (or water table) through capillary action.

caprock resistant ledge-forming layer of rock that forms flat tops of mesas, cuestas, and stripped plains.

carbohydrate sugars, starch, cellulose, etc., produced by plant photosynthesis, and used as a source of energy by plant consumers.

carbon cycle on earth, passage of carbon into atmosphere as carbon dioxide, and its return to surface to be stored in vegetation and other systems.

carbon dioxide CO_2; important atmospheric gas; absorbs long wavelength radiation and is a major factor in greenhouse effect; vented by volcanoes, plant decay and, increasingly, by human activity.

carbon 14 isotope of carbon useful in dating organic materials as old as 50,000 years.

carnivores animals that ingest other animals.

catchment, see drainage basin

cation positively charged ion which is deficient in electrons.

cation exchange replacement of one type of cation bound to the clay-humus complex by another type that is more strongly bound.

cation exchange capacity capacity of various soil colloids to acquire and retain cations.

cavern, limestone, see limestone cavern

Celsius scale formerly Centigrade scale, with 0° at freezing point of water and 100° at boiling point of water.

channel fill sediment stored within stream channel; largely removed during peak discharges.

chaparral vegetation characteristic of dry summer subtropical climate, consisting of dense shrubs that are deep-rooted, small-leaved, and generally evergreen.

chart map utilized for navigation.

chatter marks crescentic chips removed from rock abraded by glaciers.

chemical energy energy stored in bonds that join molecules of chemical compounds.

chemical load, stream substances carried in dissolved form in stream water.

chemical weathering decomposition of geological material due to chemical reactions that swell, soften, or dissolve particular minerals.

chert rock consisting of silica; occurs in thin beds and as segregations in limestone; brittle, but extremely hard and abrasive.

C **horizon** soil layer composed of weathered parent material that has not yet been significantly affected by translocation and organic modification.

cinder cone ring of pyroclastic debris produced by volcanic eruptions (also called pyroclastic cone or scoria cone).

circle of illumination boundary between sunlit and dark halves of earth; also called terminator.

circle of tangency in certain map projections, line along which a cylinder or cone is tangent to the spherical earth; line of true scale on such maps.

cirque steep-walled bowl formed by erosive action of ice at head of valley glacier; feature of alpine glaciation.

cirque glacier mass of glacial ice that is restricted to cirque basin and does not enter valley.

cirrus clouds feathery, high-altitude clouds that are composed of ice crystals.

clastic debris fragments of rock; often lithified into a new generation of sedimentary rock.

clay inorganic mineral particles smaller than 2 micrometers (0.002 mm).

clay-humus complex colloidal combination of humus and fine clay particles that behaves chemically like large molecule.

climate average long-term meteorologic conditions of a region, based on indexes such as temperature and precipitation.

climatic index measure used to distinguish one climatic type from another.

climatic optimum time, about 5,000 years ago, when the climate of Northern Europe was warmest it has been during Holocene (Recent).

climatic realms broad division of climates by Albrecht Penck: frozen realm, dry realm, and moist realm.

climax vegetation final, stabilized pattern of vegetation resulting from plant succession in region.

closed system system containing finite amounts of energy and materials with no inputs from other systems.

closed traverse a survey traverse that starts and ends at the same point.

clouds condensation forms in the atmosphere, not in contact with the ground at low elevations.

cloud seeding artificial introduction into clouds of substances that attract water, to form ice crystals or water drops that can fall as rain.

cloud streets parallel banks of clouds extending over wide area.

coal black organic rock derived from altered plant material; a basic energy resource.

coalescence model model of precipitation in which raindrops start from initial large droplet and grow by coalescence during collisions with smaller droplets.

coast edge of land area to an indefinite distance from the actual shoreline; clearly influenced by the sea.

cockpit karst tropical limestone terrain dominated by large solution pits separated by narrow-crested ridges.

cold front forward edge of mass of cold air that has moved into region previously occupied by warm air.

cold glacier, *see* polar glacier

collapsing breaker destructive type of breaker that advances as a turbulent mass of foaming water.

collision (subduction) coasts coasts in which lithospheric plate convergence and tectonic activity create strong surface relief and steep coasts.

colloid very fine material in suspension in a liquid and carrying negative charge.

colluvium deposits of unconsolidated material that has been transported by gravity, including slope debris, talus, and landslide deposits.

compass traverse simple map-making technique requiring only compass to measure directions and pacing to record distances.

composite cone, *see* stratovolcano

condensation phase change of water vapor to water droplets, in the form of fog, clouds, or dew.

condensation nuclei fine particles in atmosphere that act as collection centers for water molecules, and promote growth of water droplets.

condensed interrupted map projection map projection in which portions of Atlantic and Pacific oceans are omitted to allow more space to display landmasses.

conduction flow of heat from warmer substance in contact with colder substance by means of collisions between atomic particles.

cone karst tropical limestone terrain composed of conical hills.

confined aquifer layer of permeable, water-bearing material that is capped by a layer of impermeable rock; required for artesian flow.

confluence, stream point at which two streams join, or where tributary enters the main stream.

conformal projection map projection in which angular relationships are shown correctly; conformal projections depict shapes of small areas accurately.

conglomerate sedimentary rock composed of pebbles or cobbles in matrix of finer material.

conic projection projection of portion of globe onto tangent cone.

coniferous trees bearing cones and commonly having needleshaped leaves usually retained throughout year; adapted to moisture deficiency due to frozen ground or soils that are not moisture-retentive.

coniferous forests, northern needleleaf coniferous forest of high latitudes, evergreen except larch.

consumers in food chains, organisms that derive their energy from live plant material or animal prey.

continental drift Alfred Wegener's hypothesis that continents have moved horizontally over surface of earth (compare with plate tectonics).

continental ice sheet ice sheet covering substantial portion of a continent; e.g., Antarctic and Greenland ice sheets; ice attains depth of several km; continental glacier.

continental plate sialic portion of earth's crust that constitutes continent; floats in denser sima of ocean floor and mantle.

amount received during the wettest month of the warmest 6 months.

Cf, Df: No marked dry season occurs.

The subdivisions *Cs, Ds, Cw, Dw, Cf,* and *Df* may be further differentiated according to the seasonality of temperature by adding a third letter of notation—*a, b, c,* or *d.*

a: Average temperature of the warmest month exceeds 22°C (71.6°F).

b: Average temperature of the warmest month is less than 22°C (71.6°F), but at least 4 months have an average temperature greater than 10°C (50°F).

c: Average temperature of the warmest month is less than 22°C (71.6°F), fewer than 4 months have an average temperature greater than 10°C (50°F), and the average temperature of the coldest month is greater than −38°C (−36.4°F).

d: Same as *c,* except that the temperature of the coldest months is less than −38°C (−36.4°F).

The Köppen system of climate classification possesses additional symbols to denote special features such as frequent fog or seasonal high humidity, but they are seldom used on the global scale. A global map of the Köppen climate classification is shown in Figure 8.4 on pp. 204–205.

THORNTHWAITE'S FORMULA FOR POTENTIAL EVAPOTRANSPIRATION

The tables and graphs presented in this section make it possible to calculate potential evapotranspiration for a given month at a given location from Thornthwaite's formula, if the average monthly temperatures are known. Thornthwaite's formula is designed to be used with temperatures expressed in degrees Celsius, so temperature data on the Fahrenheit scale should be first converted to Celsius before beginning the calculation of potential evapotranspiration.

The first step in applying Thornthwaite's method is to calculate a monthly heat index, i, for each of the 12 months. The monthly heat index is defined according to the equation

$$i = \left(\frac{T}{5}\right)^{1.514}$$

where T is the long-term average temperature of the month in °C. Approximate values of the monthly heat index can be read from the graph in Figure III.3.

The sum of the twelve monthly heat indexes is the annual heat index, I. The annual heat index is representative

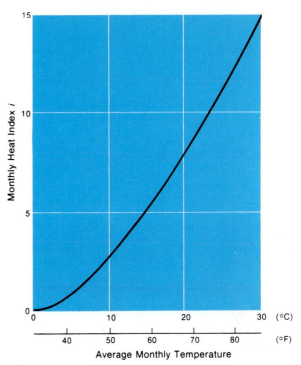

Figure III.3
Thornthwaite monthly heat index (i) as a function of average monthly temperature.

of climatic factors at a given location because it is based on long-term averages. Thornthwaite found an empirical formula that gives potential evapotranspiration, *PE,* for a given month of a particular year in terms of I. His equation for *PE,* unadjusted for duration of sunlight, is

$$\text{Unadjusted } PE \text{ (centimeters)} = 1.6 \left(\frac{10T}{I}\right)^m$$

where T is the average temperature in °C for the specific month being considered, and m is a number that depends on I. To a good approximation, m is given by the equation

$$m = (6.75 \times 10^{-7})I^3 - (7.71 \times 10^{-5})I^2 + (1.79 \times 10^{-2})I + 0.492$$

With these equations, unadjusted potential evapotranspiration can be calculated using tables of logarithms or a computer. Alternatively, approximate values of unadjusted potential evapotranspiration can be read from the nomogram in Figure III.4 with enough accuracy for most purposes, if the values for the average monthly temperature and the annual heat index are known.

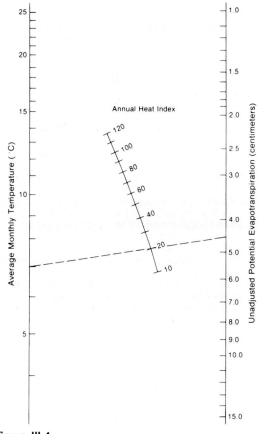

Figure III.4
Nomogram for estimation of unadjusted potential
evapotranspiration based on average monthly
temperature and annual heat index.

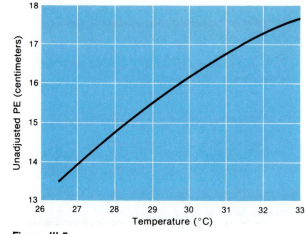

Figure III.5
Estimation of monthly unadjusted potential
evapotranspiration from average monthly temperatures
above 26.5°C.

If the average temperature of the month is below 0°C,
potential evapotranspiration is taken to be 0. If the average monthly temperature exceeds 26.5°C, unadjusted
PE is given directly in terms of temperature according to
the graph in Figure III.5.

Values of unadjusted *PE* must be corrected for the duration of daylight in order to obtain the desired final values. Unadjusted *PE* for a given month at a given location
should be multiplied by the correction factor listed in
Table III.1. The correction factors for latitude 50°N are
used for all latitudes farther to the north, and the factors
for latitude 50°S are used for all latitudes farther to the
south.

Table III.1
Daylength Correction Factors for Potential Evapotranspiration

LATITUDE	JAN	FEB	MAR	APR	MAY	JUN	JUL	AUG	SEP	OCT	NOV	DEC
50°N	0.74	0.78	1.02	1.15	1.33	1.36	1.37	1.25	1.06	0.92	0.76	0.70
40°N	0.84	0.83	1.03	1.11	1.24	1.25	1.27	1.18	1.04	0.96	0.83	0.81
30°N	0.90	0.87	1.03	1.08	1.18	1.17	1.20	1.14	1.03	0.98	0.89	0.88
20°N	0.95	0.90	1.03	1.05	1.13	1.11	1.14	1.11	1.02	1.00	0.93	0.94
10°N	1.00	0.91	1.03	1.03	1.08	1.06	1.08	1.07	1.02	1.02	0.98	0.99
0°	1.04	0.94	1.04	1.01	1.04	1.01	1.04	1.04	1.01	1.04	1.01	1.04
10°S	1.08	0.97	1.05	0.99	1.01	0.96	1.00	1.01	1.00	1.06	1.05	1.10
20°S	1.14	1.00	1.05	0.97	0.96	0.91	0.95	0.99	1.00	1.08	1.09	1.15
30°S	1.20	1.03	1.06	0.95	0.92	0.85	0.90	0.96	1.00	1.12	1.14	1.21
40°S	1.27	1.06	1.07	0.93	0.86	0.78	0.84	0.92	1.00	1.15	1.20	1.29
50°S	1.37	1.12	1.08	0.89	0.77	0.67	0.74	0.88	0.99	1.19	1.29	1.41

Source: C. Warren Thornthwaite, 1948. "An Approach Toward a Rational Classification of Climate." *Geographical Review*. 38: 55–94.

crops and their stages of growth can be distinguished by such means.

Multispectral Scanners

Another way to make use of spectral signatures is to employ several sensing devices, each fitted with a filter that allows only certain wavelengths of visible light to enter. LANDSAT, which views the earth from an altitude of nearly 1,000 km, employs four *multispectral scanners*

Figure II.20
Multispectral scanner and radar images.

  (a) Hong Kong Island appears at the lower right and Guangzhou (Canton), China, is at the top left of this LANDSAT false-color composite image sensed from an altitude of 900 km (560 miles). The color print is made from a composite of three black-and-white images, each exposed to different spectral wavelengths; the image is similar to that produced by color infrared film. Healthy vegetation stands out in red. Water color shows sediment concentration. (NASA and Earthsat Corporation)

  (b) Relief features, including linear features in the San Andreas fault zone, show crisply in this side-looking radar image of part of the San Francisco peninsula. The thin linear feature at the lower right is the 2-mile-long linear accelerator used in atomic research at Stanford University. (J. R. P. Lyon, Stanford University)

and three television cameras, each sensing the same scene in a different range, or "band," of wavelengths. The resulting images can be assigned colors and combined to produce a "false-color" image (Figure II.20a). From its orbit between the poles, LANDSAT can scan a given portion of the earth once every 18 days to detect short-term changes in the environment.

SLAR Images

Many areas in the tropics are almost continuously covered by clouds, making normal photography and multispectral scanning unsuccessful. *Side-looking airborne radar* (SLAR), which utilizes radio waves of short wavelength, can penetrate clouds and yield a high resolution picture of the surface (Figure II.20b). Radar waves emitted from the aircraft are partly reflected from the ground and detected by a receiving apparatus on the aircraft. Variations in the strength of the return signal, produced

(b)

(a)

by reflectance qualities and the orientation of reflecting surfaces, are electronically translated into a picture of the terrain. Airborne radar has been used to map the Amazon Basin, Panama, and portions of Africa and New Guinea, and has even been used to distinguish the coarseness of debris on desert alluvial fans. Although radar methods do not differentiate between types of vegetation as clearly as multispectral methods, forest, range, and agricultural land can be separately identified on large-scale radar images.

APPENDIX III
Climate Classification Systems

KÖPPEN SYSTEM OF CLIMATE CLASSIFICATION

The Köppen system of global climate classification recognizes five major climatic types, symbolized by *A, B, C, D,* and *E.* The *B* climates, which are the dry realms, are further divided into *BS* (steppe) and *BW* (desert) types; and the *E* region is divided into *ET* (tundra) and *EF* (perpetual frost) types.

Figure III.1 shows Köppen's criteria for establishing the boundaries between the *A, C,* and *D* climates and the *BS* and *BW* climates. The solid line on each graph marks the boundary between the humid and dry climates. Each of the three sets of criteria is based on average annual temperature and precipitation. Which criterion is used depends on whether precipitation occurs evenly throughout the year or primarily during the summer or the winter. Köppen considered a region to have a dry winter if at least 70 percent of the precipitation occurs during the 6 summer months, and to have a dry summer if at least 70 percent of the precipitation occurs during the 6 winter months. Regions not fitting either category are considered to have an even distribution of precipitation. Summer is interpreted as the season when the midday sun is high in the hemisphere being considered, and winter is interpreted as the low-sun season.

The classification of a place into *A, C,* or *D* on the one hand, or *BS* or *BW* on the other, can be determined graphically by plotting the average annual temperature and precipitation on the appropriate diagram. The equation for each of the boundary lines is also given in the diagrams. In each case, the upper equation is in degrees Celsius and centimeters, and the lower equation in degrees Fahrenheit and inches.

Figure III.1
Temperature and precipitation criteria for separation of *B* climates from *A, C,* and *D* climates, and *BS* from *BW* climates. The three graphs reflect the influences of precipitation seasonality. (Figures III.1–III.5 by Andrea Lindberg)

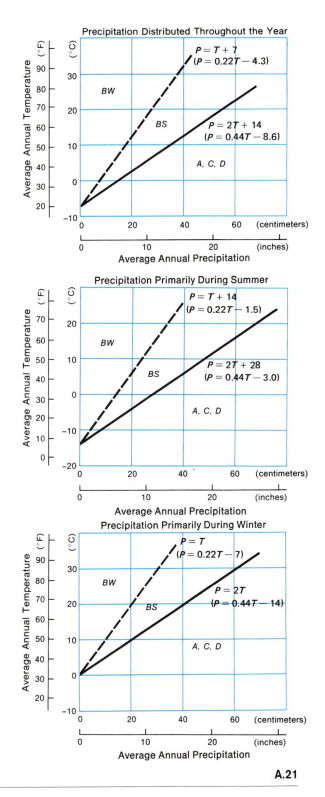

Precipitation Distributed Throughout the Year

$P = T + 7$
$(P = 0.22T - 4.3)$

BW

BS

$P = 2T + 14$
$(P = 0.44T - 8.6)$

A, C, D

Average Annual Temperature (°F) (°C)

Average Annual Precipitation

Precipitation Primarily During Summer

$P = T + 14$
$(P = 0.22T - 1.5)$

BW

BS

$P = 2T + 28$
$(P = 0.44T - 3.0)$

A, C, D

Average Annual Temperature (°F) (°C)

Average Annual Precipitation

Precipitation Primarily During Winter

$P = T$
$(P = 0.22T - 7)$

BW

BS

$P = 2T$
$(P = 0.44T - 14)$

A, C, D

Average Annual Temperature (°F) (°C)

Average Annual Precipitation

The *A, C, D, ET,* and *EF* climates are distinguished from one another according to various criteria based on temperature, with the warmest being *A* and the coldest being *EF*. The criteria are as follows:

A: Average temperature of the coldest month exceeds 18°C (64.4°F).

C: Average temperature of the warmest month exceeds 10°C (50°F). Average temperature of the coldest month lies between 18°C (64.4°F) and −3°C (26.6°F).

D: Average temperature of the warmest month exceeds 10°C (50°F). Average temperature of the coldest month is below −3°C (26.6°F).

ET: Average temperature of the warmest month lies between 10°C (50°F) and 0°C (32°F).

EF: Average temperature of the warmest month is below 0°C (32°F).

H: Unclassified highland climates.

Principal Subdivisions of *A* Climates

The *A* climates are subdivided into *Af, Am,* and *Aw* types on the basis of the amount and seasonality of precipitation. If precipitation in the driest month exceeds 6 cm (2.4 in.), the climate is classified as *Af,* as indicated in Figure III.2. *Af* regions, such as equatorial lowland rainforests, receive abundant moisture for plant growth throughout the year. If there is a winter dry period during which precipitation for the driest month is less than 6 cm, the climate is classified as *Aw* or as *Am.* The distinction between *Aw* and *Am* climates depends upon the relation between the amount of precipitation during the driest month and the annual precipitation, as Figure III.2 shows. Tropical wet and dry climates are classified as *Aw.* The *Am,* or monsoon, climate has enough annual precipitation that the moderate winter dry season does not exhaust soil moisture, and plant growth is not seriously affected.

A fourth subdivision, the *As* climate, would in principle be characterized by a summer dry season. However, *As* climates only occur locally near the equator, and the *As* classification is not significant on the global scale.

Principal Subdivisions of *BS* and *BW* Climates

The *BS* and *BW* regions are divided on a thermal basis, as specified by the additional notation *h, k,* or *k'*.

h (hot): Average annual temperature exceeds 18°C (64.4°F).

k (cold winter): Average annual temperature is less than 18°C (64.4°F) and average temperature of the warmest month exceeds 18°C.

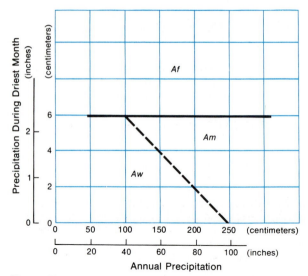

Figure III.2
Subdivisions of Köppen *A* climate based upon amount and seasonality of precipitation.

k' (cold): Average annual temperature is less than 18°C (64.4°F) and average temperature of the warmest month is less than 18°C.

The deserts of North Africa and central Australia are examples of *BWh* regions. *BWk* regions occur in the high plateaus of central Asia.

Principal Subdivisions of *C* and *D* Climates

The *C* and *D* climate regions are further differentiated according to the seasonality of precipitation. A second letter of notation, *s,* denotes regions with a dry summer (*Ds* seldom occurs); *w* denotes regions with dry winters; and *f* denotes regions with no marked dry season.

Cs, Ds: The driest month occurs during the warmest 6 months, and the amount of precipitation received during the driest month is less than one-third the amount received during the wettest month of the coldest 6 months. Also, precipitation during the driest month must be less than 4 cm (1.6 in.).

Cw, Dw: The driest month occurs during the coldest 6 months, and the amount of precipitation received during the driest month is less than one-tenth the

Figure 17.28
A submarine hydrothermal vent, or "black smoker." Such vents have been observed in the central rifts of all sea-floor spreading centers that have been photographed from deep-diving vehicles. The cloud is a plume of super-heated water discolored by sulfides of a host of metallic minerals that are precipitated in mounds around the vents. This phenomenon was first discovered in the Galápagos Rift in 1981. (Robert D. Ballard, Woods Hole Oceanographic Institution)

from "black smokers" apparently are derived from interactions between downward-percolating sea water and magma reservoirs beneath the sea-floor spreading centers. Metals dissolve in salt water hotter than 320°C (650°F) under high pressure that keeps the water from flashing to steam. These metals are precipitated when the hot water vents into the cold water in the ocean depths.

As the oceanic ridge system has an aggregate length of 70,000 km (40,000 miles), a vast but extremely difficult new area for mineral extraction becomes available. However, the rocks of oceanic spreading centers have been found on the land, as in Cyprus, which has been a source of copper for some 4,000 years. Other areas with similar rock types exist on the continents and are being evaluated anew in terms of their economic potential.

At present we have a more detailed knowledge of the surface of Mars than of the landforms of the earth's sea floors—our planet's last frontier. Future years are certain to turn up an ever-increasing list of mysteries of the deep and more problems to engage the attention of physical geographers and their colleagues in the earth sciences.

KEY TERMS

waves
currents
tides
eustatic sea level changes
tectonic sea level changes
flood tides
ebb tides
tidal range
tidal bulge
spring tide
neap tide
wind tide
storm surge
wave crest, trough
wave height, length, period

wind waves
forced waves
sea
swell
fetch
seismic sea wave (tsunami)
runup
breakers: collapsing, spilling, plunging, surging
wave refraction
shoreline angle
wavecut platform (abrasion ramp)
wave-built terrace
sea stack, arch, cave
marine terrace

swash (uprush)
backwash
beach drifting
longshore current
longshore drifting
bay head beach
pocket beach
groin
spit
bar
lagoon
tombolo
barrier island
cusp
corals
coral reefs: fringing, barrier, atoll

collision coast
subduction coast
trailing-edge coast
marginal sea coast
prograded shoreline
retrograded shoreline
coastal submergence
coastal emergence
rias
echo sounding
submarine canyon
turbidity current
guyot
seamount
oceanic fracture zone
submarine hydrothermal vent

REVIEW QUESTIONS

1. Describe the trend in sea level in the last 5,000 years. What was the trend in the preceding 5,000 years?
2. What two forces combine to produce the tides? What is the ideal tidal cycle?
3. Describe the paths of the water molecules that transmit the movement of waves in the open sea.
4. What are the relationships among wave length, wave period, and wave velocity in deep water? When does wave height begin to show a relationship to the preceding parameters?
5. Where are the principal sources of seismic sea waves? Why do seismic sea waves rise so high as they near the land?
6. What is the geomorphic importance of wave refraction?
7. What landforms are typical of irregular cliffed coasts?
8. Two things are required to create broad marine terraces. What are they?
9. Describe the mechanisms by which sediment is moved parallel to shorelines.
10. How can artificial groins affect coastal sediment budgets?
11. What is the difference between coastal spits and bars?
12. How are barrier islands important to the mainland areas behind them?
13. What are the habitat requirements of corals?
14. Distinguish between the different types of coral reefs, and explain the origin of each type.
15. What three different coastal configurations can result from coastal submergence?
16. How are submarine canyons formed, and what role do they play in submarine sediment movements?
17. How do guyots support the theory of sea-floor spreading?

APPLICATIONS

1. What large cities around the world would be submerged if the Antarctic and Greenland ice sheets melted completely? What inland cities would be transformed into ports if this happened?
2. How would the tides be affected if the center

of rotation of the earth–moon system were closer to the center of the earth? if it were between the earth and the moon?

3. In what way is a beach similar to a glacier?

4. If there is a beach near your campus, note the relationship between the slope of the beach face and the size of the particles forming the beach. Both the beach slope and the associated particle size differ from point to point as well as from day to day. What explains the relationship? Is it a constant one? Can you see any relationship between breaker type and beach slope? between beach plan (map view) and beach slope or breaker type?

5. If your department or campus has a collection of topographic maps of the United States (or other areas), locate map examples of each of the coastal depositional forms illustrated in Figure 17.18. What seems to be the source of the sediment in each case: a river or coastal erosion? Can you find a location in which the growth of a beach or spit has stopped wave erosion of a former sea cliff?

6. What accounts for the striking difference in the form of the Atlantic coast of North America north and south of Cape Cod, Massachusetts? Why is the coast of California so much less regular than the coasts of Oregon and Washington?

FURTHER READING

Adey, Walter H. "Coral Reef Morphogenesis: A Multidimensional Model." *Science,* 202 (1978): 831–837. Advanced treatment of the controls of coral reef growth, emphasizing tectonics and sea level change.

Cronan, D. S. *Underwater Minerals.* New York: Academic Press (1980), 380 pp. Synthesis of current knowledge of the formation and possible exploitation of undersea minerals, with attention to environmental and legal pitfalls.

Davis, Richard A., ed. *Coastal Sedimentary Environments.* New York: Springer-Verlag (1978), 420 pp. Chapters on individual coastal environments: deltas, bays, estuaries, salt marshes, dunes, and beaches, by experts in the respective areas. Well-illustrated and readable.

Goreau, Thomas F., Nora I. Goreau, and **Thomas J. Goreau.** "Corals and Coral Reefs." *Scientific American,* 241:2 (1979): 124–126. Excellent, well-illustrated article on all aspects of coral reefs, stressing the symbiosis between coral and photosynthetic algae that makes reef formation possible.

Inman, Douglas L., and **Birchard M. Brush.** "The Coastal Challenge." *Science,* 181 (July 6, 1973): 20–31. This article details modern findings concerning the physical processes in coastal systems, and shows how human activities have changed coastal environments, often diminishing both their utility and aesthetic qualities.

King, Cuchlaine A. M. *Beaches and Coasts,* 2nd ed. New York: St. Martin's (1972), 570 pp. King combines theory, model experiments, and field observation in detailed analyses of coastal processes and landforms.

Kuhn, Gerald G., and **Francis P. Shepard.** *Sea Cliffs, Beaches, and Coastal Valleys of San Diego County: Some Amazing Histories and Some Horrifying Implications.* Berkeley: University of California Press (1984), 193 pp. Case histories of historic coastal changes, organized by geographic location. Abundant illustrations.

Masters, P. M., and **N. C. Flemming, eds.** *Quaternary Coastlines and Marine Archaeology.* New York: Academic Press (1982), 500 pp. Interdisciplinary inquiry into human movements between continents and islands during Pleistocene low stands of the seas. Techniques of underwater archaeology.

Russell, Richard J. *River Plains and Sea Coasts.* Berkeley and Los Angeles: University of California Press (1967), 173 pp. About half of this engagingly written book is devoted to Russell's personal observations about shoreline processes. The author concentrates on low-latitude coasts.

Shepard, Francis P. *The Earth Beneath the Sea,* 2nd ed. Baltimore: Johns Hopkins Press (1967), 242 pp. This very readable nontechnical treatment of the features of the sea floor is by a leading investigator of submarine canyons. As well as presenting facts and interpretations, Shepard indicates how data are collected beneath the sea.

——— and **Harold R. Wanless.** *Our Changing Coastlines.* New York: McGraw-Hill (1971), 571 pp. This large book is a complete inventory of the coastal morphology of the United States, including Alaska and Hawaii. The text is superbly illustrated with aerial photographs.

Strahler, Arthur H. *A Geologist's View of Cape Cod.* Garden City, N.Y.: Natural History Press (1966), 115 pp. This small, nicely written book details how glacial deposition and wind and wave action have fashioned the landforms of a popular tourist area. Very well illustrated and nontechnical.

APPENDIX I
Scientific Measurements

UNITS OF MEASURE

The length, volume, mass, or temperature of an object must be expressed according to a definite system of units if the measurement is to be meaningful to others. The *metric system,* the system of units used in science, is a decimal system. Basic units of length, area, volume, and mass are multiplied or divided by 10 into larger or smaller units, greatly simplifying conversion between units of different magnitudes. A table of conversion from metric to English units follows.

Two scales are commonly used by scientists to measure temperature. On the *Celsius* (C) scale (formerly called the centigrade scale), the temperature at which pure water crystallizes, or "freezes" to ice, is taken as 0°C, and the temperature at which water "boils" into vapor is fixed at 100°C. The interval from freezing to boiling is divided into 100 equal degrees. Normal room temperature of 72° Fahrenheit (F) is equivalent to approximately 22°C, and normal body temperature of 98.6°F is equivalent to 37°C.

The second temperature scale, called the *absolute,* or *Kelvin* (K) scale, has a different zero point from the Celsius scale. The temperature of a gas is a measure of the kinetic energy of its molecules; as a gas is cooled, the molecules move more slowly and have less energy. Theoretically, there is a temperature at which all motion would cease, and this point is taken as zero on the Kelvin scale. Zero on the Kelvin scale is equivalent to approximately −273°C (−459°F), so that the freezing point of

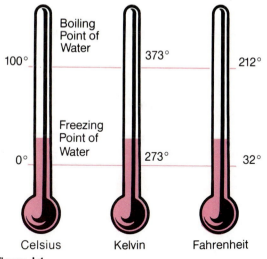

Figure I.1
Comparison of the Celsius, Kelvin, and Fahrenheit scales of temperature.

water on the Kelvin scale is 273°K (Figure I.1). Many scientific laws are simpler when expressed in absolute temperature because of its fundamental significance on a molecular level.

METRIC TO ENGLISH CONVERSIONS

Length

1 kilometer = 1,000 meters = 0.6214 mile = 3,281 feet
1 meter = 100 centimeters = 1.0936 yards = 3.281 feet = 39.37 inches
1 centimeter = 10 millimeters = 0.3937 inch
1 micron = 10^{-6} meter = 10^{-4} centimeter = 3.937×10^{-5} inch

Volume

1 cubic kilometer = 10^9 cubic meters = 0.2399 cubic mile
1 cubic meter = 10^6 cubic centimeters = 1.308 cubic yards = 35.31 cubic feet = 61,024 cubic inches

Mass

1 metric ton = 1,000 kilograms = 2,204.6 pounds
1 kilogram = 1,000 grams = 2.2046 pounds

Time

1 day = 86,400 seconds
1 year = 3.156×10^7 seconds

Speed

1 meter per second = 3.281 feet per second
1 meter per second = 3.6 kilometers per hour = 2.237 miles per hour
1 kilometer per hour = 0.62 mile per hour
1 knot = 1 nautical mile per hour = 1.151 miles per hour

Area

1 square kilometer = 10^6 square meters = 0.3861 square mile = 247.1 acres
1 square meter = 10^4 square centimeters = 1.196 square yards = 10.764 square feet = 1,550 square inches

Pressure

1 bar = 10^6 dyn/cm^2 = 10^5 pascals (newtons/m^2)
1 atmosphere = 1.01325 bars = 1,013.2 millibars = 760 millimeters of mercury = 29.92 inches of mercury

Temperatures

°C = 5/9 (°F −32°)
°F = 9/5 °C + 32°
°K = °C − 273.15°

Energy

1 calorie = 4.186 joules = 3.968×10^{-3} British Thermal Unit
1 langley = 1 calorie per square centimeter

Power

1 calorie per second = 4.186 joules per second = 4.186 watts
1 calorie per minute = 0.07 watts

Figure I.2
Scaled comparisons of Celsius to Fahrenheit and of metric to English measurement systems.

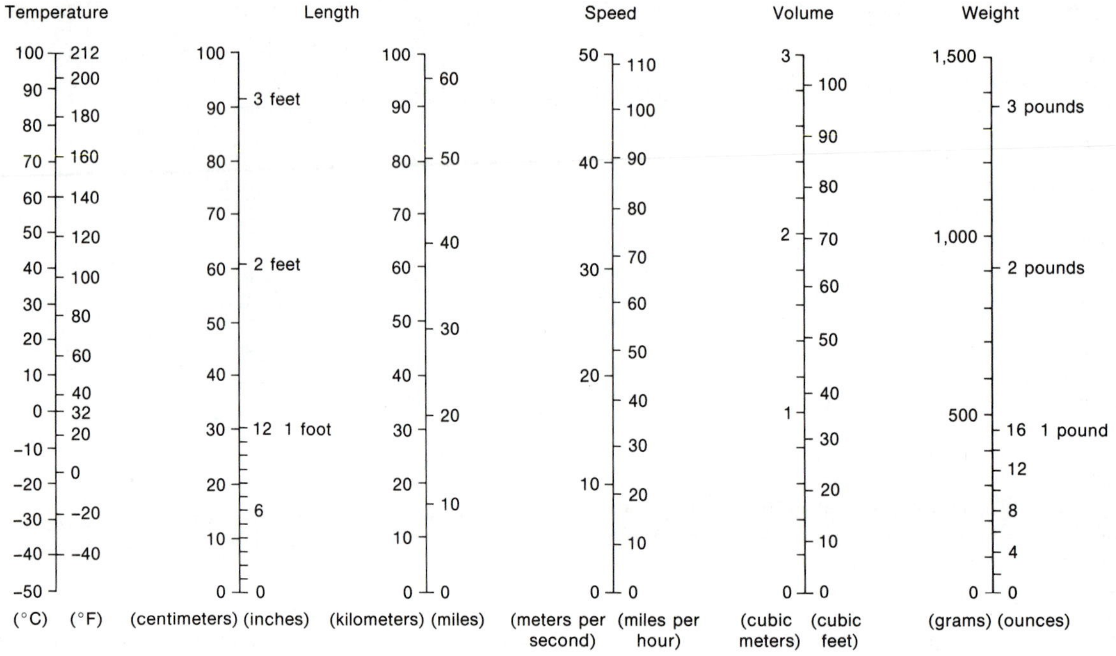

APPENDIX II
Tools of the Physical Geographer

MAPS: REPRESENTATIONS OF THE EARTH'S SURFACE

A *map* can be defined formally as a two-dimensional graphic representation of the spatial distribution of selected phenomena. A map is *planimetric;* that is, it shows *horizontal* spatial relationships on the earth. In addition to specifying location, distribution, amount, distance, direction, sizes, and shapes, maps can also represent the form of the land surface or any statistical surface based on spatial data.

Scale and Distance

A map's scale gives the relationship between length measured on the map sheet and the corresponding distance on the earth's surface. Map scale may be given in a simple *verbal statement,* such as "1 inch equals 3 miles," or it may be indicated by a *graphic scale* marked off in units of distance on the earth, as shown in Figure II.2. One advantage of a graphic scale is that, unlike a verbal scale, it remains correct if the map is copied in a larger or smaller size.

Scale is often expressed as a *representative fraction.* A representative fraction of 1/5,000 (commonly written 1:5,000) means that a distance of 1 unit on the map represents 5,000 similar units on the earth.

When a portion of the spherical earth is represented on a flat map, distortions inevitably occur. Consequently the scale of a map cannot be constant for every portion of the map. However, scale does not vary greatly on a flat map of a small region, and even the coterminous United States can be mapped in such a way that the scale does not vary by more than a few percent. Significant varia-

Figure II.1
Early map representations of the earth.
(top) This Babylonian map from 500 B.C. is one of the earliest attempts to portray the world. (The British Museum)
(bottom) Claudius Ptolemy's map of the known world—reconstructed in the fifteenth century from his detailed descriptions written in the second century—is a comparatively accurate representation that takes into account the spherical shape of the earth. (*Atlas of the Universe,* p. 13, © 1971, Mitchell-Beazley, Ltd.)

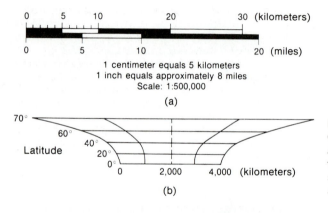

1 centimeter equals 5 kilometers
1 inch equals approximately 8 miles
Scale: 1:500,000

(a)

(b)

Figure II.2
Graphic map scales. The upper scale is applicable in all areas of a medium-scale map. The bottom scale is the type necessary on the Mercator projection, which stretches distances (in all directions) as they move away from the equator.

tions of scale occur on all world maps, however. The scale on a Mercator map of the world, for instance, is several times larger at higher latitudes than at the equator (Figure II.2).

When the representative fraction is a small number, less than 1/1,000,000, a map is called a *small-scale map.* Small-scale maps are used when a large portion of the earth's surface, such as a continent or an ocean, must be represented on a map of limited size. If a map has a representative fraction larger than 1/250,000 (1:250,000), it usually is called a *large-scale* map.

Large-scale maps of small areas are capable of showing greater detail than small-scale maps of large areas.

Location

The principal method for specifying location on the earth's surface is by the system of latitude and longitude (Figure 3.3, pp. 48–49). *Latitude,* the position of a place north or south of the equator, is expressed in angular measure relative to the center of the earth. The angle of latitude varies from 0° at the equator to 90° at the poles. *Longitude,* the position of a place east or west of a selected prime meridian, is expressed in angular measure that varies from 0° to 180° east or west of the prime meridian. The most commonly accepted prime meridian is the one on which the Greenwich Observatory in England is located, but other prime meridians have been used in the past. The framework of lines representing parallels of latitude and meridians of longitude on a map is called the *graticule* or *grid* of the map. Depending on the method chosen to construct a map, the lines of the graticule may be straight or curved, and they may or may not intersect at right angles as they do on a globe.

A more easily computed location reference system, particularly adapted to computer use, employs a rectangular grid composed of straight lines that do intersect at right angles.

The first step in constructing such a grid system is to choose a standard map projection that meets the needs of the user. (Map projections are discussed later in this appendix.) A square grid is superimposed on the map projection and numerical coordinates are assigned to the reference lines of the grid. The coordinates are expressed in units of distance from a selected origin. The grid coordinates of any location can then be read from the map as illustrated in Figure II.3. By convention, the coordinate to the east, or the *easting,* is specified first. Then the coordinate to the north, or the *northing,* is specified. The rule is to read toward the *right* and *up,* following the same order used for giving the *x* and *y* coordinates of a point on a graph. When using computers to construct maps or specify data points, rectangular coordinates may be recorded differently, as + or − values horizontally and vertically from a central point, or *right* and *down* from the top left.

Figure II.3
System of reading rectangular grid coordinates.

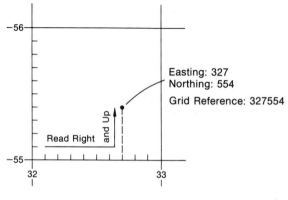

The United States National Ocean Survey (formerly the United States Coast and Geodetic Survey) has designed a rectangular grid system for each of the states, called the *State Plane Coordinate System.* The basic grid square of the state coordinates is 10,000 feet on a side; eastings and northings for the grid are listed in units of feet.

A modified grid system has been in use for many years in connection with the survey of public lands conducted by the Bureau of Land Management. The basic land unit of the survey, which was begun in the eighteenth century, is the *township,* a square plot 6 miles on a side. Townships are laid out with two sides along meridians and the other two sides along parallels of latitude. Because meridians converge toward the north, the north-south sides of the townships must jog eastward or westward every 24 miles to maintain the size of the 6-mile square (Figure II.4).

Townships are laid out with respect to a north-south *principal meridian* and an east-west *base line.* Different land surveys established thirty-one sets of principal meridians and base lines for the conterminous United States and five sets for Alaska. The location of each township in a survey region is given with respect to the point at which the principal meridian and the base line intersect. The coordinates that specify a particular township are read as the number of townships north or south of the base line; the number of townships east or west of the principal meridian is called the *range.* Townships are further subdivided into 36 squares, 1 mile on each side, which are called *sections;* sections are numbered 1 through 36 in a serpentine fashion, beginning in the upper right corner of the township (Figure II.4).

Direction

By definition, meridians of longitude on a sphere are true north-south lines that meet at the poles, and parallels of latitude are true east-west lines equal distances apart on the surface of the sphere. Because of the distortions inherent in representing the surface of a sphere on a flat sheet of paper, meridians or parallels often vary in direction across a map.

Many large-scale maps indicate the direction of *true north* by means of a star-tipped arrow or the symbol *TN* (Figure II.5). However, this direction is rarely the same as *magnetic north,* the direction in which a magnetic compass needle points. Large-scale maps usually indicate the direction of magnetic north by means of a half-headed arrow and the symbol *MN.*

The earth's magnetic field is not uniform, and the magnetic poles do not coincide with the geographic poles, so the relation between magnetic north and true north must be specified separately for each region. The difference between magnetic north and true north is known as *magnetic declination* and is expressed in degrees east or west of the true meridian of a given location. Across the con-

Figure II.4
Bureau of Land Management system of land survey; often referred to as the "township and range system."

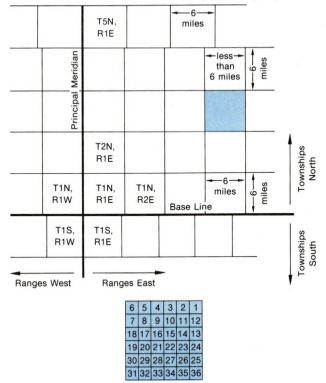

Figure II.5
Direction indicator on standard topographic maps showing direction to true north (star), magnetic or compass north (MN), and grid north (GN) for any rectangular grid present on the map.

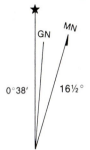

terminous United States the magnetic declination varies from 0° to as much as 25° east or west. Furthermore, the direction of magnetic north at a given location varies with time, often by as much as 1° in 20 years.

Direction on a map may also be specified as *grid north,* the northerly direction arbitrarily determined by a particular rectangular grid system, and symbolized on the map by *GN.* Grid north generally does not coincide with either magnetic north or true north. The grid north directions specified by two different grid coordinate systems usually differ from one another as well.

Directions other than north can be expressed in terms of *azimuth,* which is the angle of the desired direction measured clockwise from a chosen reference direction and expressed in degrees between 0° and 360°. According to the choice of true north, magnetic north, or grid north as a reference, the corresponding azimuths are termed *true azimuth, magnetic azimuth,* or *grid azimuth.*

MAP PROJECTIONS

A model globe is the only accurate way to represent large portions of the earth's surface because only a globe can duplicate the spherical shape of the earth. A flat piece of paper cannot be fitted closely to a sphere without wrinkling or tearing, so small-scale flat maps that represent regions hundreds or thousands of miles in extent inevitably introduce significant distortions of shape or size.

The fundamental problem of map making is to find a method of transferring a spherical surface onto a flat sheet in a way that minimizes undesirable distortions. Any method of relating position on a globe to position on a flap map is called a *map projection.* Numerous projections have been devised, each with its characteristic advantages and distortions. Because no projection is free of distortion, the choice of a projection should be made with regard to its proposed use.

The principles of projection are illustrated in Figure II.6, using the example of the gnomonic projection. This is a true geometric projection constructed by tracing the rays of light from a light source at the center of a transparent globe onto a plane that touches the globe at one point, called the *point of tangency.* Each point on the portion of the globe that is projected onto the plane constitutes a point on the map. However, only a few projections, such as the gnomonic, can be visualized geometrically. Many projections can be expressed only as a mathematical rule that relates points on a globe to points on a flat sheet.

Projections and Distortion

Maps are commonly relied upon to show correct direction and distance from one location to another and the sizes and shapes of areas. A single flat map can depict

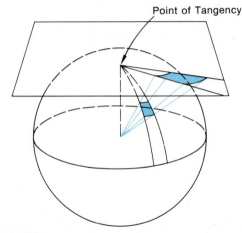

Figure I!.6
Construction of the gnomonic projection, which is projected to a tangent plane from a projection point at the earth's center. This is an example of a useful geometric projection of the azimuthal type.

one or another of these without distortion, but not all of them.

A number of projections, called *conformal* projections, have been devised so that azimuths from point to point are correct. Shapes of small areas are portrayed accurately. However, on a conformal map the scale necessarily changes from one region to another. Regions that span a large portion of the globe exhibit extreme distortion when mapped on a conformal projection. The well-known Mercator projection is conformal, but it represents areas in the higher latitudes several times larger than areas of the same size near the equator. In an obtrusive display of map ignorance, the news media commonly use this projection to portray the world.

Another important class of projections includes the *equal area,* or *equivalent,* projections. On such projections, point-to-point azimuths are incorrect, but the scale is designed to vary over the map in such a manner that the relative sizes of all areas are correct. Figure II.7 illustrates how shapes can be varied without altering their areas. Equal area projections should always be used when the geographic distributions of phenomena are displayed because regions are represented in their correct relative sizes. Many projections are neither conformal nor equivalent area but represent compromises to obtain adequate representation of shape without badly distorting size.

An impression of a projection's major properties can usually be obtained by observation of the lines of latitude and longitude. On a globe, parallels are equally spaced

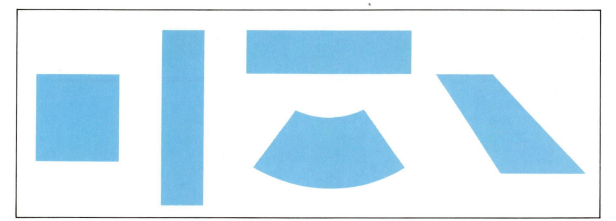

Figure II.7
Examples of the shapes that may be assumed by the
same grid cell on different map projections.

and intersect meridians at right angles. If a map shows
them intersecting at right angles, the projection may or
may not be conformal, but if they are shown intersecting
at other angles, distortion of shape is present and the
projection is not conformal. Distortions of shape or size
can be seen by comparing the map compartments
bounded by parallels and meridians to those on a globe,
recalling that the grid cells on a true globe become progressively smaller from the tropics to the poles.

The globe may be projected onto a plane, cylinder, or
cone, which are the only surfaces that can be spread flat.
Projections are conventionally classified into families: *azimuthal,* or *zenithal,* projections onto a plane; *cylindric* projections onto a cylinder; and *conic* projections onto a
cone. A fourth family of *geometrical projections* is usually
used to portray the entire globe. The projections in a
given family tend to have similar properties and similar
distortion characteristics. These are outlined in Figures
II.8 through II.11.

Interruption and Condensation of Projections

On most maps showing global distributions, the landmasses are of greater interest than the oceans. In such a
case, the projection can be *interrupted* in the oceans (Figure II.11). The projection may then be reprojected to
standard meridians through each major landmass so that
no land area is far from a meridian. Such a world map can
maintain areas in their correct proportions with only minor

shape distortion. If ocean areas are of no interest in a
particular application, portions of the Atlantic and Pacific
oceans can be omitted entirely, producing a *condensed
interrupted map.* Condensation permits the map to be at a
larger scale without using more space. Conversely, maps
concerned with oceanic phenomena on a global scale
can be interrupted in continental areas.

RELIEF PORTRAYAL

Frequently it is useful or necessary to depict land surface
form and vertical relief on maps. *Contour lines,* special
kinds of *relief shading,* and *color tints* are some of the
methods for indicating form and relief on maps. Relief
and surface form are represented most accurately by
contour lines.

Contours

A contour line on a map represents a line of constant
altitude above or below a chosen reference level, called a
datum plane, which is usually mean sea level. Consider
the hilly island in Figure II.12. The figure shows horizontal
planes at a regular vertical spacing, or *contour interval,* of
200 ft. Each horizontal plane cuts the surface of the
ground in a circle that is the contour line for that altitude;
every point on a given contour line is the same altitude
above the datum plane.

Figure II.8 Azimuthal projections.

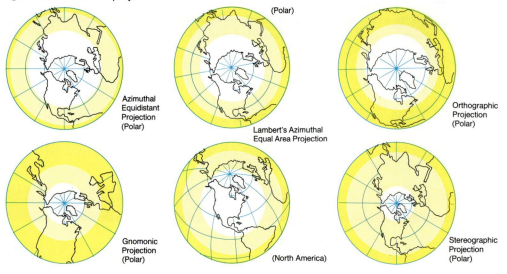

Azimuthal Equidistant Projection (Polar)

(Polar)

Lambert's Azimuthal Equal Area Projection

Orthographic Projection (Polar)

Gnomonic Projection (Polar)

(North America)

Stereographic Projection (Polar)

Table of Azimuthal Projections

PROJECTION	APPEARANCE	PROPERTIES	DISTORTION PATTERN	BEST USES
Azimuthal Equidistant (Polar)	Meridians are straight lines radiating from the pole; parallels are equally spaced circles concentric about the pole.	Scale is constant and correct along meridians. Directions from central point are correct.	Distortion increases slowly with increased distance from the center. Shapes are represented comparatively well, but areas are distorted.	Directions and distance to the center are undistorted; useful for charts of radio propagation to a given location or for showing relative distance from a given location.
Gnomonic (Polar)	Meridians are straight lines radiating from the pole; parallels are circles concentric about the pole with rapidly increasing spacing outward.	Great circles anywhere on the map are represented by straight lines.	Distortion increases rapidly with increased distance from the center. Shapes become badly distorted.	Aircraft and ship navigation. Only projection on which shortest distance between any points on the earth plots as a straight line.
Lambert's Azimuthal Equal Area (Polar)	Meridians are straight lines radiating from the pole; parallels are circles concentric about the pole with spacing decreasing outward.	The only azimuthal equal area projection. Directions from central point are correct.	Distortion increases moderately with increased distance from the center. Shapes are represented well.	Polar maps and maps of one hemisphere, especially where distributions are to be represented.
Orthographic (Polar)	Meridians are straight lines radiating from the pole; parallels are circles concentric about the pole with spacing decreasing outward.	Directions from central point are correct. Gives appearance of earth as seen from deep space.	Distortion increases moderately with increased distance from the center.	Used primarily for illustrations, shows how the earth looks from outer space.
Stereographic (Polar)	Meridians are straight lines radiating from the pole; parallels are circles concentric about the pole with spacing increasing outward.	The only azimuthal conformal projection. Directions from central point are correct.	Area distortion increases rapidly with increased distance from the center. Shapes are represented well.	Base map for the UPS grid system, poleward of latitude 80°.

(Figures II.8–II.11 by Andrea Lindberg after Arthur Robinson and Randall Sale, *Elements of Cartography*, 3rd ed., © 1953, 1960, 1969 by John Wiley & Sons, reprinted by permission)

Figure II.9 Cylindric projections.

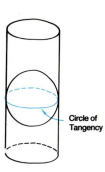

Circle of Tangency

Equirectangular Projection

Standard Mercator's Projection

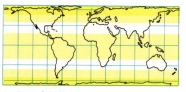

Lambert's Cylindrical
Equal Area Projection

Transverse
Mercator's Projection

Table of Cylindric Projections

PROJECTION	APPEARANCE	PROPERTIES	DISTORTION PATTERN	BEST USES
Equirectangular	Parallels and meridians are equally spaced and form a square grid. Often a parallel is chosen to be the circle of tangency.	No major properties. Can depict entire earth. Neither conformal nor equivalent.	Scale is correct along the standard parallel and along meridians, but shape and area distortion increase with increased distance from the standard parallel.	Used only for large- or moderate-scale maps of limited areas in low latitudes.
Lambert's Cylindrical Equal Area	Parallels and meridians form a rectangular grid. Employs two parallels equidistant from the equator as circles of tangency.	A global equal area projection excluding the polar regions.	Shape distortion increases with increased distance from the standard parallels. Shapes at high latitudes are severely compressed in the north-south direction.	Not widely used because of severe shape distortion at high latitudes, but satisfactory distributions in the lower latitudes.
Standard Mercator	Parallels and meridians form a rectangular grid, with the equator as the circle of tangency.	A conformal global projection. A straight line on the map is a line of constant azimuth on the earth. Excludes the polar regions.	Areas are magnified with increased distance from the equator. Shapes of small regions are represented well, but there are gross distortions of area at high latitudes.	Commonly misused for general purpose world maps. Only legitimate use is for navigation by compass headings after route is laid out on gnomonic projection.
Transverse Mercator	The circle of tangency is a meridian or portion of a meridian. Most meridians and parallels are curved.	A conformal projection. Most straight lines on the map are not lines of constant azimuth on the earth.	Areas are magnified with increased distance from the central meridian. Scale distortion is constant along lines parallel to the central meridian. Shapes of small regions are represented well.	Base map for UTM grid system, for some State Plane Coordinate grids, and for the British Ordinance Survey maps.

Figure II.10 Conic projections.

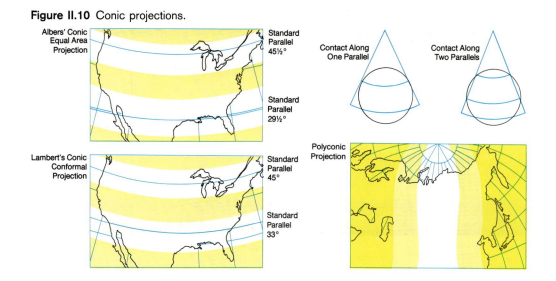

Table of Conic Projections

PROJECTION	APPEARANCE	PROPERTIES	DISTORTION PATTERN	BEST USES
Albers' Conic Equal Area	Meridians are converging straight lines, and parallels are arcs of concentric circles. Two standard parallels are used; for maps of the conterminous United States, they are 29½°N and 45½°N.	An equal area projection.	Distortion increases slowly with increased distance from the standard parallels. Scale along meridians is slightly too large between the standard parallels and somewhat too small outside them. Distortion is small in normal applications.	An excellent projection for midlatitude countries that extend primarily east-west. Usually applied over a restricted range of longitude. Distributions are shown without bias because it is an equal area projection. Chosen for the maps in the National Atlas of the United States.
Lambert's Conic Conformal	Meridians are converging straight lines, and parallels are arcs of concentric circles. Two standard parallels are used; for maps of the conterminous United States, they are often 33°N and 45°N.	A conformal projection.	Distortion increases slowly with increased distance from the standard parallels. Scale along meridians is slightly too small between the standard parallels and somewhat too large outside them. Distortion is small in normal applications, but slightly greater than for Albers' equal area projection.	Used for countries that extend primarily in the east-west direction and for some air navigation charts because straight lines on the map represent great circles.
Polyconic	Meridians are curves converging toward the poles, except for one straight central meridian. Parallels are arcs of circles, except perhaps for one parallel, but each circle has its own center. Every parallel is a standard parallel for the projection.	An excellent compromise between a conformal projection and an equal area projection.	Distortion tends to increase with increased distance east and west of the central meridian. Distortion is small when small regions are mapped.	Base for the topographic maps of the United States Geological Survey. Satisfactory for maps of small regions, particularly regions extending primarily north and south.

Figure II.11 Geometrical global projections.

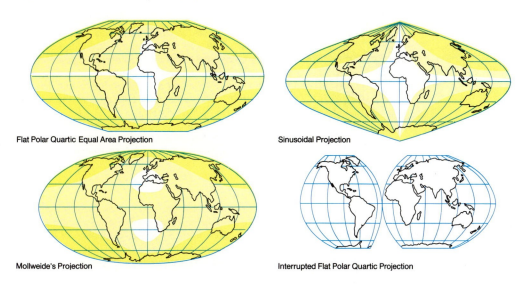

Flat Polar Quartic Equal Area Projection

Sinusoidal Projection

Mollweide's Projection

Interrupted Flat Polar Quartic Projection

Table of Geometrical Global Projections

PROJECTION	APPEARANCE	PROPERTIES	DISTORTION PATTERN	BEST USES
Flat Polar Quartic Equal Area	Parallels are straight parallel lines. Meridians are curves in general but the central meridian is straight. The poles are represented by straight lines one-third the length of the equator. Meridians converge toward the poles and are equally spaced along each parallel. The spacing between parallels decreases slightly with increased latitude. The boundary of the map is a complex curve.	An equal area projection.	Distortion is least nearest the equator and central meridian and greatest at high latitudes near the boundaries.	Generally useful as a world map and for depicting global distributions.
Mollweide	Parallels are straight parallel lines. Meridians are elliptical in general, but the central meridian is a straight line half the length of the equator. The meridians converge to a point at each pole and are equally spaced along each parallel. The spacing between parallels decreases slightly with increased latitude. The boundary of the map is an ellipse.	An equal area projection.	Distortion is least in midlatitude regions near the central meridian and greatest at high latitudes near the boundaries.	Generally useful as a world map and for depicting global distributions.
Sinusoidal (Sanson-Flamsteed)	Parallels are straight, parallel, equally spaced lines. Meridians and the boundary are sinusoidal curves. The central meridian is straight and half the length of the equator. Meridians converge to points at the poles. The length of each parallel is equal to its length on a globe of corresponding scale.	An equal area projection.	Distortion is least nearest the equator and the central meridian and greatest at high latitudes near the boundaries.	Somewhat inferior to other projections as a global map, but useful for maps of individual continents.

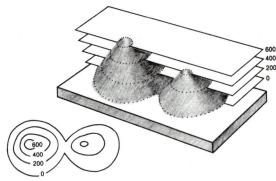

Figure II.12
Derivation of contour lines.

As shown in Figure II.13, contour lines are used on flat maps to represent vertical relief. The contour interval should be chosen to be commensurate with the nature of the landscape depicted; the contour interval will normally be larger for a mountainous region than for a gently sloping plain. The choice of contour interval also depends on the scale of the map. On a large-scale map the contour interval may be only a few feet, and the contour lines will show comparatively small changes in elevation. A small-scale map of the same region will employ a larger contour interval, and in addition the fine details of the contours will

Figure II.13
Derivation of contour lines from topography and means of constructing terrain profiles from contours. (Andrea Lindberg adapted from Whitwell Quadrangle, Tennessee, U.S. Geological Survey)

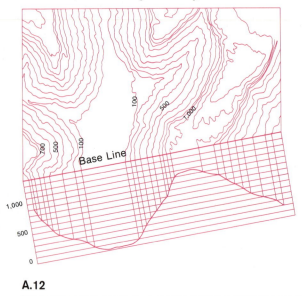

be smoothed and averaged, so that less detail will be represented.

As Figures II.13 and II.14 indicate, the horizontal spacing of contour lines can be interpreted in terms of the local slope angle. Contour lines that are close together represent steeper slopes than do contour lines that are farther apart. The convexity or concavity of a slope and the form of ridge crests and valley floors can also be inferred from the spacing and character of the contour lines.

For quantitative purposes, a *topographic profile* of the landscape along a given direction can be prepared from a contour map according to the method illustrated in Figure

Figure II.14
Vertical (top) and oblique (bottom) views of a terrain model, illustrating the relationship between contour lines and surface form. The model consists of layers representing elevation intervals of 250 feet (76 m) on Mt. Shasta, a large stratovolcano in northern California. The vertical and horizontal scales are equal. (Model by Kurt Seemann)

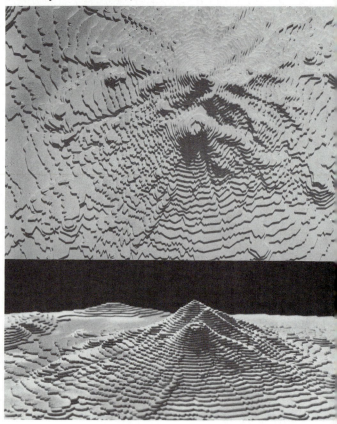

II.13. The vertical scale of a topographic profile is usually exaggerated in order to portray minor relief features more clearly.

Hachures

In addition to contour lines, a variety of qualitative or partly quantitative artistic techniques can be used to indicate relief. In the nineteenth century, slopes were depicted by straight lines called *hachures* drawn in the direction of slope descent, at right angles to contours. Hachures are drawn so that their width or spacing increases with increased slope angle, so that steep slopes appear darker than gentle slopes (Figure II.15 *left*). Expertly drawn hachures present a pleasing appearance, but they are seldom used today because they do not indicate the specific slope angle and because their construction is extremely tedious.

Shaded Relief

A modern method of symbolizing relief on a map is to add shading to give a three-dimensional impression of the terrain. Such *shaded relief* is normally drawn as though the area were illuminated from the upper left, or "northwest," corner of the map (Figure II.15 *right*). Shading can be combined with contour lines or altitude tints (see below).

Figure II.15

Relief can be depicted on maps by hachuring (left) or by shaded relief (right). (*Left*—Reprinted from Arthur Robinson and Randall Sale, *Elements of Cartography,* © 1953, 1960, 1969 by John Wiley & Sons, reprinted by permission; *right*—U.S. Geological Survey)

Altitude Tints

The use of color tints for successive ranges of altitudes, known as *altitude tinting,* is often employed on small-scale maps of large areas. Green is usually used for altitudes near sea level, with colors for higher altitudes progressing through yellow, orange, red, and brown. The green tint used for low altitudes should not be taken as indicative of vegetation cover. The progression of colors is based on human perception of "warm" colors (orange, red) as "advancing," or lifting above the page, while "cool" colors (green, blue) "recede" into the page. Hachures, shaded relief, and altitude tinting are sometimes used in combination on a contour map to depict the general character of a landscape while retaining the quantitative accuracy of contours.

Three-dimensional models with exaggerated vertical relief are mass-produced by molding thin sheets of plastic. On large-scale models, the quality of the molding is sufficiently good to represent the landscape accurately. Plastic "raised relief" maps are available for much of the United States at a scale of 1:250,000. Global relief maps are sometimes prepared from photographs of three-dimensional models, which are illuminated from the upper left to emphasize relief by light and shadow.

MAPPING IN THE UNITED STATES

Regional and city planning and the development of a country's resources are closely tied to the availability of specialized maps. In addition to maps of topography and those required for navigation, government agencies prepare maps of water, soil, geologic phenomena, timber and mineral resources, and other features. Maps are also employed to compile and report economic and demographic statistics. A map devoted to a single topic, such as the distribution of population or rainfall, is known as a *thematic* map.

The responsibility for mapping the United States is divided among many civil and military agencies, depending on the purpose of the map. Only a few examples are given here to indicate the extent of the mapping services carried on by the government.

The United States Geological Survey (USGS) publishes several series of topographic maps in different scales that cover the United States, Puerto Rico, and other territories. The USGS Topographic Maps are discussed in detail in the following section of this appendix because of their particular importance to physical geographers.

In addition to topographic maps, the Geological Survey publishes geologic and mineral investigation maps, and hydrologic maps for water-resource planning.

Maps related to navigation purposes are known as *charts.* The preparation of nautical and aeronautical

charts is carried out by several agencies. Nautical charts of foreign waters are prepared by the Naval Oceanographic Office. This office also issues small-scale bathymetric charts that portray the topography of much of the ocean floor. The National Ocean Survey prepares nautical navigation charts of United States coastal and offshore waters and some inland waterways. The harbor charts prepared by the Ocean Survey may have scales as large as 1:18,000 to show accurate locations for important features such as main channels, marker buoys, and underwater cables. The Army Corps of Engineers issues nautical charts of the Great Lakes and navigation charts for numerous rivers and inland waterways.

Aeronautical charts are produced by the National Ocean Survey and the United States Air Force Aeronautical Chart and Information Center. The scales employed vary from 1:250,000 on charts for aircraft of moderate speed to 1:2,000,000 and smaller for the needs of global jet transport. Special features of aeronautical charts include identification of airports and air navigation radio beacons, and a simplified representation that emphasizes major relief features and landmarks visible from the air.

USGS TOPOGRAPHIC MAPS

A *topographic map* is a graphic representation of a portion of the earth's surface at a scale large enough to show human works such as individual buildings. Additionally, topographic maps employ contour lines to show the configuration of the land surface. The United States has maintained a topographic map program since 1879 under the direction of the Geological Survey. The Geological Survey publishes the National Topographic Map Series, an invaluable source of information on the physical characteristics and human activities present in nearly all parts of the national territory. The maps are compiled primarily from aerial photographs, but field surveys are used to establish highly precise networks of horizontal positions and vertical elevations, and to verify details of photo interpretation.

The Topographic Series is itself made up of maps at several different scales. Each of the maps is drawn so that its sides coincide with parallels of latitude and meridians of longitude, and a similar symbolism is used in all series (Figure II.16). Topographic-map series covering most of the United States are published at scales of 1:24,000; 1:62,500; and 1:63,360 (Alaska). The entire nation is also covered by map series at scales of 1:250,000 and 1:500,000. In recent years, a number of maps using the metric system have been published at a scale of 1:100,000 (1 cm = 1 km). In addition, there are map series covering national parks, monuments, historic sites, major metropolitan areas, rivers and flood plains, Puerto Rico, other national territories, Antarctica, the moon, and Mars.

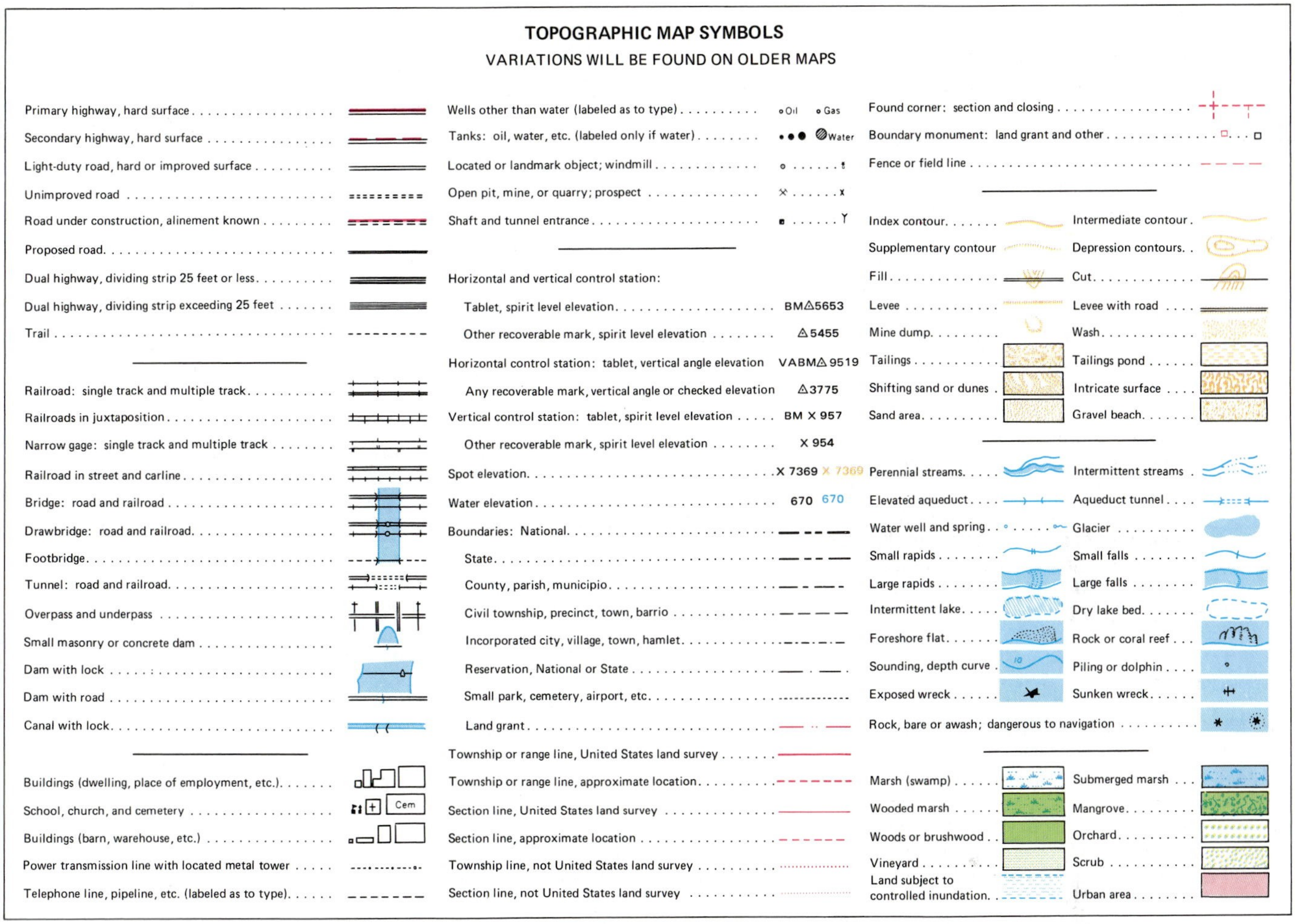

TOPOGRAPHIC MAP SYMBOLS

VARIATIONS WILL BE FOUND ON OLDER MAPS

Primary highway, hard surface	Wells other than water (labeled as to type)	○ Oil ○ Gas	Found corner: section and closing	
Secondary highway, hard surface	Tanks: oil, water, etc. (labeled only if water)	● ●● ⊘Water	Boundary monument: land grant and other	
Light-duty road, hard or improved surface	Located or landmark object; windmill	○ ●	Fence or field line	
Unimproved road	Open pit, mine, or quarry; prospect	✕ ✕		
Road under construction, alinement known	Shaft and tunnel entrance	▪ Y		
Proposed road			Index contour	Intermediate contour
Dual highway, dividing strip 25 feet or less			Supplementary contour	Depression contours
Dual highway, dividing strip exceeding 25 feet	Horizontal and vertical control station:		Fill	Cut
Trail	Tablet, spirit level elevation	BM△5653	Levee	Levee with road
	Other recoverable mark, spirit level elevation	△5455	Mine dump	Wash
Railroad: single track and multiple track	Horizontal control station: tablet, vertical angle elevation	VABM△9519	Tailings	Tailings pond
Railroads in juxtaposition	Any recoverable mark, vertical angle or checked elevation	△3775	Shifting sand or dunes	Intricate surface
Narrow gage: single track and multiple track	Vertical control station: tablet, spirit level elevation	BM X 957	Sand area	Gravel beach
Railroad in street and carline	Other recoverable mark, spirit level elevation	X 954		
Bridge: road and railroad	Spot elevation	X 7369 X 7369	Perennial streams	Intermittent streams
Drawbridge: road and railroad	Water elevation	670 670	Elevated aqueduct	Aqueduct tunnel
Footbridge	Boundaries: National		Water well and spring	Glacier
Tunnel: road and railroad	State		Small rapids	Small falls
Overpass and underpass	County, parish, municipio		Large rapids	Large falls
Small masonry or concrete dam	Civil township, precinct, town, barrio		Intermittent lake	Dry lake bed
Dam with lock	Incorporated city, village, town, hamlet		Foreshore flat	Rock or coral reef
Dam with road	Reservation, National or State		Sounding, depth curve	Piling or dolphin
Canal with lock	Small park, cemetery, airport, etc.		Exposed wreck	Sunken wreck
	Land grant		Rock, bare or awash; dangerous to navigation	
Buildings (dwelling, place of employment, etc.)	Township or range line, United States land survey			
School, church, and cemetery	Township or range line, approximate location		Marsh (swamp)	Submerged marsh
Buildings (barn, warehouse, etc.)	Section line, United States land survey		Wooded marsh	Mangrove
Power transmission line with located metal tower	Section line, approximate location		Woods or brushwood	Orchard
Telephone line, pipeline, etc. (labeled as to type)	Township line, not United States land survey		Vineyard	Scrub
	Section line, not United States land survey		Land subject to controlled inundation	Urban area

Figure II.16

Basic symbols employed on U.S. Geological Survey topographic maps at scales of 1:24,000 and 1:62,500. (U.S. Department of the Interior)

The 7½-minute quadrangle series (1:24,000) and 15-minute quadrangle series (1:62,500; 1:63,360) are large-scale maps suited to planning, engineering, and landform studies on a local scale. Such maps contain a vast amount of information.

Topographic maps employ more than 100 different symbols to depict natural and cultural features. The table of USGS topographic map symbols reproduced in Figure II.16 is not printed on the map sheets but is available separately. The colors on a topographic map are an integral part of the symbolization, with each color restricted to a particular class of features.

Black is used for all man-made objects other than major roads and urban areas. All names and labels are in black. Blue is used for all water features, including streams, lakes, marshes, canals, the oceans, and water depths indicated by contours and numbers. Brown is reserved for land surface contour lines, indications of surface type, such as loose sand or mine tailings, and certain elevation figures. Green is used for vegetation symbols: woodlands, orchards, vineyards, brush, and wooded marshland. Red is used for major roads and land survey lines. Pink indicates densely built-up urban areas in which only landmark buildings are indicated individually. Purple is used on interim maps to symbolize changes that have occurred since the previous edition.

TOOLS OF THE PHYSICAL GEOGRAPHER

REMOTE SENSING

The space age has provided new techniques for obtaining information about the earth's surface. The new methods employ special sensing devices that can acquire data at a distance from the phenomena being studied; therefore the technique is known as *remote sensing*. The principal methods used are to sense and record electromagnetic radiation of sunlight, laser light, or microwave radar that is reflected from objects on the earth, or the infrared thermal radiation directly emitted by objects (Figure II.17).

Remote sensing equipment is usually mounted in aircraft or in satellites in orbit around the earth. The advantages of remote sensing include the ability to cover vast areas in a very short time, the ability to provide data from very large to very small scale, the possibility of repetitive measurements to follow the progress of selected events, and the ability to penetrate areas inaccessible to ground survey.

The techniques of remote sensing have been greatly extended recently in an effort to obtain the information needed to manage the earth's resources more efficiently, and much of the information being gathered is of direct interest to physical geographers. The capabilities of remote sensing technology include the detection of vegetation types and their seasonal changes; the measurement of surface water distribution, soil moisture, and water temperature; the depiction of landforms and surface geologic structures; analysis of water circulation in lakes and the seas; detection of small variations in water levels in the seas; mapping variations in gravity that reveal subsurface geologic features for mineral exploration; and the monitoring of human activities that affect the environment.

Aircraft equipped with detectors of longwave thermal infrared radiation can provide images of objects on the earth's surface with intensities proportional to their temperatures. Infrared imagery methods can be employed to detect cloud temperatures, ocean currents, or the mixing of warm water from industrial effluents with cooler river water.

Application of remote sensing methods to the study of the earth has been revolutionized by using remote sensors in conjunction with satellites in orbit around the earth. Surveillance of the earth's cloud cover from weather satellites that orbit the earth at altitudes of several hundred kilometers has greatly improved the accuracy of weather predictions and the ability to track potentially damaging hurricanes and midlatitude storms.

Aerial Photography

Aerial photography, the first remote sensing method developed, is now used for most topographic mapping. Conventional aerial surveys employ the visible light reflected from the earth and detected with cameras using black-and-white film. Optical techniques allow the unavoidable distortions in a photograph taken at low or moderate altitudes to be removed so that areas close to the edges of the photograph have the same scale as areas directly below the camera. Two exposures can be used to record a stereoscopic (three-dimensional) image from which contour lines can be drawn (Figure II.18).

A single aerial photograph may cover an area of a few square km to one comprising hundreds of square km, depending on the cameras and the altitude of the aircraft. Ground features as small as 1 cm in size can be detected

Figure II.17
Remote sensing techniques utilize various portions of the electromagnetic spectrum. The visible and near-infrared regions are used for photography and television, and the infrared is used for scanning surface temperatures. Relief features are mapped by laser profiling and radar. (Calvin Woo)

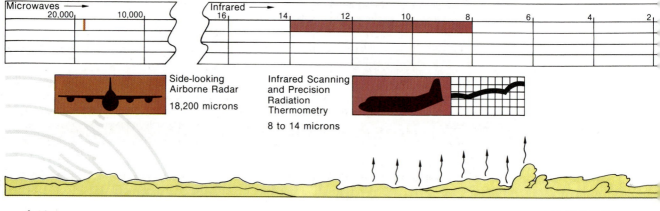

Figure II.18

Stereo pairs are used to make relief apparent in aerial photographs. When viewed with a stereoscope, relief features in this view in eastern Utah stand out as a three-dimensional model. To see the stereo effect, the left-hand photograph should be viewed with the left eye, and the right-hand photograph with the right eye. A card held between the photographs helps to separate the images. (U.S. Geological Survey)

on photographs made from aircraft flying at an altitude of a few kilometers, and satellite imagery can provide resolution of objects only a few meters in size from an altitude of several hundred kilometers. These advanced capabilities are gradually becoming available for civilian uses.

Infrared Photography

Sunlight falling upon the earth consists of electromagnetic radiation in the visible region and in the near-visible infrared region of the electromagnetic spectrum. Different objects reflect different relative proportions of each wavelength, so that the radiation reflected from an object, when analyzed in terms of wavelength, forms a means of identification known as the *spectral signature* of the object. Vegetation, for example, usually reflects proportionately more short wavelength infrared radiation than does bare soil. Furthermore, different types of vegetation can usually be distinguished by their spectral signatures. Diseased plants or lack of soil moisture can also be detected.

Color infrared film is similar to ordinary color film, except that it depicts the previously invisible infrared radiation as red, the red colors as green, and the green colors as blue (Figure II.19). However, healthy vegetation appears red on a color infrared photograph. The intensity of the color is directly proportional to the amount of chlorophyll in the vegetation. Color infrared photographs of test fields and forests have proven that different trees and

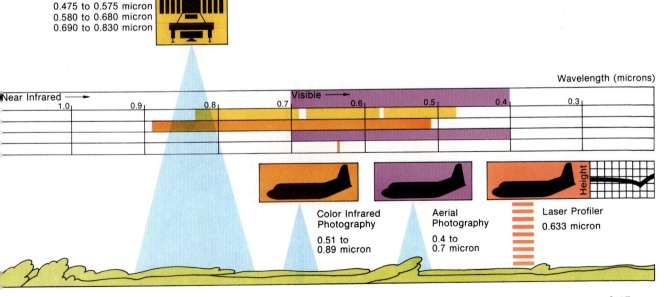

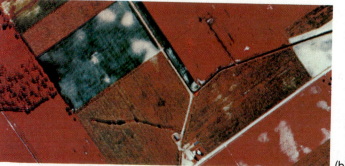

(a)

(b)

Figure II.19

Conventional and infrared color photography.

(a) The aerial view (top left) was photographed with ordinary color film.

(b) The same fields (top right) were photographed with color infrared film; healthy vegetation appears bright red. (Courtesy of the Laboratory for Applications of Remote Sensing—LARS).

(c) Vegetation on irrigated land is clearly differentiated from dry desert in this color infrared photograph of the Imperial Valley of California taken from earth orbit. The Salton Sea appears as a large dark region at the upper left. The pattern of agricultural fields is clearly distinguishable; note in particular the border between the United States and Mexico that is evident toward the bottom of the photograph. The well-irrigated crops on the United States side show as bright red, compared to the bluish hues representing bare ground or desert shrubs on the Mexican side of the border. (NASA)

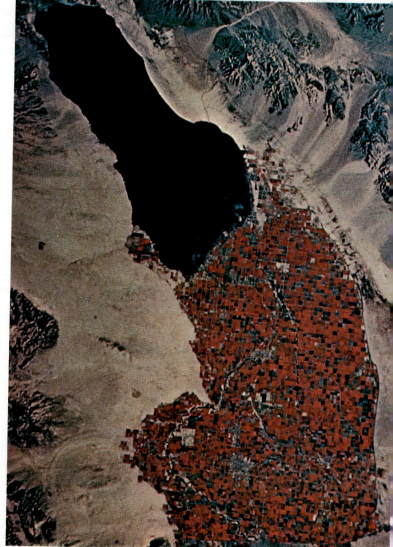

(c)

ing estuaries, called *rias,* where the sea has invaded river valleys deepened during Pleistocene low stands of the sea. The coast of North Carolina, shown in Figure 17.23, is a classic example, displaying the estuaries of several large rivers. Between rias we find cliffed headlands and sometimes sea arches and stacks. As the headlands are battered back, their debris helps produce spits that close off bays, transforming them into lagoons.

The form of submerged coasts varies according to the nature of the older subaerial (non-marine) landforms and the orientation of relief features with respect to the coastline. The rocky islets off the coasts of New England, Labrador, British Columbia, Scotland, Norway, and Chile result from submergence of glacially scoured landscapes (Figure 17.23). Other highly irregular coasts, such as those of Greece and western Turkey, are fringed by rugged islands that result from submergence of mountainous topography. In southern Greece and Turkey the line of the coast cuts across the trend of mountain ridges, resulting in a series

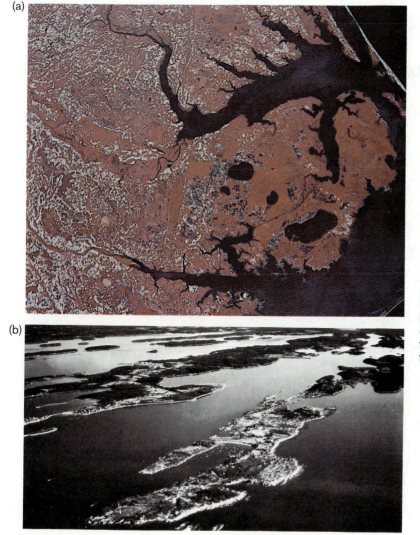

(a)

(b)

Figure 17.23
Coastal submergence by Holocene sea level rise.

(a) This LANDSAT false-color image portrays the *ria* coast of North Carolina, created by Holocene submergence of river valleys incised to the lower sea level of the late Pleistocene. At the right is Pamlico Sound, protected by a line of barrier islands that link at Cape Hatteras (Figure 17.19). The drowned valley on the north is Albemarle Sound, which is the estuary of the Roanoke and Chowan rivers. The Pamlico River creates the southern estuary. (NASA)

(b) The coast of Maine was eroded by a continental ice sheet and then submerged by postglacial rise in sea level. The long, parallel projections of land and elongated islands were continuous ridges before being submerged. Submergence of the coast drowned the valleys and exposed the ridges to wave erosion. (John S. Shelton)

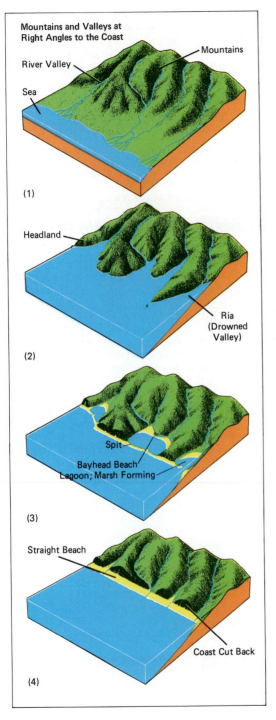

Mountains and Valleys at Right Angles to the Coast

Mountains

River Valley

Sea

(1)

Headland

Ria (Drowned Valley)

(2)

Spit

Bayhead Beach
Lagoon; Marsh Forming

(3)

Straight Beach

Coast Cut Back

(4)

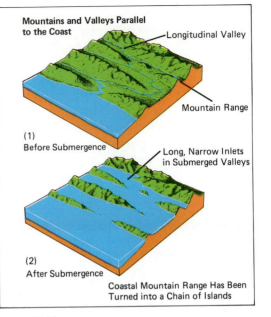

Mountains and Valleys Parallel to the Coast

Longitudinal Valley

Mountain Range

(1)
Before Submergence

Long, Narrow Inlets
in Submerged Valleys

(2)
After Submergence

Coastal Mountain Range Has Been
Turned into a Chain of Islands

Figure 17.24

Effects of submergence on coasts having varying topographic trends.

(left) Evolution of a coastline formed where valleys and ridges transverse to the coast are submerged by a rise of sea level relative to the land.

(1) The initial coastline consists of valleys and ridges behind a narrow coastal plain.

(2) The relative rise of sea level forms drowned river valleys, or *rias,* and headlands jutting into the ocean.

(3) Wave erosion cuts back the headlands and forms vertical sea cliffs. Deposition of sediment by currents builds spits and beach areas; spit growth eventually closes off the bays, forming lagoons.

(4) Continued erosion wears the coast back to a straight line bordered by sea cliffs and a narrow beach. Stream erosion will reduce the elevation of the ridges and highlands of the land area.

(above) (1) The diagram shows a coastal region where ridges and valleys are parallel to the coast.

(2) Submergence of the coastline by elevation of sea level with respect to the land leaves numerous islands oriented parallel to the coast, such as those off the Dalmatian Coast of Yugoslavia.

of peninsulas. Where the line of a submergent coast parallels mountain ridges, the coastline is more regular, with linear islands parallel to the mainland, as along the coasts of Yugoslavia and southern California (Figure 17.24).

ENIGMAS OF THE OCEAN FLOOR

While the constant change observable at shorelines makes the coastal zone one of the earth's most fascinating areas of study, the ocean floors themselves, once a mystery, are proving to have quite diversified relief, including several features that have yet to be fully explained.

The use of deep-diving vehicles to observe the landforms of the ocean floor has barely begun. Most of our information about the form of the sea floors has been gained by indirect means. The most widely used method for charting the relief of the ocean floor is *echo sounding*, a technique that was devised to track enemy submarines during World War II. Sound waves are generated at the surface, and the time it takes for the echo to return from the bottom is recorded and automatically converted to a distance measure. In this way a continuous profile of the ocean floor is traced mechanically from echo data (Figure 17.25).

Coring and dredging devices lowered from ships specially equipped for oceanographic research bring back samples of ocean floor material. Since 1968 the ongoing Deep Sea Drilling Project, utilizing the oceanographic research vessel *Glomar Challenger,* has retrieved cores from hundreds of boreholes in the deep ocean

Figure 17.25
This figure illustrates the method by which information about the character of the sea floor is obtained. The oceanographic vessel tows a sound source that creates a loud noise by an electric impulse or an explosion of gas. This acoustic signal travels outward in all directions and is reflected back from the sea floor and, if strong enough, from separate layers of sediment on the sea floor as well as from the rock beneath the sediments. The reflected signals are sensed by a towed hydrophone array. The reflected signals drive a recording device that makes a continuous trace of the type shown here in black. Such traces intentionally exaggerate the submarine relief to reveal minor details. Here we see irregular bedrock topography that is partially buried by marine sediments.

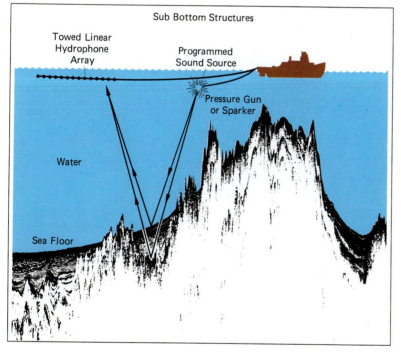

floor. Some of these were drilled in water depths exceeding 6,000 meters (20,000 ft), a very difficult procedure. While this work has done much to unveil the geologic history of the oceans, some puzzles remain. Two of these are submarine canyons and guyots.

Submarine Canyons

Canyons that rival any on the land surfaces gash the submarine continental shelves and slopes throughout the world. Their existence has been known for nearly a century. According to soundings and inspections using deep-diving vehicles, *submarine canyons* are similar in form to stream-cut canyons on the land.

They vary from broad troughs to vast steep-sided trenches with dimensions equal to those of the largest erosional features on the continents (Figure 17.26). Canyons appear off the mouths of New York's Hudson River, the Columbia River, and the Zaire (Congo) River, but

Figure 17.26
This diagram shows the *submarine canyon* that cuts across the narrow continental shelf at Monterey Bay in California. Such canyons are cut into bedrock and may attain the dimensions of the largest fluvial canyons on the land. The longshore drift process feeds sediment into the heads of the canyons, which lie quite close to the land. The canyons then funnel this sediment into the ocean depths as submarine *turbidity currents*. (Tau Rho Alpha/U.S. Geological Survey)

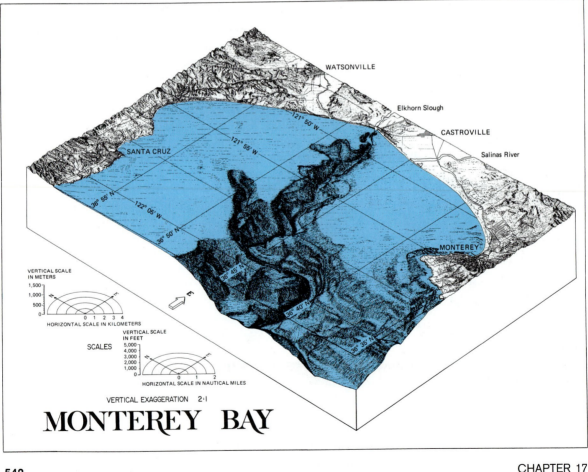

MONTEREY BAY

many others are not in the vicinity of any existing large stream.

Submarine canyons are puzzling because many extend to depths of nearly 5,000 meters (16,500 ft)—far too deep to be drowned river valleys. Some portions are excavated in soft sediments, but others are cut through hard rock in steep-walled gorges. The heads of these canyons come close enough to the land to trap sediment moving by longshore drift, diverting it from the continental shelves to the continental rises, ocean trenches, and deep ocean basins.

The most acceptable explanation for the origin of submarine canyons is that they are erosional forms created below sea level by streams of sand and silt that periodically pour down the continental slopes to the deep-sea floor. These so-called *turbidity currents* have been seen and photographed. The fact that turbidity currents occasionally rupture telegraph cables on the sea floor indicates a destructive potential that, given enough time, could account for rock-cut submarine canyons. Thick masses of alternating sandstone and shale beds that are widely encountered on the continents are thought to be the deposits of turbidity currents in ancient oceans, and are known as "turbidites." The existence of turbidites suggests that submarine canyons have been present throughout most of the earth's history.

Guyots

Another puzzling feature of the sea floor is the *guyot,* a submarine mountain with a conspicuously flat top that is far below sea level. The form was discovered during World War II and has been named after a famous Swiss-American geographer of the last century. Hundreds of submerged volcanoes, or *seamounts,* are present in the world's oceans, particularly in the Pacific. A large proportion of these are flat-topped guyots. Although guyots have the appearance of having been planed off by erosion at the surface of the ocean, their summits

are often 1,200 meters (4,000 ft) or more below the sea surface—far too deep to be affected by wave action, even during times of lowered sea level.

It has been proposed that volcanic islands formed at oceanic ridges were planed off by wave attack, and then were carried laterally down the flanks of the ridges into the ocean depths as part of the sea-floor spreading process. This hypothesis of subsidence by lateral plate motion seems to be verified by age determinations on the volcanic rocks of the guyots. Most show increasing age with increasing depth and distance from the oceanic ridge system. It is probable that many coral atolls have been built up from summits of sinking guyots.

Oceanic Fracture Zones

Before sea-floor spreading was even suspected, it was known that the submerged oceanic ridge system was offset horizontally in many places, sometimes by hundreds of kilometers. Oceanographers eventually discovered that these ridge offsets were giant fractures that extended across the ocean floors for thousands of kilometers. Some could be traced all the way from the oceanic ridge to the edges of the continents on either side of the ocean basin (Figure 17.27).

We now know that these fractures are enormous strike-slip faults that have developed as the oceans expanded from the sea-floor spreading centers (the oceanic ridge system). Many of these faults are of greater magnitude than any seen on the land. One that crosses the equatorial Atlantic Ocean from Brazil to Africa forms a trench four times the depth of Arizona's Grand Canyon. Others create great linear escarpments that interrupt the sea floor. Though invisible to us, they rank among the earth's grandest relief features.

The oceanic fracture zones are tectonically active and are centers of geothermal and earthquake activity as well as submarine volcanism. The linear fractures at right angles to ridge

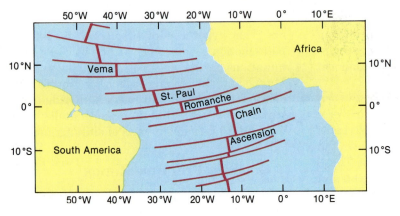

Figure 17.27
Fracture zones in the equatorial Atlantic Ocean. These fractures, often 3,000 km (2,000 miles) in length, span the breadth of the Atlantic Ocean and transform seafloor spreading to strike-slip faulting that markedly offsets the Mid-Atlantic Ridge. The larger fractures produce linear depressions and escarpments crossing the ocean floor from one side to the other. (After E. Bonatti and K. Crane, *Scientific American,* May 1984)

segments are necessary to accommodate differential rates of sea-floor spreading on the spherical surface of the earth. Thus they are a crucial element in the birth and expansion of the ocean basins.

Hydrothermal Vents

Direct access to active sea-floor spreading centers using deep-diving vehicles and indirect access through side-scanning sonar and towed cameras has recently revealed an unsuspected world of *hydrothermal vents* that jet out dark plumes of water at 350° to 400°C (650° to 750°F), with active deposition of sulfur and metallic sulfides of iron, copper, zinc, manganese, cobalt, vanadium, gold, and silver (Figure 17.28). Somewhat cooler vents support a hitherto unknown biotic community of ghostly white crabs, giant clams, and eerie clusters of 2.5 meter (8 ft) tube worms—all in total darkness under as much as 2,400 meters (8,000 ft) of ocean water. These phenomena, first discovered under the Red Sea in the 1960s, were found in the Pacific Ocean near the Galapagos Islands in 1981 and have since been located elsewhere in the Atlantic, Pacific, and Indian oceans.

The metals are dissolved in hot solutions spewed from the vents, known as "black smokers," and precipitate in the cold (2°C) seawater in yellow-orange mounds, towers, and slaggy masses. The principal activity occurs in the youngest portion of the spreading center—the axial rift between in-facing fault scarps where volcanic activity is most recent. Some metallic deposits are found in summit craters of shield volcanoes within the rift zone.

The discovery of an enormous, previously unknown source of metallic minerals is of obvious significance. Prior to this discovery geologists had attributed all mineralization to reactions between plutons and surrounding "host" rock and to subaerial weathering that releases metallic elements that form secondary concentrates on land surfaces. The metals issuing

continental polar air (cP) cold dry air derived from high latitude continental interior locations; brings clear weather, but causes rain or snow by wedging moist tropical air upward in frontal zones.

continental rise submarine region of sediment accumulation at base of continental slope.

continental shelf shallow, gently sloping undersea extensions of continents.

continental slope steeply sloping undersea region descending from continental shelf to continental rise.

continental tropical air (cT) hot, dry air originating over major desert regions in anticyclonic zones; source of dust in far-distant areas.

continentality being characterized by very large annual temperature ranges.

contour interval difference in elevation between successive contour lines on relief map.

contour line line on map joining points that have same elevation above or below reference plane.

convection vertical movement of parcels of air due to density differences.

convective cell (Hadley cell) vertical circulation system characterized by ascending warmer air and descending cooler air.

convergence process in which surface air flows in inward-converging spiral toward a low pressure center. It is forced to rise to allow for additional incoming air, frequently resulting in condensation, clouds, and precipitation.

convergent evolution adaptations of different species on widely separated continents to similar climates, resulting in similar morphological characteristics.

coral marine animal of tropical seas that builds stony exterior skeletal structure composed of calcium carbonate.

coral reef reef composed mainly of coral, of which parts have solidified into limestone.

core central portion of earth, composed of iron and nickel compounds; divided into molten outer core and solid inner core.

corestone rounded boulder, usually granitic, created by subsurface chemical weathering of jointed rock; often becomes exposed by erosion.

Coriolis force horizontal force apparent to observer on rotating earth, that tends to deflect moving objects to right in northern hemisphere but to left in southern hemisphere.

corrasion mechanical erosion caused by particles being transported by wind, flowing water, or glacial ice.

corrosion erosion of a solid substance by chemical processes.

creep slow downslope transfer of loose rock and soil due to gravitational energy.

crescentic dunes, see barchan

crevasse deep crack in glacier, usually oblique or transverse to direction of glacier flow; also localized overflow of floodwater in lower Mississippi River Valley.

crust outermost shell of earth composed of solid rock material called sial and sima; lithosphere.

crustal plate (lithospheric plate) individual segment of earth's crust, bounded by oceanic ridges, subduction zones, or transform faults. Interaction of separate crustal plates is basis for global tectonics.

cryergic processes geomorphic processes related to low temperatures, especially in tundra climate.

crystalline rock igneous and metamorphic rock composed of a variety of visible mineral crystals.

crystalline shield surface expanse of very ancient crystalline rocks that form the nuclei of all continents.

cuesta one-sided plateau capped by resistant layer of sedimentary rock; composed of gently dipping back slope and terminated by erosional escarpment.

cumulonimbus clouds (thunderheads) towering, anvil-headed clouds associated with thunderstorms, hail, and occasionally tornadoes.

cumulonimbus mammatus uncommon cloud type covering extensive area, with downward bulges on lower surface; often associated with severe storms and tornadoes.

cumulus clouds puffy white clouds that form locally on rising columns of air.

currents, ocean persistent surface drift of water driven by prevailing winds; at depth, water may move in opposite direction.

cusp triangular projection, point, or promontory.

cuspate foreland coastal projection formed where differently oriented spits or barrier islands have become linked.

cuspate spit coastal spits that link in a V-shaped configuration, with the point extending seaward; smaller than a cuspate foreland.

cycle of erosion, see Davisian cycle of erosion

cyclogenesis formation of cyclonic circulations.

cyclones converging airflows around central region of low pressure; ascending, large-scale vortex of air.

cylindric projection map projection in which globe is projected onto tangent cylinder.

d

datum plane chosen reference level for establishing relative elevation, such as mean sea level.

Davisian cycle of erosion evolutionary sequence of landforms devised by William Morris Davis to describe midlatitude landscapes shaped by fluvial processes.

debris flow flow of water-saturated rock debris often including large boulders; destructive in settled areas.

decay food chain food chain that begins with dead organic matter and goes on to microorganisms and to their predators, bacteria and fungi. These decomposers break down organic material, returning minerals to soil.

deciduous shedding of leaves during specific season to avoid moisture loss or cell damage; applied to plants, particularly trees.

decomposers in food chains, organisms that derive their energy from dead plant and animal tissues.

deep-water waves waves in water that is deeper than half the wave length; their speed is independent of water depth.

deflation, aeolian process of removal of fine rock debris by wind.

degradation, fluvial prolonged scouring (vertical incision) by streams.

delta alluvial deposits laid down by stream at its mouth in lake or sea.

dendritic treelike branching pattern; applied particularly to drainage patterns in which tributaries enter main stream at acute angles.

dendrochronology, see tree ring analysis

denudation   lowering of land surface by all processes of erosion.

denudation rate   rate of lowering of the land surface by erosion; usually measured as mm, cm, or feet per 1,000 years.

deposition   laying down of sediment by transporting agent, such as water, wind, or glacial ice.

deranged drainage   type of drainage that shows no geometrical pattern or constant direction and usually includes numbers of lakes.

desalinization   removal of salt from water to make water usable.

desert   area lacking continuous vegetative cover, sometimes entirely barren, usually resulting from extreme moisture deficit.

desert pavement (desert armor)   surface veneer of stones left as a lag deposit as finer material is removed by water runoff or aeolian deflation.

desert varnish   thin coating of clay and dark-colored mineral oxides that forms on exposed rock surfaces in desert regions; principally manganese and iron oxides.

desilication   removal of silica in solution as part of laterization process in humid tropics.

dew   moisture that condenses on surfaces at dew point temperature.

dew point   temperature at which relative humidity becomes 100 percent.

differential weathering and erosion   local variation in decay and removal of rock as a consequence of variation in rock properties.

dike, volcanic   solidified sheetlike intrusion of igneous rock that cuts through older rocks; usually vertical or steeply inclined.

dike ridges   wall-like ridges made by exposure of resistant dikes of igneous rock as a result of differential erosion.

dip   angle from horizontal made by dipping plane such as fault or bed of sedimentary rock.

dip-slip fault   fault in which relative displacement of blocks is parallel to direction of fault dip.

discharge, stream   volume of water transported by stream past given point during standard time interval; measured in cubic meters per second or cubic feet per second.

dissolved load (chemical load)   material carried in solution by stream.

distributaries   separate channels produced by downstream division of stream flow, characteristic of deltas.

doldrums   latitudinal zone centered on equator, in which trade winds converge into low-pressure trough of generally light, variable winds and calms; intertropical convergence zone (ITC).

doline,   see sinkhole

dolomite   sedimentary rock; a magnesium-rich limestone; also a magnesium carbonate mineral.

dome   circular or elliptical uplift in area of sedimentary rock, produced by vertical tectonic stress, often related to volcanic intrusion.

double planation   subsurface chemical decomposition of rock at weathering front, and surface removal of weathered material by erosion; generally applied to humid tropics.

downslope gravitational stress   portion of gravitational force on a particle on a slope, which tends to pull the particle downward parallel to the slope.

draa   large sand mountains composed of superimposed sand dunes.

drainage basin   area that supplies water discharged by given stream channel.

drainage density   total number or total length of stream channels per unit of area.

drainage net   system of stream channels in drainage basin.

drainage pattern   geometrical arrangement of stream channels in drainage net; often reflects type and arrangement of bedrock units within drainage basin.

drainage texture   relative term referring to spacing of stream channels in area; usual contrast is between coarse texture (few channels) and fine texture (many channels).

drift, glacial   term referring to material of all types deposited by glacial processes; includes till, ice-contact stratified drift, and outwash.

drift plain   in a glaciated region, a plain composed of unspecified types of glacial drift (till, ice-contact stratified drift, outwash).

drizzle   form of rainfall that consists of smaller than average raindrops.

drumlin   streamlined elongated hill composed of glacial deposits; long axis parallels direction of glacier's flow.

dry adiabatic rate   rate of adiabatic cooling of air that is not saturated and where condensation is not occurring; its value is 10°C for every km of increasing altitude.

dry climates   climates in which precipitation is inhibited because of atmospheric subsidence, stable lapse rates, or distance from oceanic moisture sources.

dry realm   one of Penck's three realms of climate, where precipitation is less than potential evapotranspiration; characterized by soil moisture deficiency.

dry summer subtropical climate   climatic region along western margins of continents in subtropical latitudes; characterized by wet winters and very dry summers.

dune backslope   slope of sand dune that receives full impact of wind, up which sand moves to dune crest.

dune slip face   steep downwind face of dune, which is always at angle of repose of dry sand, about 34°.

duricrust   rocklike crust developed by soil-forming processes on flat surfaces over long periods of time; includes laterite, calcrete, and silcrete.

dynamic equilibrium   pertains to systems that oscillate about an average condition; compare with metastable equilibrium.

e

earthflow   bulge of collapsed soil that has been pushed out and downslope by slump.

earthquake   violent disturbance of earth's surface produced by sudden displacement of subsurface rock masses.

easterly waves   extended weak, low-pressure trough that drifts westward in region of trade winds; often accompanied by thunderstorms.

ebb tides   daily periods of falling sea level along coasts; usually occur twice every 24 hours and 50 minutes.

echo sounding   determination of water depth by reflected sound waves.

ecological niche   specific environment exploited by an organism in an ecosystem.

ecology   study of the interactions between living organisms and their environment.

ecosystem community of plants and animals generally in equilibrium with inputs of energy and materials in their particular environment.

ecotone boundary or zone of transition between adjacent ecosystems; applicable at various scales.

edaphic factor influence of soil conditions upon vegetation.

eddy current of local circulation, which causes turbulent motion of gases or liquids.

E horizon eluvial soil horizon; light-colored horizon in soil showing maximum loss of material to translocation processes; below A horizon and above B horizon.

electromagnetic radiation energy which is transferred in wave form through space without attenuation.

electromagnetic spectrum various kinds of electromagnetic radiation, arranged according to wavelength; electromagnetic spectrum includes radio waves, infrared radiation, visible light, ultraviolet radiation, x-rays, and gamma rays.

element simple substance composed of atoms having the same atomic number, may exist in varying isotopes.

El Niño periodic oceanic phenomenon in which warm equatorial water replaces cold upwelling water in eastern Pacific, causing die-off of marine organisms and weather changes even in distant areas.

eluvial layer E horizon of soil from which material has been translocated in solid and dissolved form.

eluviation flushing of fine particles or dissolved substances to lower levels in soil by downward percolating water.

emergent coast coast along which land has risen relative to sea so that submarine relief features are exposed subaerially.

end moraine accumulation of material deposited at terminal position of glacier; terminal moraine.

energy capacity to do work, or to produce heat or motion.

energy budget balance between energy inputs and outputs in a system, taking account of energy storage and transformations.

energy conversion efficiency amount of energy converted to specific form relative to total energy supplied.

energy flow diagram schematic diagram that emphasizes transfer of energy from one part of system to another or from one system to another.

energy flow pyramid pyramidal structure of all ecosystems resulting from energy loss (heat, work, respiration) at each level of the food chain.

entisols soils with poor horizonation due to youth of soil, rapid erosion as soil forms, waterlogging, or human interference.

entrainment process of putting stationary loose particles into motion in a moving medium; erosion.

environmental lapse rate vertical distribution of temperature in segment of atmosphere.

eon in geologic time scale, broadest division, recognizing Precambrian and Phanerozoic eons.

epicenter, earthquake point on the earth's surface that is directly above, and closest to, the underground focus (point of energy release) of an earthquake.

epiphytes plants that grow on other plants, particularly in rain forests; they extract moisture from air and nutrients from plant debris lodged in tree branches.

epochs in geologic time scale, subdivision of a geologic period (e.g., Pleistocene epoch of Quaternary period).

equal area projection map projection in which each region depicted has same area as it would have on globe of same scale; angles are distorted, especially near margins.

equator imaginary line cutting the earth into northern and southern hemispheres; surface trace of plane cutting axis of earth rotation perpendicularly midway between poles; 0° latitude.

equatorial trough zone of low atmospheric pressure situated near equator; intertropical convergence zone.

equilibrium state of stability in which inputs and outputs of energy and material are balanced so as to maintain system in unchanged state.

equilibrium line, glacier line that separates portion of a glacier having a net gain of material from that having a net loss of material.

equilibrium slope slope that tends to maintain just sufficient steepness to equate erosional stress to material resistance.

equinox instant, occurring twice each year, at which sun is vertically overhead at equator; on this day, hours of sunlight and darkness are equal everywhere on earth.

equivalent projection, see equal area projection

era broad division of geologic time dominated by distinctive life forms; (e.g., Mesozoic, Cenozoic).

erg (geomorphology) vast accumulation of sand in desert region; (physics) unit of energy or work.

erosion various processes by which rock debris is detached and transported.

escarpment steep face at edge of plateau, cuesta, fault block, or region of high local relief.

esker ridge of sand and gravel, sometimes many km long, deposited by streams beneath or within glacier.

estuary enlarged mouth of river resulting from coastal submergence, where fresh water and sea water mix under pulse of tides.

eustatic changes, sea level sea-level changes related to the volume of liquid water in the ocean basins; affected by storage of water in glaciers.

evaporation change of substance's physical state from liquid to vapor phase.

evapotranspiration transfer of water vapor to atmosphere by combined action of evaporation from soil and transpiration from vegetation. (See also potential evapotranspiration.)

evapotranspirometer device used to measure potential evapotranspiration; it consists of tank containing well-watered vegetation in which water is carefully measured.

evergreen vegetation vegetation that retains its foliage year-around: e.g., rainforests, coniferous forests, chaparral.

exfoliation physical (mechanical) weathering in which rock layers separate in sheets; common in granite.

extrusive igneous rock molten material that solidifies into volcanic rock at the land surface; lava.

eye, of hurricane, see hurricane eye

f

factors of soil formation factors causing variation of soil characteristics, including parent material, climate, site, organisms, and time.

fault fracture or fracture zone in earth's crust along which relative displacement of rocks has occurred.

fault block rock mass, bounded by faults on at least two opposing sides that moves vertically as a unit.

fault creep very slow (cm per year) horizontal displacement along strike-slip fault; no seismic (earthquake) activity associated.

fault scarp escarpment formed by faulting; usually modified by erosion.

felsenmeer "sea of rocks"; extensive area covered by large, irregular, frost-shattered rocks; characteristic of flat mountain summits in cold climates.

ferrallitization soil enrichment in iron resulting from loss of other substances by leaching in a tropical climate; part of laterization process.

ferricrete (laterite) bricklike crust formed when concentrates of iron and aluminum oxides in soil of tropical regions are exposed to air.

fetch open water distance across which wind blows in generally constant direction; factor in wave height.

field capacity maximum amount of soil moisture that can be retained after excess has drained away.

filling (fluvial) elevation of stream bed by deposition of stream-transported sediment.

firn compacted granular snow; intermediate form of glacier ice, formed from snow during first year after its accumulation. Also called névé.

firn basins glacial cirque in which firn (névé) is gradually transformed into glacial ice.

firn line line on surface of glacier between accumulated firn (white) and glacier ice (gray-blue).

first order of relief largest scale of relief features on planet; on earth, continents and ocean basins.

first-order stream smallest distinct channels in headwater area of a drainage net; channels receiving no tributaries.

fjord steep-walled glacial trough that has been invaded by sea, producing deep inlet; found along mountainous coasts in high latitudes.

flood basalts vast plateaulike accumulations of basalt, with flow superimposed on flow to thicknesses of hundreds to 1,000 or more meters (e.g., Columbia Plateau).

flood tides periods of rising sea level along coast caused by tides; usually two periods every 24 hours and 50 minutes.

floodplain flat plain bordering stream, composed of sediments deposited by stream.

fluted moraine morainal deposit that has been striated by advancing glacial ice.

fluvial deposition deposition of sediment carried by a stream.

fluvial processes processes associated with channelled flows of running water; stream processes.

fluvial terrace former valley floor standing above current floodplain as a result of stream incision.

focus, earthquake (hypocenter) subsurface point of rock rupture creating shock waves that produce earthquake.

fog cloud of water droplets formed at or near ground.

fold bend in rock strata, caused by horizontal compression of earth's crust.

fold limb, see limb of fold

folding wrinkling of sedimentary rock strata by crustal compression.

foliation alignment of minerals in metamorphic rocks resulting from plastic behavior, or "flow," of rock material under pressure deep within the earth's crust.

food chain series of organisms having interrelated feeding habits, each in turn serving as food for the next; represents energy transfer through environment.

food webs interconnected food chains.

footwall mass of rock below inclined fault plane.

forced waves ocean waves that are raised directly by wind stress on water surface, and that have sharp peaks and broad troughs.

foreshore lower shore zone of beach, between normal levels of high tide and low tide.

fossil remains or mineral replacements of plants or animals that have been preserved by natural processes, often by incorporation into sedimentary rock.

fossil record history of life forms recorded as fossils in rock strata, with earliest forms in deepest layers.

friction force that acts to retard any object in motion on earth or in the atmosphere.

friction cracks crescent-shaped tension fractures produced where rocks transported at base of glaciers press upon solid bedrock.

fringing reef coral reef which is built out laterally from shore.

front zone between two air masses that have different temperatures and humidities.

frost condensation of water vapor into solid phase when dewpoint temperature is below freezing.

frost weathering fragmentation of rock due to stress produced by expansion of water as it passes from its liquid to its solid state.

frozen realm one of A. Penck's three climatic realms in which potential evapotranspiration equals zero; characterized by year-around subfreezing temperatures.

fusion change from a solid to a liquid through the agency of heat; melting.

g

galaxy discrete aggregation of dust, gases, and thousands of millions of stars, widely separated from other galaxies; disclike, globular, or diffuse in form.

gelifluction form of mass wasting in which thawed, wet soil moves downslope in series of distinct lobes; occurs in tundra regions; solifluction.

gelifluction lobe lobate mass of soil that has moved downslope over permanently frozen ground in tundra regions; solifluction lobe.

genera catagories of closely related life forms that are further subdivided into species.

general circulation global patterns of atmospheric circulation that are persistent features.

geographical cycle, see Davisian cycle of erosion

geologic structure nature and arrangement of materials of the earth's crust.

geomorphology study of the processes that create landforms.

geostrophic wind wind blowing at right angles to pressure gradient and parallel to isobars; above the friction layer of the atmosphere.

geosyncline extended trough in earth's crust that fills with sediment as trough subsides; early stage in development of most mountain systems.

geothermal energy energy derived from earth's interior heat.

glacial drift general term for all types of material deposited directly or indirectly from glaciers.

glacial erractics glacially transported boulders that differ in composition from bedrock where they are lodged.

glacial horn, see horn peak

glacial outwash sediments deposited by meltwater streams issuing from glaciers.

glacial stairway irregular longitudinal valley profiles caused by uneven excavation of rock by valley glaciers.

glacial surge temporary rapid advance of glacier margin due to various causes, mainly excessive water at base.

glacial till unstratified and unsorted rock debris that is transported and deposited by glaciers.

glacial trough former stream valley that has been deepened, oversteepened, and smoothed by glacial erosion.

glacier mass of ice flowing under gravitational force; formed by recrystallization of snow that accumulates over periods of tens to hundreds of years.

glacier mass balance balance between snow accumulation and ablation of ice and snow of glacier. *See also* negative mass balance; positive mass balance.

gleization pedogenic regime found where soil is waterlogged. Soils produced have low pH and are blue-gray in color.

gley soils wetland soils, colored black to blue-gray by reduced iron compounds.

gneiss metamorphic rock produced by alteration of other rock types by heat, pressure, and chemical action; generally shows wavy bands of segregated minerals.

gnomonic projection special-purpose map projection used in navigation; only projection on which all great circles are straight lines.

Gondwanaland hypothetical supercontinent formed in southern hemisphere by the breakup of Pangaea.

graben portion of earth's crust that has subsided downward along faults on at least two sides.

gradation process by which projections on earth's surface are worn down and depressions are filled in by movement of fragmented rock material from high to low elevations.

graded stream stream that has reached condition of equilibrium in which there is no trend toward either filling or scouring.

gradient rate at which quantity varies with distance; vertical descent of stream per unit of horizontal distance.

gram calorie, *see* calorie

granite common plutonic igneous rock type found on continents.

graphic scale scale bar on a map, used to estimate distances or measure them accurately.

graticule network of lines of latitude and longitude on map.

gravel sediment particles of moderate size; larger in diameter than 2 mm.

gravitational energy energy related to elevation of mass; also called potential energy.

gravity attractive force exerted by earth due to its mass.

grazing food chain food chain that begins with green plants and goes on to herbivores and then to carnivores.

Great Basin area 700 km (400 mi.) wide, centered in Nevada, being stretched by tension, creating classic fault-block landscape.

great circle any circle on surface of earth whose center coincides with center of earth; shortest distance between two points on earth's surface is arc of great circle.

greenhouse effect atmospheric warming by transmission of incoming shortwave solar radiation and trapping of outgoing longwave terrestrial radiation.

grid azimuth azimuth measured with respect to grid north.

grid north reference direction determined by grid system.

grid system location reference system on a map; latitude and longitude system or other grid.

groin barrier constructed perpendicular to shoreline to induce sand deposition.

gross productivity rate at which plant converts light to chemical energy.

groundmass in igneous rocks, the matrix of fine particles in which larger crystals (phenocrysts) are embedded.

ground moraine debris laid down in sheet beneath glacier.

groundwater water in all openings in soil, sediment, and rock masses in saturated zone beneath water table.

Gulf Stream powerful northeast-flowing warm oceanic current in western Atlantic, off coast of southeastern United States.

guyot flat-topped undersea mountain; seamount.

gypsum calcium sulfate, a mineral concentrated in thick beds where lakes or inland seas have evaporated.

gyre large-scale, spiral circulation pattern, particularly in oceans.

h

hachuring method of indicating relief on maps by shading with short lines (hachures) parallel to slope direction; the steeper the slope, the thicker the hachures.

Hadley cell large-scale atmospheric convection cell originally deduced to explain global circulation pattern.

hailstones form of precipitation consisting of layered ice balls formed by repeated vertical cycling in clouds.

half-life time required for half of an initial amount of radioactive isotopes of atoms to decay.

halite rock salt, $NaCl$ (sodium chloride).

hanging valley valley that has floor noticeably higher than level of valley or shore it intersects.

hanging wall mass of rock above inclined fault plane.

hardpan relatively dense or impermeable layer in soil; may be composed of clay or a cemented duricrust.

headland cape or promontory that juts into sea; usually high and prominent.

heat capacity ability of substance to absorb heat in relation to its volume.

heat energy energy associated with internal random motion of molecules in matter.

heat island local region where temperature is generally higher than that of surrounding areas; large cities are often heat islands.

herbivores animals that subsist on plants for food.

highland ice cap unconfined ice cap through which rock peaks may protrude; much smaller than a continental glacier.

histosols soils composed primarily of plant material; they develop in waterlogged environments where lack of oxygen causes organic matter to decompose slowly.

hoar frost feathery crystals of ice that form on cold surfaces when dew point is below freezing point of water.

hogback ridge narrow, sharp-crested ridge with two steep slopes, formed by resistant rock strata tilted at steep angle.

Holocene present (or Recent) geologic epoch; beginning approximately 10,000 years ago.

homoclinal ridge ridge composed of inclined layer of resistant sedimentary rock.

homoclinal structure geologic structure in which sedimentary rocks dip in a particular direction; fold limbs are homoclines.

homoclinal valley valley eroded into inclined layer of erodible rock.

horizon, soil distinctive horizontal layer in soils, produced by pedogenic processes.

horn peak faceted pyramidal peak formed at intersection of three or more cirques.

horse latitudes belts of descending warm, dry air, fair weather, and weak surface wind in regions of subtropical anticyclones.

horst vertically uplifted mountain block bounded by faults on both sides.

hot spots, volcanic points of persistent subcrustal magmatic activity; movement of lithospheric plates across fixed hot spots leaves a "track" of volcanoes that are progressively older with distance from the hot spot (e.g., Hawaiian Islands).

hot springs outflows of groundwater heated by a near-surface body of molten magma.

human geography geography of human activity on the earth, present and past.

humid continental climate the climatic regions in higher middle latitudes of northern hemisphere where continents are broad; characterized by marked seasonal temperature differences. Precipitation is greatest during summer in these regions.

humid subtropical climate climatic regions on eastern sides of continents at subtropical latitudes; characterized by heavy precipitation throughout year.

humidity amount of water vapor in air. *See also* absolute humidity; relative humidity; specific humidity.

humus semisoluble chemical complex in soil, produced by bacterial transformation of decaying organic matter; major element in soil fertility.

hurricane tropical cyclone that forms in Caribbean region or near subtropical east or west coasts of North America.

hurricane eye cloud-free area of descending adiabatically heated air in the center of a tropical cyclone.

hydrocarbons compounds containing carbon and hydrogen; petroleum and natural gas.

hydrograph, stream graph plotting stream discharge through time.

hydrologic cycle processes by which water passes from surface of earth to atmosphere and back to earth.

hydrology study of water on or below surface of land.

hydrophytes plants that grow in water.

hydrosphere realm of water; one of the earth's major systems along with the atmosphere, lithosphere, and biosphere.

hydrothermal vents, submarine points along sea-floor spreading centers where metal-bearing hot solutions are vented into sea water; "black smokers."

hygroscopic tendency of substance to take up water vapor.

hypocenter, earthquake, *see* focus, earthquake

hypothetical continent model used to explain how delivery systems for energy and moisture determine regional climate.

i

ice age popular term for period when continental glaciers are present in middle latitudes.

ice cap, *see* highland ice cap

ice-contact stratified drift sorted, bedded, and deformed deposits of glacial debris originally deposited in contact with glacial ice.

ice crystal model of rainfall, *see* Bergeron ice crystal model

icefall steep and heavily crevassed portion of valley glacier.

ice-forming nuclei fine particles in atmosphere that promote freezing of supercooled water droplets.

ice sheet extensive and nearly flat expanse of ice that spreads in all directions from accumulation area in high latitude region; continental glacier.

ice-stagnation topography disorganized topography resulting from decay of ice laden with varying amounts of rock debris. *See also* kame and kettle topography.

ice wedge downward tapering mass of ice that forms in ground that contracts and cracks during freezing in tundra environment; wedges form networks.

ice-wedge polygon area of soil bounded by wedges of ice that forms and persists under the surface; occurs in tundra regions.

Icelandic Low persistent low pressure center between Iceland and Greenland, most pronounced in northern-hemisphere winter; steers winter storms in North Atlantic Ocean.

igneous rock rock formed by solidification of molten magma; usually exhibits crystalline structure.

Illinoian glaciation glaciation occurring more than 100,000 years ago; deposits cover southern Illinois.

illuvial layer, soil B horizon in soil where fine particles and soluble substances arrive after translocation from A and E horizons.

illuviation deposition in soil B or C horizon of material translocated from higher portion of soil.

inceptisols young soils with weak profile development that are present in humid regions on recent alluvium, glacial or aeolian deposits, or volcanic ash.

inert gas any member of family of essentially unreactive chemical elements, including helium, neon, argon, krypton, and xenon; noble gas.

infiltration movement of water from surface into ground.

infiltration rate or capacity maximum rate at which given soil can absorb water.

infrared radiation portion of electromagnetic spectrum between radio waves and visible light; about 0.7 to 1,000 micrometers; thermal radiation.

inselberg "island mountain"; isolated, steep-sided summit rising abruptly from surface of low relief; formed by erosion.

insolation solar radiation received at the earth's surface.

instability, atmospheric situation in which warm parcels of air rise convectively to great altitudes, permitting development of cumulus or cumulonimbus clouds.

interception process whereby leaves catch and hold raindrops falling on plant cover.

interglacial periods intervals within the Pleistocene epoch during which ice sheets withdrew from Europe and North America, causing sea level to rise close to its present position.

interlobate moraine morainic accumulation formed where two adjacent lobes of continental ice sheet brought debris into central dumping ground between them.

interruption, of projection to permit land areas on world maps to be more accurately represented, the maps are "split open'' in oceans. If seas are the object of interest, land areas may be interrupted (or split).

intersection in surveying, technique for specifying locations by measuring azimuths from both ends of known base line.

intertropical convergence zone (ITC) zone of low pressure near equator where trade winds converge; shifts seasonally.

intrusion, igneous molten magma that has invaded older rock and solidified below surface.

intrusive igneous rock coarse-grained rock that solidified from a molten condition below the land surface.

inversion, temperature condition in which air at lower altitudes is cooler than air immediately above, so that vertical motion of atmosphere is impeded.

inversion of relief development of topographic high on prior structural or topographic low, by process of differential erosion.

ion electrically charged atom; atom with excess or deficiency of electrons.

ionize dissociation of compound into positively and negatively charged atoms.

island arc line of islands formed by volcanic activity on convex side of oceanic subduction zone (e.g., Japan).

isobar line on a weather map connecting points of equal atmospheric pressure.

isostacy concept that earth's crust is in hydrostatic equilibrium with continents ''floating'' in denser material of mantle.

isostatic compensation vertical uplift or subsidence of crust to compensate for change in buoyancy resulting from transfer of load as a consequence of erosion and deposition, or growth and decay of ice sheets or large lakes.

isotherm line connecting points of equal temperature.

isotopes atoms of chemical element having constant number of protons, but varying numbers of neutrons in atomic nucleus.

j

jet stream narrow stream of rapidly moving air, occurring at altitudes between 10 and 15 km.

jetty long, artificial barrier extended outward from coast to modify tidal currents or stream flow; used to stabilize or deepen channel and to prevent deposition.

joint one of a set of systematic fractures that normally divide rock masses, with no displacement along fractures.

jokulhlaup ''glacier burst''; any large flood of water from a glacier; related to ice-dammed lakes and subglacial volcanic eruptions.

k

kame hill that originated as mass of sand and gravel deposited against glacial ice by glacial meltwater.

kame and kettle topography topography of small hills (kames) and closed depressions (kettles) resulting from stagnation and decay of glacial ice.

kame moraine end moraine composed of kame and kettle topography.

kame terrace terrace composed of ice-contact stratified drift accumulated along margins of mass of stagnating glacial ice in valley.

Kansan glaciation ancient Pleistocene glaciation in North America leaving glacial drift over eastern Kansas; under younger drift elsewhere.

karst topography topography developed as consequence of surface and subsurface solution of limestone bedrock; including sinkholes, caverns, underground streams, and other distinctive landforms.

karst towers, see tower karst

Kelvin scale of temperature scale of temperature fundamental for physical processes; its zero point, called absolute zero, is approximately $-273°C$ $(-459°F)$.

kettle, glacial depression formed by melting of massive ice block embedded in glacial drift plain.

kettle lake lake in glacial kettle on drift plain or end moraine.

kinetic energy energy arising from motion of objects; also called energy of motion.

knickpoint sharp vertical break in longitudinal profile of stream.

Köppen climate classification system the most widely used descriptive climatic classification; originally devised to accord with boundaries of global vegetation formations.

l

laccolith roughly lens-shaped intrusion of magma that pushes overlying rock upward to form dome at surface.

ladang, see swidden

lagoon shallow body of water entirely or nearly cut off from sea by a spit or coral reef.

lahar mudflow or debris flow related to volcanic activity or composed of previously erupted volcanic material.

laminar flow streamline flow in which fluid moves without turbulence; speed and direction of flow at given point are steady.

land breeze local circulation that blows from land toward sea, usually at night, in response to thermal contrast.

landforms relief features of earth's surface that may be observed in single view; for example, mountain ridge rather than mountain range; produced mainly by erosional and depositional processes.

LANDSAT series of NASA satellites in oblique orbit about 1,000 km above the earth's surface, passing over most points on earth every 18 days; records images of the earth's surface in varying spectral bands.

landslide sudden downslope motion of relatively dry mass of soil or rock.

langley unit of energy flux defined as 1 calorie per square centimeter.

lapilli particles coarser than sand ejected during a volcanic eruption.

lapse rate, average global average rate of temperature decrease upward in the atmosphere; approximately 6.5°C per km (3.6°F per 1,000 feet).

lapse rate, environmental measured rate of temperature change upward in the atmosphere at a location at a given time; highly variable.

large-scale map map that portrays a small portion of the earth's surface at a large size.

latent heat energy stored or released when substance changes phase.

latent heat of condensation heat that is released when 1 gram of gaseous vapor condenses to liquid at same temperature; numerically equal to latent heat of vaporization at that temperature.

latent heat of fusion   energy required to change the state of 1 gram of substance from a solid to a liquid at same temperature.

latent heat of sublimation   energy required to change state of 1 gram of substance from solid to gaseous vapor at same temperature.

latent heat of vaporization   energy required to change state of 1 gram of substance from liquid to gaseous vapor at same temperature.

lateral moraine   ridgelike deposits of material formed along margins of valley glacier.

lateral planation   process of valley floor widening by lateral displacement of stream meanders that erode valley walls on outside of each bend.

laterite   surface residue of iron and aluminum oxides and hydroxides that forms on latosols or lateritic soils in humid tropics and savannas; bricklike duricrust when dried.

laterization (ferrallitization, desilication)   pedogenic regime that produces red iron-rich soils found in wet portions of tropics; involves solution and removal of bases and silica.

latitude   angular distance north or south of equator on earth's surface measured in degrees; lines of constant latitude are called parallels.

latosols   soils of warm humid rainforest; they are deeply weathered and are colored red by iron compounds; oxisols and ultisols.

Laurasia   ancient landmass that became separated into North America, Greenland, and Eurasia.

lava   fluid or semifluid magma that flows out at surface during volcanic eruptions and eventually solidifies as extrusive igneous rock.

leaching   downward translocation of soluble substances dissolved by percolating soil moisture.

legume   plant having particular seed structure, which also hosts nitrogen-fixing bacteria (attached to roots).

levee, natural,   *see* natural levee

liana   climbing vine supported by tree in wet forest.

limb of fold   inclined rock strata on either side of fold axis.

limestone   sedimentary rock composed of calcium carbonate of organic origin; soluble in slightly acid groundwater, producing sinkholes and caverns.

limestone cavern   subsurface void produced by solution of limestone at joint and bedding plane intersections; forms just below water table; exposed by drop in water table.

limestone solution   processes of carbonation and solution that dissolve limestone.

linear erosion   downward stream incision resulting in valley deepening.

lithification   process by which unconsolidated material is converted to solid rock.

lithosphere   outer layer of rigid rock on earth, consisting of crust and upper mantle, and including crustal plates.

lithospheric plate(s)   major horizontal divisions of earth's crust, separated by sea floor spreading centers, subduction zones, and transform faults; in motion away from sea floor spreading centers.

Little Ice Age   period extending from 1550 to 1850, in which average temperatures in much of Europe were low enough to allow advance of glaciers.

llanos   savannas of South America.

load, stream,   *see* stream load

loam   soil that contains a mixture of sand, silt, and clay.

local water budget   computation system accounting for principal components of water distribution at a locality, including precipitation, soil moisture, evapotranspiration, and runoff.

loess   dust deposited from turbid atmosphere, often forms deep, semi-consolidated beds.

longitude   angular distance east or west on earth's surface, measured in degrees from specific prime meridian. Lines of constant longitude are called meridians.

longitudinal dunes   parallel, linear dunes that make striped pattern on earth's surface; produced by strong winds from two directions.

longitudinal profile   graphic plot of vertical elevation of stream bed vs. horizontal distance along channel.

longshore current   current parallel to shore, produced by approach of waves at angle to shore.

longshore drifting   movement of marine sediment parallel to shoreline in zone of wave break.

longwave radiation   radiation from relatively low temperature sources, with wavelengths greater than 4 micrometers; terrestrial radiation.

m

magma   molten rock-forming material in earth's interior, which reaches earth's surface as lava during volcanic eruptions.

magnetic azimuth   azimuth measured from magnetic north.

magnetic declination   difference between azimuths to magnetic north and true north; varies with location.

magnetic north   northerly direction indicated by magnetic compass.

magnetic reversals   changes in polarity of earth's magnetic field, frozen into volcanic rocks and sediments containing magnetic minerals; determines whether magnetic compass would point north or south.

magnetic stripes   pattern of normal and reversed magnetic polarity frozen into sea-floor lavas of different ages.

mammals   mostly hair-covered vertebrate animals, born alive and initially nourished by mother's milk. Evolutionary advance beyond reptiles.

mantle   portion of earth between core and crust, composed of high-density rock material; molten near its outer margin, which permits lateral displacement of crust above it.

mantle convection   slow ''boiling'' motions in asthenosphere that creates ocean ridges, causes sea-floor spreading, and drags crustal plates toward subduction zones.

map   two-dimensional representation of all or part of earth's surface showing selected natural or human phenomena; preferably constructed on definite projection, with specified scale.

map projection,   *see* projection, map

map series   set of maps that share common properties, such as scale and design.

maquis (pronounced mah-KEE)   vegetation that consists of drought-adapted scrub, occurring in dry-summer regions near Mediterranean Sea.

marble   limestone that has been metamorphosed by heat and pressure.

marginal sea coasts   coasts such as those of Gulf of Mexico that are separated from oceans by island chains and that have broad continental shelves and low submarine relief similar to trailing-edge coasts.

marine climate climatic region along western margins of continents in higher latitudes; characterized by moist maritime air and precipitation throughout year.

marine terrace former marine abrasion platform or wavecut bench that has been uplifted tectonically to form terrace.

maritime polar air (mP) cool air with high moisture content; major source of precipitation when lifted orographically, by convection, or along fronts.

maritime tropical air (mT) warm air having a very high moisture content; major source of rainfall when lifted by convection or interaction with cooler non-tropical air.

marsupial early type of mammal distinguished by birth at very immature stage of development; matures in pouch on mother's stomach.

mass balance, glacier, see glacier mass balance

mass wasting downslope movement of loose rock or soil under direct influence of gravity.

material budget balance between inputs, outputs, storage, and transformation of materials in a system.

mature stage descriptive term applied to landscape that is all in slope, being thoroughly dissected by streams with no original surfaces preserved and no permanent new surfaces developed; in Davis cycle, intermediate stage of landform evolution.

meander one of series of looping, sinuous curves in course of stream.

meander scar meander that has filled with sediment after being abandoned by stream; a former oxbow lake.

meander wavelength distance between similar points on adjacent stream meanders.

meandering channel stream channel that exhibits regular pattern of serpentine curves; normal tendency of stream channels.

mechanical weathering any physical process of weathering; commonly, freezing of water or salt crystallization.

Mediterranean climate dry summer subtropical climate with winter rainfall.

megarelief large-scale relief features of the continents and ocean basins produced by plate tectonic processes.

meltwater water produced by melting of glacial ice; seen in meltwater streams issuing from glacier margins.

Mercator projection cylindric map projection devised for use in plotting compass courses for navigation; inappropriate for general use because of extreme distortion of scale except near equator.

meridians of longitude lines connecting points of equal angular distance east or west of the plane of the prime meridian at Greenwich, England.

mesa flat-topped erosional remnant of sedimentary rock or lava, bounded by cliffs on all sides.

mesophytes plants that occur where water supply is neither scanty nor excessive.

mesosphere portion of the outer atmosphere, above the stratosphere, in which temperature decreases upward; approximately 50 to 80 km above earth's surface.

Mesozoic the fourth of the five great geologic eras; the era of giant reptiles.

metamorphic rock rock formed by recrystallization of igneous or sedimentary rock as result of heat, pressure, or chemical action.

metastable equilibrium pertains to systems that must periodically readjust to a different equilibrium condition; compare with dynamic equilibrium.

meteorite solid body of stone, iron, nickel or other substance that reaches the earth from elsewhere in the solar system; range from dust to boulder size; creates impact craters on planets and their moons.

meteorology study of weather, especially of physical processes in atmosphere.

metric system decimal system of measurements used in science, based on the meter, the liter, and the kilogram.

microclimate small-scale variations in temperature and moisture conditions close to the ground.

micrometer metric unit used to measure wavelength, equal to one millionth of a meter; same as micron.

micron, see micrometer

Mid-Atlantic Ridge major component of oceanic ridge system; important sea-floor spreading center.

mid-bay spit spit extending into bay between bay mouth and head of bay.

Milky Way "home" galaxy of the earth and its solar system.

millibar metric unit of pressure used in meteorology, equal to approximately 0.75 millimeters on mercury barometer; standard atmospheric pressure is 1,013.25 millibars.

milpa cultivation, see swidden

mineral recurring natural combination of chemical elements in solid form; particles composing rocks.

mixed tropical forests and shrublands tropical forests that are adapted to seasonal drought; less luxuriant and rich in species than tropical rain forests; often thorny and dry-season deciduous.

mixing ratio weight of water vapor contained in given sample of moist air compared to weight of air alone; usually expressed in grams of water per kilogram of dry air.

model representation of a system, used to explain or to predict the effects of changing conditions.

Modified Mercalli scale scale of earthquake intensity based on visible effects and damage produced.

Mohorovičić discontinuity, Moho, M discontinuity boundary, 10 to 70 kilometers (6 to 40 miles) below the earth's surface, between regions in which seismic waves move with different speeds; taken to be boundary between crust and mantle.

moist adiabatic rate rate of adiabatic cooling in saturated air, with condensation occurring; varies with condensation rate; between 5°C and 10°C per km of elevation.

moist realm one of A. Penck's three climate realms, the realm of forests, where precipitation exceeds potential evapotranspiration.

moisture deficit in local water budget, difference between potential and actual evapotranspiration.

moisture index measure of deficiency or surplus of precipitation compared to potential evapotranspiration.

moisture surplus in local water budget, available moisture in excess of that required by evapotranspiration and soil moisture recharge.

mollisols soils with dark, humus-rich A horizons and with high base saturation; largely developed under natural grass cover; best of all agricultural soils.

monadnock isolated summit in erosional landscape of low relief; in Davis cycle of erosion, landform of old age.

monocline flexure in sedimentary rocks, such that flat-lying strata curve upward and then level off at a higher altitude.

monsoon circulation wind system that reverses direction seasonally in response to pressure differential between continents and oceans.

moraine landform composed of unsorted rock debris deposited directly by glacier.

mountain wind current of cold air that drains from mountains to valleys, usually at night.

mudflow downslope flow of water-saturated mud; usually seen in areas of sparse vegetation; lacks coarse fragments of debris flows.

mudstone sedimentary rock composed of fine silt particles.

multispectral scanners remote sensing devices that record reflectance of surfaces in different wavelengths of electromagnetic spectrum.

Munsell color notation system, soil classification of soil color in terms of hue, value, and chroma, for purpose of precise soil description.

Mycenaeans ancient inhabitants of Greece, pre-dating era of classic Greek civilization; at peak, approximately 13th century B.C.

n

natural arch opening through narrow wall of rock; produced by collapse of portion of rock face.

natural bridge bridge of rock left where adjacent stream meanders wear through rock wall formerly separating them.

natural levee silt embankment adjacent to stream channel; produced by overbank deposition of fine material during stream floods.

natural selection proliferation of species that has advantages enabling it to survive stress of environment.

natural vegetation climax vegetation in region, especially where human influence is minimal.

neap tide tide having minimum tidal ranges during month; occurs when sun and moon make 90° angle with earth.

Nebraskan glaciation extremely ancient glacial deposits in midwestern United States, exposed on the land surface in eastern Nebraska; age probably exceeds 1 million years.

needle ice ice formed under uppermost soil particles on damp ground subject to freezing temperature at night; growth lifts surface soil millimeters to centimeters.

negative feedback process whereby disturbance of system that is in equilibrium triggers changes that tend to restore original system.

negative mass balance, glacier condition where annual loss of glacial ice by ablation processes exceeds input of new snow and ice in accumulation zone, causing glacial thinning and retreat of ice margin.

neoglacial moraine ridges composed of coarse rock rubble, found mainly in cirques, which mark glacial readvance during past 5,000 years.

net productivity net rate at which plant stores energy, exclusive of energy it uses for respiration.

net radiation balance between incoming shortwave radiation and outgoing longwave radiation; net gain or loss of radiative energy at the earth's surface.

névé, *see* firn

nimbostratus clouds thick, dark layer of stratus clouds that produce precipitation.

nitrogen fixation conversion of gaseous nitrogen by soil bacteria to forms that can be utilized by plant life.

nor'easters strong, blustery winds from the northeast that precede passage of cyclonic storms in New England.

normal fault fault in which hanging wall, or block above fault plane, moves downward in relation to footwall.

North Atlantic Drift northeastward continuation of the Gulf Stream ocean current, bringing warm water to northern Europe.

northern coniferous forests, *see* coniferous forests, northern

nuclear energy energy binding particles in atomic nuclei.

nuée ardente type of volcanic eruption producing glowing cloud illuminated by incandescent ash and scoria flowing down slopes of erupting volcano-like fluid, burning everything in its path.

nutrient cycle mutually beneficial cycling (uptake, conversion, and return) of nutrients between organisms and their environment.

o

oblique slip motion along fault in direction oblique to both strike and dip of fault.

obsidian glassy extrusive igneous rock; indicates cooling too rapid for crystal formation.

occluded front overtaking of warm front by cold front in cyclonic disturbance, with air mass ahead of cold front and behind warm front being cut off (occluded) from contact with ground.

ocean basins first-order relief feature; along with the continents, one of the fundamental features of the earth's surface form; formed by sea-floor spreading process; floored by basaltic lava.

ocean currents, *see* currents, ocean

oceanic circulation system of wind-driven surface currents, and deeper countercurrents related to salinity (density) variations.

oceanic fracture zone strike-slip fault perpendicular to oceanic ridge; related to sea-floor spreading on spherical surface of earth.

oceanic ridge rifted high welt on ocean floor where new crustal material is being produced by submarine volcanism.

ocean trench linear depression on ocean floor where one crustal plate is descending beneath another in process of subduction.

offshore part of shore zone from low tide seaward to where erosion and deposition by waves does not occur.

offshore bar material that has been peeled away from beach by high energy waves and deposited in zone of wave break; may be a seasonal phenomenon.

O horizon layer of undecomposed plant debris or raw humus at soil surface.

old age term used to describe landscape eroded to plain of low relief with broad river valleys and isolated hills, or monadnocks; in the Davis cycle, advanced stage of landscape evolution.

open systems systems that continuously or periodically receive energy or materials from other systems.

open traverse in surveying, a traverse that starts and ends at different locations.

orbit circular or elliptical path of moving body such as planet or moon.

orders of relief classification of relief features of earth's surface into three orders of magnitude.

ore mineral from which economically valuable metal may be obtained by refining processes.

organic activity (in soil formation) all biotic (plant and animal) influences on soil.

organic evolution concept that organisms evolve toward more efficient forms by survival and reproduction of those individuals best adapted to their environment.

orogeny in geologic time, a phase of large-scale crustal compression resulting in strong deformation and mountain building.

orographic fog fog formed when warm, moist air is forced up mountain slopes and chilled to its dew point.

orographic lifting occurs when horizontally moving air is forced to rise to higher altitudes over mountain barrier.

outer core outer portion of earth's core that has liquid properties according to seismic analysis.

outlet glacier glacier that carries ice from an unconfined icecap outward through a confined valley.

outputs, earth's systems energy or materials transformed in one system that become an input to another system.

outwash plain plain composed of sediment carried out of glacier and deposited by meltwater streams.

overland flow rainwater that has accumulated at surface and begins to drain down available slopes.

overthrust nearly horizontal fault carrying slices of rock thousands of meters thick over top of adjacent rock masses to distances of tens of kilometers.

oxbow lake crescent-shaped lake formed where stream meander has been cut off from main stream.

oxidation chemical process in which oxygen ions in water combine with other ions in compound.

oxides compounds of oxygen with other elements.

oxisols thoroughly leached soils of wet tropics; characterized by subsurface horizon consisting of residue of clay and iron and aluminum oxides and hydroxides, with virtually all bases removed.

ozone gas composed of three oxygen atoms; forms layer in atmosphere that absorbs much dangerous ultraviolet radiation from sun.

p

Pacific Ring of Fire andesitic volcanoes encircling Pacific Ocean; created by subduction in ocean trenches.

pahoehoe lava smooth-skinned, but often wrinkled, form of lava made fluid by its content of dissolved gases.

paired fluvial terraces terraces at same elevation on both sides of stream; indicates periodic stream aggradation or incision.

paleomagnetism record of ancient magnetism in volcanic rocks.

Paleozoic era third of the five great geologic eras, from the first complex life forms to the early dinosaurs.

Pangaea hypothetical supercontinent thought to have contained all earth's landmasses approximately 200 million years ago.

parabolic sand dune sand dune shaped like parabola, with its concave side facing wind and its horns pointed into wind; blowout dune.

parallel of latitude line connecting all points on the earth's surface of equal angular distance north or south of the plane of the equator.

parallel-pinnate, *see* pinnate drainage pattern

parent material raw material from which soil is formed; weathered rock or unconsolidated sediments.

paternoster lake one of a series of lakes connected by a stream in a glacially eroded valley; resembling the beads of a rosary.

patterned ground division of ground surface into polygons, caused by contraction and cracking of the ground at subfreezing temperatures; characteristic of landscapes in tundra climates.

PE, *see* potential evapotranspiration

peat partly decomposed plant matter, formed in waterlogged soils.

ped(s) grains of soil joined together in clumps, the shape and size of which determine soil structure.

pedalfers general class of soils in which aluminum and iron compounds are dominant due to leaching of bases in humid regions.

pediment extensive gently sloping erosion surface created by backwearing of hillslope or escarpment; best seen in arid regions.

pedocals general class of soils in which calcium compounds are abundant; occur in regions of moisture deficit.

pedogenic regimes systems of soil-forming processes, each of which produces distinct soil types that reflect variations in biotic activity and energy and water budgets.

pedology scientific study of soils.

peneplain "almost a plain": lowland erosional plain of little relief; in Davis system, end result of erosion cycle.

perched water table body of groundwater lying above normal water table as a result of presence of an impermeable subsurface layer in the form of a saucer.

percolation passage of water downward to lower levels in soil.

perennial plants endure seasonal climatic fluctuations and persist throughout year for a period of several years; often die back in cold or dry season, but resprout in favorable season.

periglacial block field, *see* felsenmeer

periglacial climate nearglacial climate; climate of tundra vegetation areas; characterized by extreme cold, but with insufficient precipitation to produce glaciation.

perihelion point in the earth's orbit when it is closest to the sun; early January.

period in geologic time scale, interval of about 50 million years characterized by certain dominant life forms that differ in significant ways from those of preceding and following periods (e.g., Cretaceous, Permian, Devonian, Cambrian).

permafrost permanently frozen ground below active layer at depth of few feet in cold tundra climates.

permeability capacity to transmit water through interconnected pores.

permeable containing interconnected pores through which water can move.

pH measure of soil acidity or alkalinity; specifically, measure of hydrogen ion concentration in soil moisture.

Phanerozoic eon the last 700 million years, during which multicellular life has been present on the earth.

phase physical form of substance that may be solid, liquid, or gas, depending upon its temperature.

phenocrysts   visible crystals embedded in the groundmass of an igneous rock.

photosynthesis   process by which green plants use the energy of light to produce stored chemical energy from water and carbon dioxide.

phyla   primary subdivisions of the organisms of the biosphere.

physical geography   study of the natural processes that act together to produce the earth's physical environments; interaction between humans and environmental processes.

physical model   physical replica of environmental process or system, constructed at greatly reduced scale.

physiographic diagram   bird's-eye view diagram showing landforms and geological structure; not an accurate map due to relief displacement.

piedmont glacier   large expanse of glacial ice formed where one or more valley glaciers spread out in open lowland.

piezometric surface   level to which groundwater rises in well sunk into confined artesian aquifer.

pillow basalts   basaltic lava erupted into or beneath body of water, resulting in pillowlike blobs of solidified lava.

pingo   domed hill formed when soil cover is pushed up by lens-shaped mass of ice; occurs in tundra regions.

pinnate drainage pattern   featherlike drainage pattern composed of many closely spaced parallel tributaries feeding into larger streams in long, straight valleys.

piracy,   *see* stream piracy

pitted outwash plain   plain with scattered depressions occupied by lakes or marshes, resulting from the deposition of glacial outwash around isolated masses of glacial ice that subsequently melted.

plane table   simple survey apparatus that consists of horizontal drawing board supported by stand or tripod.

planetesimals   smaller astronomical bodies that coalesced to form planets.

planimetric map   map on which all points represented have same angular relationships to other points as they do on the earth.

plankton   microscopic animals and plants floating in waters of seas, lakes, ponds, and streams; base of aquatic food chains.

plant community   association of vegetation types that are in stable interaction with one another and with the environment.

plant succession   sequence of plant communities occupying a locality through time, leading toward climax vegetation.

plate tectonics   theory that attributes crustal deformation, such as mountain uplift and volcanism, to movement and interaction of lithospheric plates, resulting from sea-floor spreading.

plateau   surface of low relief occurring at high elevation; usually descends abruptly to lower regions and is cut by streams in canyons.

playa   lake bed that is dry for much of the year; basin in which rainwater accumulates and soon evaporates; feature of arid landscapes.

Pleistocene   epoch of geologic time, about 1.8 million years in duration, that ended 10,000 years ago.

plucking   process of glacial erosion in which material is pulled from glacial bed by being frozen into moving ice.

plug dome   type of volcano formed where viscous magma pushes to surface and forms irregular dome rather than producing lava flows.

plunging breaker   forward collapse of steep wave in surf zone producing a cylindrical ''curl.''

pluton   large individual intrusion of magma of uniform composition; numbers of plutons together form a batholith.

plutonic rock   coarse-grained igneous rock that crystallizes from a molten condition many kilometers below the earth's surface.

pocket beach   small beach on rocky coast; usually confined by cliffs.

podzolization   pedogenic regime that involves breakdown of clay and eluviation of iron, other cations, and humus, leaving silica behind as residue; creates infertile acid soils.

point bar   alluvial sediment deposit on inner (concave) side of stream meander; coarse in texture (sand, gravel, cobbles).

polar climatic region   climatic region found only in northern hemisphere where wide continents extend poleward of latitude 60°N; characterized by very cold winters and short summers with meager precipitation.

polar easterlies   surface air that flows out from polar highs in both hemispheres and is deflected westward by Coriolis force.

polar front   zone where cold air from high latitudes meets warm air from tropics.

polar glacier   glacier in which ice temperature is considerably below pressure melting point; a polar glacier is frozen to its bed and moves only by internal deformation.

polar high   high-pressure area in the atmosphere formed by subsidence of upper air over polar region.

polar outbreaks   areas where polar front bulges toward equator, allowing polar air to penetrate to subtropical latitudes.

pore space   space between particles composing rock and soil, through which water passes and in which it is stored.

positive mass balance, glacier   annual input of new snow and ice in glacier accumulation zone exceeds loss of ice in ablation zone, causing glacier to thicken and ice margin to advance.

potential energy   energy related to altitude and mass, resulting from the potential of gravity to pull the object downward; also the energy stored by compression or tension on an elastic body.

potential evapotranspiration (PE)   rate at which water would be lost to atmosphere from dense, homogeneous vegetation cover supplied with unlimited amount of soil moisture.

potential photosynthesis   maximum value of plant's net productivity, given sufficient water and nutrients.

prairie   midlatitude grassland between forests and deserts; usually refers to tall-grass prairie, but sometimes used for short-grass steppe.

Precambrian eon   includes about 75 percent of the earth's history; the earth's first 4 billion years, when only the simplest life forms existed.

precipitation   coalescence of water droplets into raindrops, snowflakes, or hailstones that are large enough to fall to the ground.

predator   animal that utilizes other animals for food.

pressure, gas or fluid   force exerted on unit area by molecular collisions; *see also* atmospheric pressure.

pressure gradient force   in the atmosphere, force caused by horizontal differences in pressure across surface; force tends to push parcel of air from higher to lower pressure.

pressure melting, glacier   melting of basal ice of glacier due to increased pressure on upstream side of obstacle on glacier bed.

primary producers in food chains, photosynthetic plants that convert solar energy, water, and carbon dioxide gas to organic energy in the form of plant tissue.

prime meridian reference meridian for specifying longitude; meridian passing through observatory in Greenwich, England is now used as the prime meridian by nearly all countries.

principal meridian north-south reference lines for U.S. Bureau of Land Management land surveys.

profile, see soil profile

prograded shoreline where shoreline has moved seaward by growth of coastal depositional landforms such as deltas and filled lagoons.

projection, map any method for transforming positions on earth's spherical surface to flat map.

pumice frothy, bubble-filled form of solidified lava.

pyroclastic ejecta solid fragments of volcanic rock explosively ejected from a volcanic vent.

q

quarrying in glacial erosion, see plucking

quartzite brittle, siliceous rock derived from metamorphosis of sandstone; sometimes applied to silica-cemented quartz sandstone.

Quaternary period Pleistocene and Holocene time; the earth's most-recent 1.8 million years.

r

radar, side-looking airborne airborne device used to create images of earth terrain by reflection of pulsed radio waves from topographic surfaces.

radial drainage drainage pattern indicating presence of isolated high area from which streams flow outward in several directions.

radiant energy, solar energy from sun that heats the earth's atmosphere and solid surface.

radiation (physics) transmission of energy through space by particles or waves; (surveying) technique for specifying location by measuring azimuth and distance from origin.

radiation fog fog that forms when moist air near surface is cooled to its dew point by contact with ground that has been cooled by radiation heat loss.

radioactive decay spontaneous transformation of one atomic nucleus into another, accompanied by emission of subatomic particles or gamma rays.

radioactive isotopes atoms of chemical element, having constant number of protons, but varying number of neutrons, and thereby varying masses, that decay spontaneously into another element at predictable rate; used to measure geologic time.

radioactivity instability in nucleus of atom, causing nucleus to change form spontaneously, releasing energy.

radiometric dating use of known rate of radioactive decay for dating of minerals and rocks.

rainforest, temperate wet forests of certain midlatitude marine climate coasts; similar in many ways to tropical rainforests, but lack equal species diversity.

rainforest, tropical forests of equatorial tropics developed under conditions of year-around warmth and moisture. Distinctive for species diversity and quantity of biomass.

rainmaking, see cloud seeding

rainshadow dry area resulting from persistent descent of air on the lee side of a mountain range.

rainsplash displacement of soil particles by impact of raindrops.

range in Bureau of Land Management survey system, position of a township numbered east or west from a principal meridian.

rating curve graph relating stream gauge height to stream discharge.

reach given segment of stream.

realms of climate, see climatic realms

Recent epoch, see Holocene

recessional moraine morainic loop behind terminal moraine that marks pause in retreat of margin of glacier.

recharge area surface area supplying water to subsurface aquifer.

rectangular drainage drainage pattern that reflects strong jointing of resistant bedrock, with streams incising along joint planes.

recurved spit spit that curves toward the land at its free end.

refraction, see wave refraction

regolith cover of loose rock fragments over solid bedrock; produced by rock weathering.

rejuvenation in Davis cycle of erosion concept, return to youthful stage from more advanced stage as consequence of tectonic uplift or other new source of energy.

relative humidity ratio of amount of water vapor present in quantity of air compared to amount that could be held by same air if it were saturated, usually expressed as a percentage.

relaxation time time required for a system to reestablish equilibrium after a significant disturbance has occurred.

relict landform surviving landform that developed in the past under climates or conditions different from those of the present.

relief vertical difference in elevation between high and low places on a surface.

relief inversion erosional landform that is the reverse of the geologic structure or the initial surface form.

relief texture intricacy of land surface, measured as number or length of fluvial valleys per unit of area; similar to drainage density.

remote sensing study of earth's surface from distance using various forms of electromagnetic radiation.

representative fraction method of specifying scale of map in terms of ratio between distance on map and corresponding distance on earth, expressed in same units of measure.

respiration process by which organisms oxidize nutrients to produce energy and expel waste products such as carbon dioxide (animals) or oxygen (plants).

response time time required for system to begin to change in response to changed inputs of energy or material.

retrograded shoreline retreating shoreline resulting from coastal erosion.

reverse dip slip block resting on inclined fault plane rides up and over block below fault plane; result of compressive stress on crust.

reverse fault, see thrust fault

R horizon   unweathered bedrock at base of soil profiles.

rhyolite   siliceous extrusive igneous rock; the extrusive equivalent of granite.

rhyolitic tuffs   extrusive igneous rock composed of rhyolitic volcanic tephra (ash) welded into a solid mass by heat; evidence of extremely violent volcanic eruption.

ria   former stream valley that is drowned by submergence of coast, forming long inlet.

Richter scale   logarithmic scale of energy release during earthquakes; earthquake magnitude scale.

riffle   shallow section of stream where bed load sediment has been deposited.

rift topography   landforms associated with strike-slip faults.

rift valley   linear depression where earth's crust is separating, in some cases allowing molten rock to well up; analogous to spreading center.

right lateral strike slip   horizontal fault displacement, with block on far side of fault shifted to right.

rill   very small surface drainage channel, only a few centimeters deep, but sharply incised.

rill erosion   erosion of slopes by closely spaced, sharply incised shallow channels.

ripples, sand   micro-topographic phenomena of sand dunes, river beds, tidal estuaries, in which sand moves as small ridges transverse to direction of wind or water flow.

rock   consolidated naturally occurring crustal materials composed of mineral crystals or fragments of older rocks; both organic and inorganic origin.

rock cycle   cycling of geological materials from igneous rock to sedimentary and metamorphic rock, and their melting and reconstitution as new igneous rock.

rock salt   rock-like mass of salt occurring in beds, derived from the evaporation of large salt lakes or inland seas in the geologic past.

rockslide   downslope slide of relatively dry rock masses; transport can achieve very high velocity and great horizontal distance.

Rossby waves   undulations in flow of upper level westerly winds in higher midlatitudes; develop in a cyclic manner.

runoff   water that enters streams as groundwater outflow and excess precipitation that flows over land surface; measure of average depth of water that flows from drainage basin during specific time period.

runup   height on the shore reached by a very large wave, such as a seismic sea wave.

S

sag pond   pond in local depression formed as a consequence of movement of strike-slip fault.

salinization   pedogenic regime in which deposit of salt is left in or on soil by evaporation of saline water above water table.

salt   any of several soluble chemical compounds in which hydrogen atoms of an acid have been replaced by metal atoms; sodium chloride (NaCl).

saltation   bouncing motion executed by particles driven by wind or flowing water; particle is repeatedly lifted from solid surface into air or water as it moves forward.

San Andreas Fault (California)   best-known example of an active strike-slip fault posing a major earthquake hazard to large population centers.

sand   particles with diameters between 50 and 2,000 micrometers (0.05 to 2 mm).

sand budget, coastal   balance between sand arrivals and losses on beaches and spits, that determines whether landform grows, remains stable, or shrinks.

sand creep, aeolian   forward migration of surface sand particles, driven by the impact of saltating sand grains in dune area.

sand dune   mound or hill of loose windblown sand; usually steepest on the downwind side.

sand ripples,   see ripples, sand

sandstone   common sedimentary rock produced by lithification of sand deposits.

saturation   condition in which air at given temperature contains its maximum capacity of water vapor.

saturation vapor pressure   pressure exerted by water molecules in air that contains maximum amount of water vapor possible at given temperature.

savanna   tropical grassland with scattered trees, lying between tropical rainforests and hot deserts.

scale, map   relationship between distance measured on map and corresponding distance on earth's surface; may be expressed verbally, numerically, or graphically.

scarp, cuesta   steep erosional slope that cuts across bedding, as opposed to back slope that corresponds to dip of strata forming cuesta.

scarp, fault   steep slope created by uplift of crustal block along fault; normally somewhat eroded.

scattering   process in which energy of electromagnetic radiation is distributed in various directions by reflection from microscopic particles.

schist   flaky metamorphic rock produced by alteration of shale or lava by heat, pressure, and chemical action.

scoria   solidified form of liquid lava fragments thrown into air during volcanic eruptions; pyroclastic ejecta; volcanic cinders.

scoria cone   volcanic cone composed of pyroclastic ejecta.

scour   removal of material from channel bed by stream flow.

scree   see talus

sea arch   opening worn through headland or offshore rock by wave action.

sea breeze   wind that blows from sea toward land, usually during day, as a consequence of thermal contrasts.

sea cave   hollow cut in shoreline rock by wave action.

sea cliff   shoreline cliff produced by wave action.

sea floor spreading   production of new crust by upwelling of molten material at oceanic ridges combined with lateral motion of sea floor away from ridge systems.

sea level   average elevation of sea surface, taking into account tidal effects.

seamount   submerged volcano, found particularly in Pacific Ocean; compare *guyot*.

sea stack   small rocky island or pillar that has been separated from coastal rocks by wave erosion.

second order of relief   global relief features of intermediate scale, such as mountain ranges, interior lowlands, ocean ridges, and ocean trenches.

second-order stream   larger stream channel formed at the confluence of two smaller, unbranched first-order streams.

secondary circulations   migratory pressure and wind systems that control day-to-day weather phenomena; midlatitude cyclones, tropical hurricanes, and wave disturbances.

section one of the 36 subdivisions of a Bureau of Land Management township; approximately 1 mile square, containing 640 acres.

sediment rock debris or organic matter deposited in beds by water, wind, or ice.

sedimentary rock rock formed by consolidation and lithification of beds of sediment.

seiche changes in lake or sea surface elevations resulting from resonance induced by wind, contrasts in atmospheric pressure, or earthquake tremors.

seif dunes knife-edged longitudinal sand dunes of extreme desert environments.

seismic sea wave sea wave of great length generated by submarine earthquake or volcanic eruption; it rises to great heights in shallow water; tsunami.

seismograph any of various instruments for measuring and recording vibrations of earthquakes.

semiarid region transitional zone between arid region and either humid tropical or humid midlatitude region.

sensible heat heat created by molecular movement that can be measured by a thermometer.

settling velocity steady speed eventually attained by object falling through gas or liquid.

7th Approximation classification of soils U.S. Comprehensive Soil Classification System of Department of Agriculture.

shaded relief use of shading on map to give impression of relative height and steepness.

shale common sedimentary rock produced by lithification of deposits of clay-size particles.

shear horizontal movement of adjacent crustal segments, producing strike-slip faults or folds.

shear stress force applied obliquely against surface, as by moving air, water, or ice; force causing detachment of particles in most forms of erosion.

sheet flow, water slow, laminar flow of water in thin films on smooth surfaces; rarely seen on natural slopes.

shield volcano volcano having nearly level summit and gentle outward slopes; produced by nonexplosive lava eruptions from central vent.

shoreline angle point where sea cliff and wave-cut platform meet; usually at high-tide level.

short-grass prairie area of short grass between tall-grass prairie and steppe consisting of bunch grass.

short-wave radiation radiation from a high temperature source, with wavelengths shorter than 4 micrometers; such as visible and ultraviolet radiation emitted by sun.

sial rock types that form continental crust, consisting primarily of compounds of silicon and aluminum; less dense than sima.

side-looking airborne radar, *see* radar, side-looking airborne

silcrete soil duricrust composed of silica.

silicates most-common rock-forming minerals; basic building block is silica tetrahedron (SiO_4).

silicic volcanism generally explosive volcanism producing minor amounts of lava and large volumes of fine-grained tephra (volcanic ash).

sill, volcanic sheetlike intrusion of magma between predominantly horizontal rock strata.

silt particles from 2 to 50 micrometers in diameter.

siltstone sedimentary rock composed of silt-sized particles.

sima rocks constituting oceanic crust, composed mainly of compounds of silicon and magnesium; denser than sial.

sinkhole surface depression resulting from solution and removal of subsurface limestone; karst feature; doline.

sinuosity, stream degree of departure from straight course; measured as stream length divided by valley length.

slash-and-burn agriculture shifting agriculture often practiced in tropical forests; called "swidden" in Southeast Asia and "milpa" in Latin America.

slate metamorphic rock produced by alteration of shale; fine-grained and easily split into sheets.

slickensides rock surface smoothed and striated by abrasion of adjacent rock mass slipping past along a fault.

slip face, dune, *see* dune slip face

slopewash removal of soil particles from sloping surfaces by turbulent water flowing in rills.

slump local form of mass wasting or gravitational transfer in which mass of soil tilts backward in rotational movement as it moves downward.

small-scale map map that shows a large area of the earth's surface at a small size.

snout, glacier downslope terminus of glacier.

snowline level above which some winter snow can persist throughout year.

soil complex of fine rock particles and organic material serving as medium for plant growth, and differentiated from its parent material by actions of various genetic and environmental influences.

soil creep slow downslope movement of rock and soil caused by gravity.

soil horizons layers that distinguish a soil from weathered regolith; produced by organic activity and downward percolation of water.

soil management practices utilized to retard soil erosion and to maintain soil fertility.

soil moisture water in soil that is bound loosely enough to be available for absorption by plant roots.

soil moisture recharge component in Thornthwaite water budget; portion of precipitation in excess of potential evapotranspiration that compensates for soil moisture deficit; must be satisfied before runoff is generated in water budget computation.

soil orders in 7th Approximation Soil Classification system, broad global categories that can be used to construct small-scale maps of soil types.

soil profile sequence of horizontal layers resulting from soil-forming processes of translocation and organic activity.

soil series basic local unit of U.S. soil classification schemes; named after geographic location in which present.

soil structure shape and size of peds of soil.

soil texture distribution of particle sizes in mineral portions of soils.

solar constant rate at which solar radiant energy is received at top of earth's atmosphere on surface perpendicular to sun's rays; approximately 1.94 langleys (calories per sq. cm) per minute.

solar energy, *see* solar radiant energy

solar radiant energy energy transferred from sun by electromagnetic radiation.

solar radiation spectrum of electromagnetic radiation emitted by sun.

solar system the sun, its nine planets and their satellites (moons), asteroids, comets, and meteorites.

sole, of glacier base of glacial ice in contact with the ground.

solid load bed load and suspended load of stream.

solifluction, *see* gelifluction

solstices times of year, about June 21 and December 22, when vertical rays of midday sun strike earth farthest from equator; longest daylight period in summer hemisphere, shortest daylight period in winter hemisphere.

solum *A, E,* and *B* soil horizons together; soil generated by pedogenic processes.

solution process of dissolving soluble substance.

sorted polygons networks of rock circles with centers of silty material seen in tundra regions; produced by frost cracking of ground combined with sorting action produced by repeated freeze and thaw.

species subdivision of plant and animal kingdoms below genus category; a species is a group of similar organisms that are able to breed with one another.

specific gravity ratio of weight of given volume of substance to weight of same volume of water.

specific heat amount of heat energy required to raise temperature of 1 gram of a substance by 1°C.

specific humidity number of grams of water vapor per kg of air.

spectral signature characteristic distribution of wavelengths reflected by a substance; can be used to distinguish different types of vegetation, soils, and land use.

spilling breaker toppling of wave with steep face, so that crest slides downward as wave advances.

spit linear marine sediment accumulation that is attached to land at one or both ends.

spodosols soils in which leached and eluviated light-colored *E* horizon overlies illuvial *B* horizon colored by iron or aluminum compounds or relocated organic carbon; infertile soils, being both acidic and unretentive of moisture.

spot height elevation of high point indicated on map.

spreading center boundary between crustal plates where new crust is being formed by upwelling of molten rock.

spring tide tide having greatest tidal range during month; occurs when sun aligns with earth and moon, at time of new and full moon.

spur projection in mountain base line.

stability, atmospheric tendency of air to resist vertical motion due to stable density stratification; situation in which temperature inversion inhibits vertical movement of air.

stable lapse rate rate of upward temperature decrease less than moist adiabatic rate, so that rising air is cooler than surrounding atmosphere, resisting upward motion.

stalactites hanging icicle-like depositional features in limestone caverns.

stalagmites pillar-like depositional forms rising from floors of limestone caverns.

standard parallel east-west reference lines for U.S. Bureau of Land Management land surveys.

standing crop (biomass) amount of stored chemical energy present in plants at any given time.

star dunes large sand dunes with peaked summits and spiralling arms found in extreme deserts.

stationary front surface air mass front in which boundary between tropical and polar air masses remains at about same location for a day or more.

stemflow passage of water along plant branches and stems to ground.

steppe semiarid midlatitude grasslands, extensive in central and northern Asia and in central North America; grass is short and grows in bunches.

stereoscopic image 3-dimensional image constructed mentally by viewing two separate 2-dimensional images recorded from different points.

stock small body of intrusive igneous rock that cuts through older rocks.

stomata microscopic slits, principally in leaves of green plants, through which carbon dioxide enters plant and water vapor leaves it.

stone garlands, rings, stripes patterns formed by sorting of stones because of frost action; characteristic of tundra regions.

storm surge high water along coast produced by combination of large waves and onshore winds related to cyclonic storm.

stoss and lee topography asymmetric forms produced by glacial erosion of bedrock; stoss side (facing ice advance) is abraded, lee side (facing down-glacier) is plucked.

straight channels, stream stream channels that do not meander and are not braided; result of geological controls.

stranglers vine-like tropical-forest plants that begin as epiphytes with aerial roots that eventually envelop and destroy host tree by intercepting its nutrients.

strata layers of sedimentary rock (singular: stratum).

stratified drift glacial debris deposited by meltwater; shows bedding and size-sorting.

stratosphere portion of atmosphere above troposphere; temperature increases with altitude.

stratovolcano cone-shaped volcano that consists of layers of both volcanic ash and lava; composite cone.

stratus clouds low clouds that form in uniform gray layer, often extending from horizon to horizon; common in winter.

stream discharge, *see* discharge, stream

stream gradient rate of stream descent per unit of horizontal distance.

stream hydrograph, *see* hydrograph, stream

stream load total transport of solid and dissolved materials by stream as bed load, suspended load, and dissolved load.

stream order numerical classification of streams according to their position in the hierarchy of a drainage net.

stream patterns geometrical arrangement of streams in a drainage net, often revealing underlying geologic structure or tectonic deformation.

stream piracy natural diversion of water out of one stream channel and into another due to impingement of lower channel into area drained by higher channel.

streamflow measured flow of water at a point in a natural channel; expressed as stream *discharge* in cu meters or cu feet per second.

strike compass direction of line of intersection between horizontal plane and plane of fault or inclined stratum; perpendicular to dip.

strike-slip fault fault in which relative displacement is horizontal, parallel to direction of fault strike.

stripped plain plain produced by wide exposure of resistant layer of sedimentary rock from which more erodible layers have been removed.

structure, geologic, *see* geologic structure

subarctic climate climatic region poleward of humid continental climates; characterized by cold winters and mild summers, during which most precipitation falls.

subduction descent of a lithospheric plate into earth's mantle, producing oceanic trenches and volcanism on margin of overriding plate.

subduction coast coast affected by active or recent subduction-related processes; generally mountainous, with complex geology reflecting strong crustal compression.

subhumid climates category in Thornthwaite climatic classification in which precipitation and potential evapotranspiration are about equal; corresponds to grasslands.

sublimation, *see* latent heat of sublimation

submarine canyon deep canyon cut into continental shelf and slope and extending thousands of meters below sea level; occurs frequently off mouths of major rivers.

submergent coast coast along which relief features produced by subaerial erosion have been drowned by rise of sea level or subsidence of land below sea level.

subpolar low atmospheric low-pressure area at boundary of polar easterlies, marked by convergence of polar and tropical air at surface.

subtropical anticyclones regions of atmospheric subsidence, high pressure, and divergent winds; centered over oceans in subtropical latitudes.

subtropical highs, *see* subtropical anticyclones

subtropics regions bordering tropics; approximately between latitude 35° and 25° north and south of the equator.

succession in plant community, process by which vegetation changes from first colonizers after disturbance to final climax vegetation type; vegetation effects in each stage make next stage possible.

summer solstice in northern hemisphere, the day of the year when the noon sun is directly overhead at the Tropic of Cancer, latitude 23½° N; occurs about June 21; southern hemisphere summer solstice is about December 21, with the sun overhead at the Tropic of Capricorn, 23½° S.

sun center of the solar system; a typical star consisting of a nuclear furnace in which hydrogen is converted to helium, releasing the energy received on earth as heat, visible light, and other electromagnetic radiation.

sunspot magnetic disturbance of solar surface that appears as a dark spot; appears to disturb weather patterns on earth.

supercooled refers to water that remains liquid at temperatures below the normal freezing point.

superimposed stream stream that originated on young rocks or sediment covering older more complex geologic structures eventually exposed and cut by stream incision.

supersaturated air air containing more than its normal saturation amount of water vapor.

surface currents, ocean, *see* currents, ocean

surface tension force of molecular attraction at surface of liquids that acts much like cohesive membrane on liquid.

surge, glacier, *see* glacial surge

surging breaker movement of very low-energy waves onto beach without breaking in surf zone.

surplus, *see* moisture surplus

suspended load material carried in buoyant suspension within stream of water or air.

sustainable yield maximum yield that can be obtained from ecosystem without depleting average population.

swash pulse of water that rushes up beach after wave breaks offshore; also called uprush.

swell regular pattern of round-crested oscillatory water waves, caused by distant disturbances and prevailing winds.

swidden traditional method of agriculture in rainforests in which a section of forest is cleared by cutting and burning, crops are planted for a few years, and the land is then abandoned and allowed to revert to forest; also known as milpa and "slash and burn" cultivation.

synclinal mountain topographic high, composed of resistant rock folded into a structural trough; result of relief inversion.

synclinal valley topographic valley that coincides with structural trough.

syncline troughlike fold in rock strata, from which limbs of fold rise upward.

system any collection of interacting processes and objects. Open systems receive energy and material from outside; closed systems are self-contained.

t

table mountain flat-topped mountain capped by lava that filled ancient stream valley cut in erodible rock that has subsequently been removed; a type of relief inversion.

taiga subarctic coniferous-forest regions of northern hemisphere.

tall-grass prairies generally open plains of vast extent covered by grass that is a meter or more in height.

talus cone or embankment of loose rock fragments that have accumulated at foot of cliff by sliding or falling from above; scree.

tarn small lake or pond in rock basin excavated by alpine glacier.

tectonic activity motions of the earth's crust resulting from the earth's internal energy.

tectonic sea level change change in sea level resulting from change in shape and capacity of ocean basins.

tectonism crustal motions that create specific arrangements of rock masses that provide framework for landscape development.

temperate glacier glacier in which ice temperature is near pressure melting point, so that motion by basal slip occurs easily.

temperature effect produced by energy of random movements of molecules or ions; hotness or coldness measured with respect to a datum such as freezing temperature of water (°C) or arbitrary zero (°F) or absolute zero (Kelvin scale).

temperature inversion upward increase in temperature in the atmosphere, reversing the normal upward decrease in temperature; where warm air overlies cold air.

temperature regimes characteristic temperatures and temperature changes over a given interval of time at a particular location.

tension stress that tends to pull material apart.

tephra collective term for all fragmented volcanic material ejected into the air during volcanic eruptions: volcanic dust, ash, cinders, lapilli, scoria, pumice, bombs, and blocks.

terminal moraine end moraine that marks terminus of major ice advance.

terrace, fluvial, *see* fluvial terrace

terrace, marine, *see* marine terrace

terrace, wave-built, *see* wave-built terrace

terranes, geologic fragments of older continents transported long distances by sea-floor spreading and eventually accreted to existing continents.

textural *B* horizon   clay-rich horizon in soil.

thalweg   line of greatest depth along stream channel.

thaw lakes   water-filled circular or elliptical depressions formed by thawing of permafrost.

thematic map   map devoted to single topic such as temperature, precipitation, or vegetation type.

thermal   parcel of warm air rising by convection.

thermal convection   vertical movement of air due to density differences.

thermogram   color image showing temperature variations of surfaces and solid bodies recorded by a sensing device.

thermoisopleth diagram   diagram that shows hourly temperature at station for every day of the year.

thermosphere   uppermost portion of the atmosphere, composed of scattered subatomic particles; approximately 90 km above the surface.

thin sections, rock   paper-thin slice of rock mounted in glass for microscope examination of mineralogy.

third order of relief   landforms; relief features visible in single view, such as hills, valleys, cliffs, beaches, etc.; erosional and depositional forms.

third-order stream   larger stream formed by the confluence of two second-order streams.

Thornthwaite climatic classification system   system of classifying climate according to water budget parameters.

threshold   the point at which a system becomes so unbalanced that its processes begin to change to restore system equilibrium.

throughfall   water that reaches soil after initial interception by vegetation.

throughflow   moisture that moves laterally downslope between soil particles.

thrust fault (reverse fault)   fault type in which block lying above fault plane has moved upward in relation to block below fault plane; result of compression.

tidal bulge   mass of ocean water raised by tide-producing forces; slight deformation of the solid earth also occurs; rotation of earth through tidal bulges on opposite sides of earth produces twice daily rise and fall of sea along most coasts.

tidal range   difference in water elevations at high and low tides at a location.

tidal wave,   *see* seismic sea wave

tides   rhythmic rise and fall of oceans and bodies of water extending from them, caused by centrifugal force and gravitational effects of moon and sun.

till   unsorted rock debris deposited by glacier.

till plain   surface of low relief that is composed of glacial till.

tilted block   crustal block, bounded by faults on at least two sides, that has rotated, rising on one side and sinking on the other side.

tombolo   deposit of marine sediment that ties island to mainland or to another island.

topographic map   large-scale map that shows both physical and cultural features of earth's surface; normally portrays terrain by use of contour lines.

topographic profile   precise vertical cross section of terrain; normally constructed from contour lines.

tor   projection of jointed rock, often granite, that rises from a hillslope or ridgecrest.

tornado   highly localized column of rapidly rotating air creating a dark "funnel," associated with intense thunderstorm activity.

tower karst   vertical pillars of limestone, tens to hundreds of meters high, created by solution of jointed limestone under wet tropical conditions.

township   basic land unit of U.S. Bureau of Land Management land survey; approximately 6 miles on a side (36 square miles).

traction   rolling or sliding transport of particles.

trade winds   persistent northeasterly (northern hemisphere) and southeasterly (southern hemisphere) surface winds between horse latitudes and equator.

trailing-edge coasts   coasts that are low-lying and lacking in tectonic activity due to their position on trailing edges of lithospheric plates.

transform faults   faults that occur when sea-floor spreading is transformed to strike-slip faulting by series of fractures.

translocation   downward displacement of solid substances and dissolved substances in soil due to action of percolating water.

transpiration   transfer of water vapor from soil through roots to atmosphere through stomata of leaves.

transverse dunes   dunes having crestlines perpendicular to the dominant wind.

transverse streams   streams that cut across linear geologic structures such as anticlines and fault blocks.

traverse   survey that extends outward from an initial base line or base point, along a linear route.

tree ring analysis   analysis of number and size of annual-growth rings to determine age of wood and history of climatic fluctuations at site.

trellis drainage   stream pattern composed of parallel and diagonal elements where inclined layers of sedimentary rock are exposed at the surface.

triangular facets   blunt ends of mountain spurs that terminate along a line, indicating the presence of a dissected fault scarp.

trophic level   position of an organism in the energy flow pyramid of an ecosystem.

Tropic of Cancer   northernmost parallel of latitude (approximately 23½° N) at which sun reaches zenith (about June 21).

Tropic of Capricorn   southernmost parallel of latitude (approximately 23½° S) at which sun reaches zenith (about December 21).

tropical cyclone   intense cyclonic storm in which wind exceeds 120 km per hour; hurricane or typhoon.

tropical rainforest,   *see* rainforest, tropical

tropical savanna,   *see* savanna

tropical secondary circulations   transient weather phenomena of tropics: easterly waves and tropical cyclones, including hurricanes (typhoons).

tropical wet and dry climate   climatic region located between tropical wet zone and drier subtropical areas; characterized by precipitation in summer and relative drought in winter.

tropical wet climate   climatic region near equator (between about 15° N and S) that lacks cold season and in which precipitation is frequent year around.

tropics   regions lying between (or near) Tropics of Cancer and Capricorn.

tropopause level at which upward temperature decrease in troposphere is succeeded by temperature increase in stratosphere.

troposphere portion of atmosphere nearest earth's surface in which most weather phenomena occur; temperature decreases as altitude increases.

true azimuth azimuth measured from true north.

true north true direction to North Pole, as measured along local meridian.

tsunami, *see* seismic sea wave

tuff rock made up of fine volcanic tephra (ash), usually welded by heat.

tundra vegetation in cold, short-summer regions at high latitudes and high altitudes, consisting of low shrubs, grasses, flowering herbs, mosses, and lichens.

turbidity, atmospheric opacity of atmosphere to radiant energy due to presence of dust, smoke, and other particulate matter.

turbidity current suspension of sand and silt that can flow with considerable force down submarine slopes; possible agent of undersea erosion producing submarine canyons.

turbulent flow irregular flow marked by eddies; speed and direction of flow at given point are constantly changing.

typhoon tropical cyclone that forms in western Pacific.

u

ultisols soils leached of bases, occurring in warm, wet climates. Relatively speaking, they contain a high proportion of iron and aluminum oxides in *E* horizon; are poor in humus and low in agricultural productivity.

ultraviolet radiation part of electromagnetic spectrum that spans range of wavelength from 0.4 to 0.01 micrometers, with wavelengths shorter than violet light; ultraviolet light.

unconfined aquifer water-bearing material below the water table; water is not under pressure, and must be pumped to surface.

unconsolidated sediment sediment in which individual particles are not bound together by a cementing substance that creates rocklike material.

underfit stream stream that is abnormally small in relation to the geometry of its valley.

U.S. Comprehensive Soil Classification System, 7th Approximation non-genetic, purely morphological classification of soil types in current use by North American soil scientists.

unpaired terraces fluvial terraces that are not at the same elevation on the two sides of a stream; indicate lateral stream planation during continuous, slow stream incision.

unstable lapse rate rate of upward temperature decrease greater than dry adiabatic rate, so that rising air is warmer than surrounding atmosphere, and continues to rise.

upper air ridge area of high pressure in upper troposphere; causes jet stream to swing toward poles.

upper air trough area of low pressure in upper troposphere; causes jet stream to swing toward equator.

upper atmosphere usually refers to atmosphere at the level of jet streams, where atmospheric pressure is about half that at the surface.

uprush, wave, *see* swash

upwelling upward movement of deep, cool ocean waters along west coasts of continents in subtropical and lower middle latitudes.

urban heat island, *see* heat island

v

valley depression formed by erosional activity of running water, assisted by mass wasting processes.

valley glacier river of ice flowing in preexisting valley; alpine glacier.

valley train outwash plain confined by valley walls in area of glaciation.

valley wind wind that blows upslope through mountain valley on clear days in response to thermal contrasts.

vaporization change in physical state from liquid to vapor; *see also* latent heat of vaporization.

vapor pressure pressure exerted by water molecules in given sample of moist air.

vapor pressure gradient rate at which vapor pressure varies from point to point.

varves paired annual layers of coarse and fine sediment deposited in lakes near glacier margins; used in estimating dates, and rates of glacial retreat.

ventifact rock outcrop, boulder, or cobble showing abrasion by wind-driven sandblast or dust polishing; seen in deserts and tundra regions.

vernal equinox springtime date when sun is directly overhead at noon at the equator; about March 21 in northern hemisphere; about September 21 in southern hemisphere.

vertisols clay soils in which horizon development is impeded by churning effects of repeated changes in volume due to alternating wetting and drying.

visible radiation part of electromagnetic spectrum corresponding to visible light, spanning range of wavelengths from 0.7 to 0.4 micrometers; visible light.

volcanic ash, *see* scoria

volcanic bomb large clot of fluid magma ejected into air during volcanic eruptions, solidifies in air to shape resembling a football (due to spin).

volcanic neck pinnacle composed of igneous rock originally emplaced in vent feeding magma to volcano; exposed by erosional removal of volcano.

volcanism intrusion of molten magma into earth's crust, and extrusion of tephra and molten lava at earth's surface.

volcano opening in earth's surface where magmatic gases, scoria, and lava are erupted.

w

warm front forward edge of warm air mass that has moved into region formerly occupied by colder air mass.

washboard moraines transverse ridges with rippled till surfaces a few meters high and tens of hundreds of meters apart; created by advancing ice.

water budget computation of the hydrologic cycle at a specific location for a specific period of time; *see also* water budget equation.

water budget climate classification of climates in terms of results of annual water budget computations.

water budget equation relation that expresses distribution of precipitation between runoff, soil moisture storage, and evapotranspiration.

water gap erosional gap in ridge of resistant rock, through which a stream passes.

water table level below land surface at which all openings are saturated with water.

water vapor gaseous phase of water.

waterspout tornado funnel over coastal waters and seas.

wave break collapse of water wave resulting from increasing wave height and steepness as wave velocity is slowed by friction with bottom of water body.

wave-built terrace submarine deposit of sand that skirts abrasion platform and extends further seaward.

wave crest highest point of water wave.

wave-cut bench, *see* wave-cut platform

wave-cut notch indentation at base of actively retreating sea cliffs; carved by wave impact and wave-tossed rock shrapnel.

wave-cut platform nearly flat surface that extends seaward from base of sea cliff; created by wave erosion; intertidal platform.

wave height vertical distance between crests and troughs of water waves.

wave length distance between successive crests or successive troughs of water waves.

wave period time required for successive wave crests to pass same point.

wave refraction change in direction of wave travel, caused by friction with bottom of water body.

wave trough lowest part of a water wave, lying between successive crests.

wavelength, of radiation distance, measured in micrometers, between similar points on wave forms in electromagnetic spectrum.

weather momentary condition of the atmosphere at a locality, including temperature, pressure, wind, cloud cover, humidity, and precipitation.

weather map map portraying meteorological conditions by use of pressure isobars, wind speed symbols, fronts, and air mass types.

weathered mantle blanket of fine particles, often several meters deep, that covers solid bedrock and that is formed by physical and chemical weathering of bedrock.

weathering fragmentation of rock materials in place at or near the earth's surface by both chemical and physical processes.

weathering front surface, below the ground, that separates weathered from unweathered rock.

westerlies warm air flowing poleward near surface; this flow is deflected toward the east to become the southwesterlies of the northern hemisphere and the northwesterlies of the southern hemisphere.

whaleback dunes large, linear dunes with broadly rounded crests; usually longitudinal dunes.

wilting point depletion of soil moisture to point at which it can no longer be extracted by plants, causing plants to begin to wilt.

wind gap conspicuous notch in level crest of homoclinal ridge, indicating position of a former stream channel transverse to the ridge.

wind tide elevation of water along a coast as a consequence of strong onshore winds related to a cyclonic storm; storm surge.

wind waves, *see* forced waves

wineglass valley valley form seen along fault scarps in mountainous deserts; narrow stream gorge cut below wider upper part of valley cross section; reflects recent stream rejuvenation due to upward movement of fault block.

winged headland cliffed headland with sand or gravel spits extending outward on either side.

winter solstice in either hemisphere the time when the midday sun is lowest in the sky, being directly overhead at latitude 23½° in the other hemisphere; the northern hemisphere winter solstice occurs about December 22.

Wisconsinan glaciation continental glaciation in North America between about 70,000 and 10,000 years ago; the last great glaciation.

x

xerophytes plants that are structurally adapted to survive extreme, prolonged drought; desert shrubs.

y

yardangs sharp-crested wind-scoured ridges between deflation furrows in deserts; usually carved in erodible material.

yield amount of energy stored during growing season in desired portion of crop, such as the fruit.

youthful stage descriptive term applied to initial stage in cycle of erosion proposed by W. M. Davis; surface of low relief that is trenched by streams in sharply incised valleys or canyons.

z

zenith point in sky directly overhead.

zonal soils soils with clearly distinguishable horizons, which are associated with specific zonal climates and vegetation types.

Index

oxygen, 11, 13, 14, 234, 333
oxygen isotopes, 482, 484
  ratios of, and climatic change, 484
  record of, marine, 506
ozone, 14, 52, 56
  formation of, 14
  layer, in atmosphere, 14, 53, 54, 71

Pacific Coast (U.S.), 431, 526, 527
  forests of, 275
Pacific high, 88
Pacific Northwest (U.S.), 455
Pacific Ocean, 96, 192, 325, 327, 328,
    329, 347, 456, 463, 521, 535,
    537, 543, 544
Pacific plate, 451
Pacific Ring of Fire, 456
Pakistan, 305, 469
Palaeoarctic zoogeographic realm,
    229, 230
Paleo-Indian hunters, 273
Panama, 227
Pangaea, 325, 329
parallel-pinnate drainage, 402
parallels of latitude, 49, A.5, A.6
parent material, of soil, 295
particle size classification, 290
pastoral nomads, 435
Patagonia, 453, 521
paternoster lakes, 497–498
patterned ground, of tundra, 481
peak flow, of stream, 420
peat, 297, 305
pedalfer, 312
pediment, 391–392
pedocal, 312
pedogenic regimes, 299
peds, soil, 291
peneplain, 377, 378, 391
penguins, 227
Pennsylvania, 296, 441, 444, 468
    Harrisburg, 469
    Philadelphia, 413
    Pittsburgh, 270
    Susquehanna River, 468, 469
perched water table, 173
perennial plants, 252
periglacial block field, 387
periglacial climates, landforms of,
    386
periglacial geomorphic systems, 389
periglacial landforms, relict, 388
perihelion, 46, 50, 216
permafrost, 369, 386, 480
Persian Gulf, 330
Peru, 98, 99, 198, 279, 384, 393
  Andes Mountains, 279
Peru–Chile trench, 521
pesticides, 314
petroleum, 14, 330, 339, 340, 350
  and overthrusts, 353
pH:
  of precipitation, 117
  of soil, 293
phase change, 28, 286
  of water, 28, 29
phenocrysts, 336
Philippine Archipelago, 345
Philippines, 456

phosphorus, 236, 289
photosynthesis, 13, 14, 25, 35, 62,
    224, 234, 235, 237, 252
phyla, 224
physical geography, 3, 5
physical systems, 32
physiographic diagram, 442
Piedmont Region, 413
piezometric surface, 173
pine, 270, 275
  barrens, 296
  forest, 298
pingo, 386, 387, 481
piracy, stream, 420, 469
pitted outwash plain, 504
placental mammals, 227
planetesimals, 7, 10
planets, 7, 9, 223
  formation of, 7
  gaseous, 9
  moons of, 7, 9, 23
  terrestrial, 9
planimetric map, A.3
plankton, 98
plant and animal:
  dispersal, 227
  interactions, 224–226
  kingdoms, 224
plant-animal interaction, 224–226
plants:
  annual, 252
  communities of, 252, 253
  deciduous, 251
  drought-evading, 264
  drought-resisting, 264
  effect on landforms, 15
  evergreen, 251, 269
  green, 13, 14
  herbaceous annual, 265
  hydrophytes, 252
  leaves of, 108, 234, 251
  litter of, 297
  marine, 13
  mesophytes, 252
  needleleaf evergreen, 269
  pathogens of, 314
  perennial, 252
  respiration in, 236, 240
  roots of, 251, 313, 383
  succession of, 253–254
  xerophytes, 252, 264–265
  yield of, 237
plate tectonics, 217, 325, 326, 328,
    329, 343, 346–354, 361, 450,
    453, 458, 462, 484, 536, 543–
    544
  and climatic change, 217, 484
  and megarelief, 329
platinum, 332
platypus, 227, 228
Pleistocene epoch, 477
  glaciation in, 217, 254, 477–484
    indirect effect of, 478
  sea-level changes, 420, 479, 483,
    515, 538
plowing, 314
plug dome, 454–455
pluton, 335, 544
plutonic injection, 462

plutonic rock, 335, 336, 341, 462
Podocarpus, 269
podzolization, 300, 301, 303
point bar, 416, 418, 419, 423
point bar deposit, 423
point of tangency, map projection,
    A.6
Poland, 478
polar bear, 227
polar high-pressure cell, 86, 198
polar outbreaks, 87, 151
pollution:
  atmospheric, recorded in glaciers,
    482
  groundwater, 175
pools, in stream bed, 413, 418
pools and riffles, 414
Portugal, 266
potassium/argon age determination,
    18
potential energy, 25, 404, 419
potential evapotranspiration *(PE)*,
    178, 199, 309, A.23
  computation of, A.23
  daylength correction factors,
    A.24
  estimation of, 178
  map (U.S.), 179
  seasonality of, 179
  unadjusted, A.23
prairie:
  short-grass, 270, 272, 435
  tall-grass, 270, 273
precipitation, 104, 122, 126, 128
  interception of, 170
  pH of, 293
precipitation distribution:
  global, 128–130
  United States, 130–131
precipitation models:
  coalescence, 116
  ice-crystal, 117, 120
prediction:
  of earthquakes, 465
  of volcanic eruptions, 465
  of weather, 93
preglacial stream valley, 502
pressure:
  of a gas, 31, 32
  and surface winds, 83
  and upper level winds, 83
  vapor, 107, 110, 111, 120
  and winds, on a uniform earth,
    84
pressure cells, semipermanent, 87
pressure gradient force, 80, 81, 83
pressure melting, of glacier, 487
primary producer, 231, 234
prime meridian, 49
principal meridian, A.5
profile:
  of stream, 412
  terrain, A.12
prograded coast, 538
protons, 7
Ptolemy, Claudius, A.3
Pueblo Indians, 363
pumice, 451
pyroclastic debris, 334, 450